CONTACT MECHANICS IN TRIBOLOGY

SOLID MECHANICS AND ITS APPLICATIONS
Volume 61

Series Editor: G.M.L. GLADWELL
Solid Mechanics Division, Faculty of Engineering
University of Waterloo
Waterloo, Ontario, Canada N2L 3G1

Aims and Scope of the Series

The fundamental questions arising in mechanics are: *Why?*, *How?*, and *How much?* The aim of this series is to provide lucid accounts written by authoritative researchers giving vision and insight in answering these questions on the subject of mechanics as it relates to solids.

The scope of the series covers the entire spectrum of solid mechanics. Thus it includes the foundation of mechanics; variational formulations; computational mechanics; statics, kinematics and dynamics of rigid and elastic bodies; vibrations of solids and structures; dynamical systems and chaos; the theories of elasticity, plasticity and viscoelasticity; composite materials; rods, beams, shells and membranes; structural control and stability; soils, rocks and geomechanics; fracture; tribology; experimental mechanics; biomechanics and machine design.

The median level of presentation is the first year graduate student. Some texts are monographs defining the current state of the field; others are accessible to final year undergraduates; but essentially the emphasis is on readability and clarity.

For a list of related mechanics titles, see final pages.

Contact Mechanics in Tribology

by

I. G. GORYACHEVA

Institute for Problems in Mechanics,
Russian Academy of Sciences,
Moscow, Russia

SPRINGER-SCIENCE+BUSINESS MEDIA, B.V.

A C.I.P. Catalogue record for this book is available from the Library of Congress.

ISBN 978-90-481-5102-8 ISBN 978-94-015-9048-8 (eBook)
DOI 10.1007/978-94-015-9048-8

Printed on acid-free paper

Originally published by Kluwer Academic Publishers in 1998
Softcover reprint of the hardcover 1st edition 1998

In memory of my teacher

Professor L.A. Galin

Contents

Preface

Tribology is the science of friction, lubrication and wear of moving components. Results obtained from tribology are used to reduce energy losses in friction processes, to reduce material losses due to wear, and to increase the service life of components.

Contact Mechanics plays an important role in Tribology. Contact Mechanics studies the stress and strain states of bodies in contact; it is contact that leads to friction interaction and wear. This book investigates a variety of contact problems: discrete contact of rough surfaces, the effect of imperfect elasticity and mechanical inhomogeneity of contacting bodies, models of friction and wear, changes in contact characteristics during the wear process, etc.

The results presented in this book were obtained during my work at the Institute for Problems in Mechanics of the Russian Academy of Sciences. The first steps of this research were carried out under the supervision of Professor L.A.Galin who taught me and showed me the beauty of scientific research and solutions. Some of the problems included in the book were investigated together with my colleagues Dr.M.N.Dobychin, Dr.O.G.Chekina, Dr.I.A.Soldatenkov, and Dr.E.V.Torskaya from the Laboratory of Friction and Wear (IPM RAS) and Prof. F.Sadeghi from Purdue University (West Lafayette, USA). I would like to express my thanks to them. I am very grateful to Professor G.M.L.Gladwell who edited my book, helped me to improve the text and inspired me to this very interesting and hard work. Finally, I would like to thank Ekaterina and Alexandre Goryachev for their help in preparation of this manuscript.

I hope that this book will be useful for specialists in both contact mechanics and tribology and will stimulate new research in this field.

Irina Goryacheva
Moscow, Russia
December 1997

Chapter 1

Introduction

Tribology deals with the processes and phenomena which occur in friction interaction of solids.

The subject of tribology is *the friction contact* that is the region of interaction of bodies in contact.

Various processes of physical (including mechanical, electrical, magnetic and heat), chemical and biological nature occur at the friction contact. *Friction force*, i.e. resistance to the relative displacement of bodies, is one of the main manifestations of these processes. It is well-known that one third of the world energy resources is now spent on overcoming friction forces.

Lubrication of surfaces is the most efficient method for reducing friction. Various greases, liquid and solid lubricants are used for friction components, depending on the environmental conditions, materials of surfaces and types of motion.

Wear of contacting surfaces is the other manifestation of the processes occurring in contact interaction. Wear is a progressive loss of material from surfaces due to its fracture in friction interaction showing up in gradual change of the dimensions and shape of the contacting bodies. The precision of machines is impaired by wear, sometimes the wear leads to the machine failure. Thus the study of wear and its reasons, and elaboration of methods for improvement of wear resistance are important problems of tribology.

These discussions point to the other definition of tribology as *the science of friction, lubrication and wear of materials.* The history of tribology is presented in the monograph by Dowson (1978). The monographs by Bowden and Tabor (1950, 1964), Kragelsky (1965), Rabinowicz (1965), Kostetsky (1970), Moore (1975), Kragelsky, Dobychin and Kombalov (1982), Hutchings (1992), Singer and Pollock (1992), Chichinadze (1995), etc., the handbooks by Peterson and Winer (1980), Bhushan and Gupta (1991), etc. are devoted to fundamental and applied investigations in tribology.

Tribology can be considered as an applied science since the diminishing of the energy losses and deleterious effects of friction and wear on the environment, and the increase of machine life are the main purposes of tribological investigation.

However, deep understanding of the nature of friction and wear is the only reliable way to the successful solution of these problems. The increasing interest in fundamental problems of tribology confirms this conclusion.

Tribology has evolved on the basis of mechanics, physics, chemistry and other sciences. However, the results obtained in these fields cannot be applied directly. Tribological processes are complicated and interconnected involving multiple scales and hierarchical levels, and must be considered using results of different scientific disciplines simultaneously.

One of the main roles in the study of friction interaction belongs to mechanics.

1.1 Friction contact from the standpoint of mechanics

Stress concentrations near contact regions affect all processes occurring in friction interaction. High contact pressures and sliding velocities cause heating at contact zones, and substantial changes of properties of the surface layer; they also stimulate chemical reactions, resulting in the formation of secondary compounds and structures, and accelerate the mutual diffusion. The subsurface layer is subjected to high strains due to mechanical and thermal action that lead to crack initiation and growth, and finally to surface or subsurface fracture.

Mechanics of solids, in particular contact mechanics and fracture mechanics, is a powerful tool for the investigation of basic tribological problems. Contact mechanics investigates the stress-strain state near the contact region of bodies as a function of their shapes, material properties and loading conditions. Fracture mechanics is used to evaluate specific conditions which lead to the junction failure.

The first investigation in the field of contact mechanics was made by Hertz (1882) who analyzed the stresses in the contact of two elastic solids. Hertz's theory was initially intended to study the possible influence of elastic deformation on Newton's optical interference fringes in the gap between two glass lenses. This theory provided a basis for solution of many tribological problems. It led to methods for the calculation of the real contact area of rough surfaces and the contact stiffness of junctions, to the investigation of rolling and sliding contact, wear of cams and gears, to estimation of the limiting loads for rolling bearings, etc.

However, it is well known that the Hertz theory is based on some assumptions which idealize the properties of contacting bodies and the contact conditions. Among other things, it is assumed that

- the contacting bodies are elastic, homogeneous and isotropic;
- the strains are small;
- the surfaces are smooth and non-conforming;
- the surface shape does not change in time;

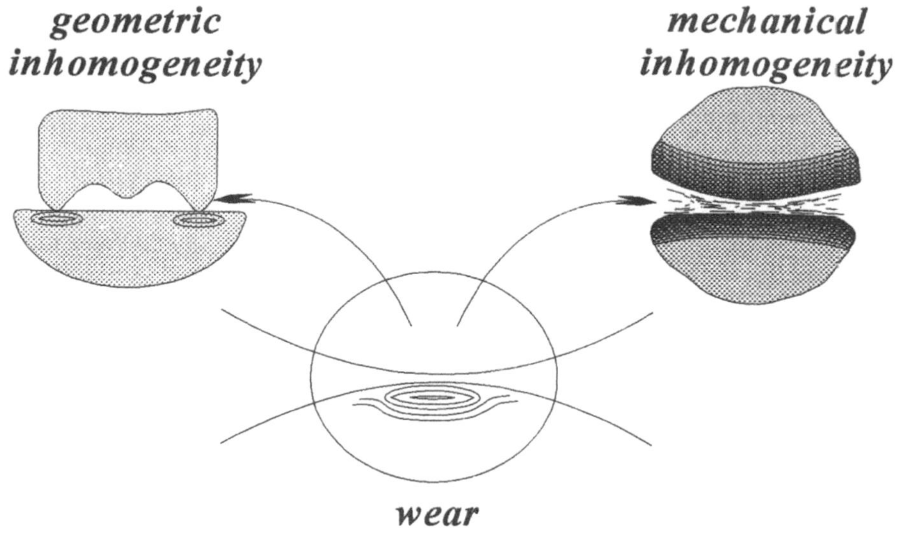

Figure 1.1: Scheme of contact of elastic bodies with geometric (a) and mechanical (b) inhomogeneities.

– the contact is frictionless.

These assumptions are often unwarranted in tribological problems. It is known for instance that, in contact interaction, stresses increase in a thin surface layer, the thickness of which is comparable with the size of contact region. Fig. 1.1 illustrates the scheme of contact and stresses near the surface. Due to the high stresses, cracks initiate and grow in this layer; this leads to particle detachment from the surface (wear). Thus, the properties of a thin surface layer play an important role in the subsurface stress and wear analysis.

Due to the surface treatment (heating, mechanical treatment, coating, etc.) the surface layer has different kinds of inhomogeneity. These significantly influence the stress distribution and wear in contact interaction.

Geometric inhomogeneity, such as macrodeviations, waviness or roughness (see Fig. 1.1(a)), which is a deviation of the surface geometry from the design shape, leads to discreteness of the contact between solid surfaces. Geometric inhomogeneity influences the contact characteristics (real pressure distribution, real contact area, etc.) and internal stresses in the surface layer. Due to the roughness of contacting bodies, the subsurface layer is highly and nonuniformly loaded, so there is a nonuniform internal stress distribution within this layer. These peculiarities of the stress field govern the type of fracture of the surface layer.

Mechanical inhomogeneity of materials of contacting bodies (see Fig. 1.1(b)) also arises due to different kinds of surface treatment, application of coatings and solid lubricants or in operating. Specifically, the mechanical properties of the surface layer are different from the bulk material. In spite of the small thickness of this layer, its characteristics can significantly influence the friction and wear processes.

The intermediate medium between the contacting bodies (third body) also influences the stress distribution in subsurface layers, e.g. application of a thin film of lubricant essentially decreases the friction and wear of surfaces.

These properties of friction contact prove that special problems of contact mechanics (contact problems) must be formulated to describe the phenomena important for tribological needs: problems which include complicated boundary conditions, the properties of the intermediate medium, surface inhomogeneity and so on.

1.2 Previous studies and the book outline

Contact mechanics has evolved from the consideration of simple idealised contact configurations to the analysis of complicated models of contacting bodies and boundary conditions.

The following fields of contact mechanics are well developed:

- contact problems with friction;

- contact problems for layered and inhomogeneous elastic bodies;

- contact problems for anisotropic elastic bodies;

- thermoelastic contact problems;

- contact problems for viscoelastic and elasto-plastic bodies.

These fields of contact mechanics have been considered in monographs by Staierman (1949), Muskhelishvili (1949), Galin (1953, 1976b, 1980, 1982), Ling (1973), Vorovich, Aleksandrov and Babeshko (1974), Rvachev and Protsenko (1977), Gladwell (1980), Popov (1982), Aleksandrov and Mhitaryan (1983), Mossakovsky, Kachalovskaya and Golikova (1985), Johnson (1987), Goryacheva and Dobychin (1988), Kalker (1990), etc.

The gap between contact mechanics and tribology has been narrowed; they have the same subject of investigation, i.e. friction contact. Contact problem formulations now include specific properties of friction contact such as surface microstructure friction and adhesion, shape variation of contacting bodies during the wear process, surface inhomogeneity, etc.

1.2.1 Surface microstructure

To take into account the surface microstructure, such as roughness or waviness, Staierman (1949) proposed a model of a combined foundation. Surface displacement of this foundation under loading was represented as a sum of the elastic displacement of the body with given macroshape and an additional displacement due to the surface microstructure. This model became a basis for investigation of the contact of rough bodies which was further developed for nonlinear models of combined foundation and for various surface shapes of contacting bodies. Based on this approach, we can calculate the nominal (averaged) contact characteristics (nominal pressure and nominal contact area).

Another way of looking at the problem of the contact of rough bodies was developed by Archard (1951), Goodman (1954) Greenwood and Williamson (1966), Greenwood and Tripp (1967), Demkin (1970), Hisakado (1969, 1970), Rudzit (1975), Hughes and White (1980), Thomas (1982), Kagami, Yamada and Hatazawa (1983), Sviridenok, Chijik and Petrokovets (1990), Majumdar and Bhushan (1990, 1991), etc. They considered models of discrete contact of bodies with surface microgeometry which made it possible to calculate such important characteristics of the rough body contact as the real contact pressure and the real contact area. Note that the most of the discrete contact models include the assumption that the stress-strain state near each contact spot is determined only by the load applied to this contact spot, i.e. these models neglect the interaction between contact spots. This assumption is valid only for small loads when the density of contact spots is not too high.

Generally, the problem of the discrete contact of rough bodies is a three-dimensional boundary problem of contact mechanics for a system of contact spots comprising the real contact area. This problem is discussed in detail in Chapter 2 where the contact problem for bodies with surface microgeometry is formulated as a multiple contact problem for elastic bodies.

1.2.2 Friction

The other important property of contact interaction is the friction between contacting bodies. In classical formulation of contact problems, friction is introduced phenomenologically by a definite relation (friction law) between the tangential and normal stresses in the contact zone.

The method of complex variables developed by Muskhelishvili (1949), Galin (1953), Kalandiya (1975) is mainly used to determine the stress distribution for the 2-D contact problems in the presence of friction. The linear form of the friction law is normally used in the problem formulations.

If a tangential force T applied to the body satisfies the inequality $T < \mu P$, where P is the normal force and μ is the friction coefficient, then partial slip occurs; this is characterized by the existence of slip and stick zones within the contact region. The friction is *static friction*. In slip zones the linear relation between the normal (p) and tangential (τ) stresses is usually used, i.e. $\tau = \mu p$. In stick zones the displacements of contacting bodies at each point are equal. Contact problems with partial slip in contact region were considered by Mindlin (1949), Galin (1945, 1953), Lur'e (1955), Spence (1973), Keer and Goodman (1976), Mossakovsky and Petrov (1976), Mossakovsky, Kachalovskaya and Samarsky (1986), Goldstein and Spector (1986), etc. The solution of the problems includes the determination of the positions and sizes of stick and slip zones for given loading conditions. In particular, it is shown that the area of stick zones decreases and tends to zero, if $T \to \mu P$.

If $T = \mu P$, there is limiting friction, and the condition of full slip occurs in the contact region. This case is also called *sliding friction*. Axisymmetric contact problems with limiting friction were investigated by Mindlin (1949), Lur'e (1955), Muki (1960), Westman (1965), Hamilton and Goodman (1966), Korovchinsky (1967), Gladwell (1980), etc. In most cases the assumption was made that the tangential stress in the contact region does not influence the contact pressure distribution. This assumption is valid for a small value of the parameter $\varepsilon = \mu\vartheta^*$, where

$$\vartheta^* = \frac{[(1-2\nu_1)/G_1] - [(1-2\nu_2)/G_2]}{2\,[(1-\nu_1)/G_1] + [(1-\nu_2)/G_2]},$$

$$G_i = \frac{E_i}{1+\nu_i}, \quad i = 1, 2.$$

For contacting bodies of identical material, and also for the case $\nu_1 = \nu_2 = \dfrac{1}{2}$, $\varepsilon = 0$, the assumption is true.

3-D contact problems with limiting friction (taking into account the influence of the tangential stress on the normal stress within the contact region) were investigated in Kravchuk (1980, 1981), Galin and Goryacheva (1983), Mossakovsky, Kachalovskaya and Samarsky (1986).

Chapter 3 presents some solutions of contact problems in the 2-D and 3-D formulations with limiting friction which include the influence of the tangential stress on the contact pressure distribution and on the size and the position of contact region.

Amontons' friction law $\tau = \mu p$, where τ and p are the tangential and normal contact stresses, is mainly used in formulation of the contact conditions in slip zones. From the standpoint of the molecular-mechanical theory of friction, Amontons' law takes into account only the mechanical component of friction force arising from the deformation of asperities of rough contacting bodies. Deryagin (1934), Bowden and Tabor (1950), Kragelsky (1965) showed that adhesion plays a key part in the friction force formation. Taking into account adhesion gives rise to Coulomb's law $\tau = \tau_0 + \mu p$. Chapter 3 also describes some results which follow from the solution of the contact problems with Coulomb's law.

1.2.3 Imperfect elasticity

Many phenomena taking place in friction interaction cannot be explained on the basis of elastic bodies. Specifically, they are the dependence of the friction force on the temperature and velocity, self-oscillations during a friction process, etc. Thus, more complicated models taking into account imperfect elasticity of contacting bodies must be used in the analysis.

Among such investigations, there is the contact problem for a rigid cylinder rolling over a viscoelastic foundation considered by Ishlinsky (1940). The author used the one-dimensional Kelvin–Voigt model to describe the relation between the normal stress σ_y and the deformation ϵ_y of the foundation:

$$\sigma_y = E\epsilon_y + HT_\epsilon\dot{\epsilon}_y,$$

where E, H and T_ϵ are characteristics of the viscoelastic body. The results showed that the dependence of the friction force T on the rolling speed V had a nonmonotone character: for low speed it was described by

$$T = \frac{HT_\epsilon VP}{ER},$$

while for high speed

$$T = \frac{2}{3}\sqrt{\frac{P^3 l_0}{HT_\epsilon VRl}},$$

where l and R are the length and the radius of the cylinder, and l_0 is a characteristic length.

It is interesting to note, that if these two asymptotic formulae had been obtained earlier, they might have brought an end to the discussion raised between Dupuit (1837) and Morin (1853) in the nineteenth century concerning the dependence of the friction force on the radius of the roller. Dupuit suggested that $T \sim R^{-1/2}$, and Morin thought that $T \sim R^{-1}$. Ishlinsky's formulae support both suggestions.

More complicated and also more realistic models of viscoelastic bodies are based on the mechanics of solids. The methods of solution of some contact problems for viscoelastic solids have been presented in May, Morris and Atack (1959), Lee and Radok (1960), Hunter (1960, 1961), Morland (1962, 1967, 1968), Galin and Shmatkova (1968), Ting (1968), Braat and Kalker (1993), etc. and also in monographs by Il'ushin and Pobedrya (1970), Ling (1973), Rabotnov (1977), etc. Some problems for inelastic solids concerning normal, sliding, and rolling contact and impact are discussed by Johnson (1987).

The analysis of the contact problem solutions taking into account inelastic properties of solids and friction allows the establishment of the dependence of the contact characteristics on the mechanical properties of bodies and the contact conditions. It also makes possible to determine the conditions that allow us to use the simplified models.

Some rolling and sliding contact problems for viscoelastic bodies are also presented in Chapter 3. The solutions of these problems are used to calculate the mechanical component of friction force and to analyze its dependence on the sliding velocity.

1.2.4 Inhomogeneous bodies

Since specific surface properties of contacting bodies considerably influence the stress distribution near the contact region and the friction force, the solution of contact problems for bodies with elastic parameters which vary with depth is of great interest for tribology. A review of early works devoted to investigation of contact problems for inhomogeneous elastic bodies may be found in Galin (1976b) and Gladwell (1980). Most of these works are concerned with the special forms of the functions describing the dependence of elastic moduli (the Young modulus and the Poisson ratio) on the depth.

Different kinds of coatings and surface modification are widely used in friction components to decrease friction, to increase the wear resistance and to prevent seizure between contacting surfaces. The lifetime of coatings and their tribological characteristics (friction coefficient, wear resistance, etc.) depend on the mechanical properties of coatings, their thickness and structure and on the interface adhesion. It is important for tribologists to choose the optimal mechanical and geometrical characteristics of coatings for any particular type of junctions.

Contact mechanics of layered bodies can help to solve this problem. Many researches in this field are reviewed in monographs by Nikishin and Shapiro (1970, 1973), Vorovich, Aleksandrov and Babeshko (1974), Aleksandrov and Mhitaryan (1983). They give solutions of plane and axisymmetric contact problems for an elastic layer bonded to or lying without friction on an elastic or rigid foundation.

Considerable attention has been focussed recently on the production of thin coatings, the thickness of which is commensurable with the typical size and distance between asperities. This initiated the investigation of contact problems for layered bodies with rough surfaces. Problems of this kind are considered in detail in Chapter 4. The effect of the boundary conditions at the interface between the coating and the substrate is also analyzed in that chapter. This analysis elucidates the influence of the interface adhesion on the internal stresses and the fracture of coating.

Chapter 4 also includes contact problems for viscoelastic layered elastic bodies. Solving these problems for rolling or sliding elastic indenter with smooth or rough surface is very important for studying the dependence of the friction force on the speed for junctions operating in the boundary lubrication condition or in the presence of solid lubricants.

What is the influence of a thin viscoelastic layer on the stress distribution in the lubricated contact of two elastic rollers? This question is also discussed in Chapter 4 where the model of lubricated contact includes equations from hydrodynamics, viscoelasticity and elasticity. This model allows us to analyze the dependence of the friction coefficient on speed for variable mechanical and geometrical characteristics of the surface layer.

1.2.5 Surface fracture

Investigation of the contact problems taking into account friction, microstructure, presence of surface layers and an intermediate medium allows us to determine contact and internal stresses in a thin subsurface layer, where the cracks initiate. Such analysis becomes a basis for prediction of the surface fracture process (wear) in friction interaction.

The methods and models of fracture mechanics are most commonly used to model the fracture of surface layer in friction process. However, modelling of fracture in tribology has specific features. First, to predict the type of wear, we must know both bulk and surface strength characteristics of materials. Secondly, detachment of one wear particle from the surface does not mean the failure of the junction; the volume of wear particles detached from the surface during the life of junction may be considerable. The surface fracture process changes surface properties (the shape of the surface and its microgeometry, mechanical properties, damage characteristics, etc.). The variable surface properties influence, in turn, the wear process. Some problems of contact fracture mechanics are discussed in monographs by Marchenko (1979), Waterhause (1981), Kolesnikov and Morozov (1989), Hills and Nowell (1994), and Menčik (1996) and in papers by Miller, Keer and Cheng (1985), Hattori et al. (1988), Waterhause (1992), Liu and Farris (1994), Szolwinski and Farris (1996), etc.

The models of delamination of the surface layer and wear particle detachment in friction of rough surfaces are presented in Chapter 5. They are based on the theory of fatigue damage accumulation in cyclic loading.

1.2.6 Wear contact problems

Wear of surfaces leads to the continuous irreversible changes of the surface macroshape in time. Consideration of these changes requires new contact problem formulations and solution methods. All contact characteristics (pressure distribution, shape variation, size and position of contact region, approach of bodies) are unknown functions of time in this case. Calculation of the wear process for different junctions is a necessary condition for design of long-life machines.

The first formulation of the wear contact problem suggested by Pronikov (1957) (see also Pronikov, 1978) did not take into account the deformation of contacting bodies; the contact conditions included only the irreversible surface displacements due to wear.

The contact problem for elastic bodies taking into account the surface shape variations during the wear process was formulated by Korovchinsky (1971). In this work the displacements of the surface due to wear are supposed to be commensurable with the elastic displacements. At any instant of time, the shape of the surface is determined by wear at each point, and simultaneously influences the contact pressure. The wear rate at each point of the contact region at any instant of time is, in turn, a function of the contact pressure at this point. Thus, all functions (pressure distribution, wear and elastic displacements of the surface, etc.) in the wear contact problem are time-dependent and interconnected.

The system of equations governing wear contact problems includes a wear equation which can be found experimentally or can be obtained by modelling the wear process (an example of such model is presented in the Chapter 4).

After the fundamental works by Galin (1976a, 1977, 1980), wear contact problems were intensively investigated in Russia. The methods of solution of the 2-D and 3-D wear contact problems for the contacting bodies of different shape (half-space, strip, beam, parabolic indenter, cylinder, etc.), for various models of deformable bodies (elastic, viscoelastic, etc.) under different loading conditions and a type of motion were presented in Aleksandrov and Kovalenko (1978, 1982), Goryacheva (1979a, 1980, 1987, 1989), Bogatin, Morov and Chersky (1983), Teply (1983), Soldatenkov (1985, 1987), Galakhov and Usov (1990), etc.

Some of these problems are discussed in Chapters 6 – 8 of this book. These chapters include general formulations of the wear contact problems and methods for their solution, the analysis of such particular problems as wear of thin coatings, wear of bodies with variable wear coefficient, wear of discrete contact, etc. Some applications of the methods to the analysis of the wear kinetics of components (plane journal bearing, wheel and rail, abrasive and cutting tools, etc.) are also presented there. The results can be used to predict the lifetime of these components and to optimize the wear process.

The close connection between tribology and contact mechanics has led to new fields in contact mechanics. These fields are the theoretical basis for further investigations in tribology and in the modelling of the phenomena that occur in friction interaction. Some of them are discussed in the chapters that follow.

Chapter 2

Mechanics of Discrete Contact

2.1 Multiple contact problem

2.1.1 Surface macro- and micro- geometry

Contact problems in the classical formulation are posed for topographically smooth surfaces; this ensures that the contact region will be continuous.

In fact, contact between solid surfaces is *discrete* (discontinuous) due to deviations of the surface geometry from the design shape (*macrogeometry*). So a real contact region consists of contact spots, the total area of which (*the real contact area*) is a small fraction of the *nominal contact area* which is the minimal connected region enclosing all the contact spots. The size and arrangement of the contact spots depend on contact interaction conditions (load, kind of motion, etc.), materials, surface macrogeometry and the deviations from it.

These deviations (*asperities*) have various sizes and shapes. Their heights vary within wide limits: from a fraction of a nanometer (for example, the surface deviations of magnetic disks, Majumdar and Bhushan, 1990) to several millimeters. Depending on the scale, they are called macrodeviations, waviness or roughness. For example, *macrodeviations* are characterized by a small height and asperities with gentle slopes; they are caused by an imperfect calibration of an instrument, its wear, etc. *Waviness* is used to describe surface conditions which lie between macrodeviations and roughness. For waviness, the ratio of the distance between asperities (asperity pitch) to the height of an asperity is usually more than 40 (Sviridenok, Chijik and Petrokovets, 1990). *Roughness* is defined as a conglomeration of asperities with a small pitch relative to the base length. It forms *a surface microgeometry* which has a complex statistical character. It is usually a result of the surface treatment. Microgeometry of a surface can also be created artificially to provide the optimal conditions for frictional components to operate. Surfaces with artificial microgeometry are widely applied in devices used for processing and storing information (Sviridenok and Chijik, 1992).

To obtain the complete information on the microshape deviations, various methods of surface topography measurement are used; they may, or may not, involve contact. Devices such as profilometers, optical interferometers, tunnel and atomic-pound microscopes make it possible to describe the microgeometry of a given element of the surface, and to determine its roughness characteristics: the mean height, and the mean curvature of asperities, the number of asperities per unit area of the surface, etc.

Surface deviations from macroshape influence contact characteristics (real pressure distribution, real contact area, etc.) and internal stresses in subsurface layers. To estimate these effects, it is necessary to solve *a multiple contact problem*, that is a boundary problem in the mechanics of solids for a system of contact spots comprising a real contact area.

2.1.2 Problem formulation

We consider a contact interaction of a deformable half-space and a counter- body, the shape of which is described by the function $z = -F(x,y)$ in the system of coordinates connected with the half-space (the plane Oxy coincides with the half-space surface in the undeformed state, and the z-axis is directed into the half-space). After deformation a finite number N or an infinity of contact spots ω_i occur at the surface $z = 0$ of the half-space within the nominal contact region Ω. If $N \to \infty$, the region Ω coincides with the plane $z = 0$.

The real contact pressure $p_i(x,y)$ acts at each contact spot $(x;y) \in \omega_i$. We assume here that tangential stresses are negligibly small. The contact pressure provides the displacement of the half-space surface along the z-axis. This displacement $u_z(x,y)$ depends on the pressure $p_i(x,y)$ applied to all contact spots

$$u_z = A[p_1, p_2, \ldots, p_N]\,. \tag{2.1}$$

The operator A is determined by the model of the deformable bodies in contact. For the contact between a rigid body with a rough surface and an elastic half-space, the relation (2.1) is

$$u_z(x,y) = \frac{1-\nu^2}{\pi E}\sum_{i=1}^{N}\iint\limits_{\omega_i}\frac{p_i(x',y')\,dx'dy'}{\sqrt{(x'-x)^2+(y'-y)^2}}, \tag{2.2}$$

where E and ν are the Young's modulus and Poisson's ratio of the half-space, respectively.

The contact condition must be satisfied within each contact spot ω_i

$$u_z(x,y) = D - F(x,y), \quad (x,y) \in \omega_i, \tag{2.3}$$

where D is the displacement of the rigid body along the z-axis. If D is not given in advance, but the total load P, applied to the bodies and directed along the z-axis is known, we add to Eqs. (2.2) and (2.3) the equilibrium equation

$$\sum_{i=1}^{N} \iint_{\omega_i} p_i(x,y)dxdy = P. \tag{2.4}$$

The system of equations (2.2), (2.3) and (2.4) can be used to determine the real contact pressure $p_i(x,y)$ within the contact spots ω_i. However, the solution of this multiple contact problem is very complicated, even if we know the sizes and the arrangement of contact spots. In the general case we must determine also the number N, and the positions and shapes of the contact spots ω_i for any value of load P. For a differentiable function $F(x,y)$ we can use the condition $p_i(x,y)\Big|_{x,y\,\in\,\partial\omega_i} = 0$ to determine the region ω_i of an individual contact.

2.1.3 Previous studies

The contact problem formulated in § 2.1.2 can be solved numerically. In this case the faithfulness of the stress-strain state so determined depends on the accuracy of the numerical procedure. A computer simulation has been used to solve a contact problem for a rough body and an elastic homogeneous half-space (3-D state) in Seabra and Berthe, (1987) and for coated elastic half-plane (2-D state) in Sainsot, Leroy and Villechase, (1990) and in Cole and Sales, (1991). In these studies a function $F(x,y)$ was obtained experimentally (for example, in the 2-D contact problem, the surface profile was determined by stylus profilometry).

It is worth noting that there is little point in developing the exact solution of the multiple contact problem formulated in § 2.1.2, because the function $F(x,y)$ is usually determined approximately by measurements of some small surface element before deformation. There are basic constraints on the accuracy of measurements of a surface microgeometry by different devices. The function $F(x,y)$ may vary from element to element. In addition, the function $F(x,y)$ can change during contact interaction (for example, in a wear process). Not only do such numerical solutions consume computer time, but they are not universal. A solution for one set of contact characteristics and environment (load, temperature, etc.) cannot be used for another set.

For these reasons, the multiple contact problem for rough surfaces is usually investigated in a simplified formulation. First of all, some model of a real rough surface is considered. The model and the real surface are assumed to be adequate if some chosen characteristics of the real surface coincide with the corresponding characteristics of the model one.

The theory of random functions is widely used to model a rough surface (Sviridenok, Chijik and Petrokovets, 1990). This theory is used to determine the parameters needed to calculate contact characteristics. It was developed by Nayak (1971) for an isotropic surface and by Semenyuk and Sirenko (1980a, b, c), Semenyuk (1986a, b) for anisotropic surfaces.

Fractal geometry seems to be appropriate for rough surface modelling, because of the property of self-similarity of surface microgeometry. Majumdar and Bhushan (1990, 1991) showed experimentally that many rough surfaces have a fractal geometry, and they developed a procedure for determining fractal dimensions of rough surfaces.

It is traditional for tribology to model a rough surface as a system of asperities of a regular shape, the space distribution of which reflects the distribution of material in the surface rough layer. Researchers use various shapes of asperities in their models. A complete list of asperity shapes, with their advantages and disadvantages, is given in Kragelsky, Dobychin and Kombalov (1982). The shape of each asperity is determined by a number of parameters: a sphere by its radius, an ellipsoid by the lengths of its axes. These parameters are calculated from the measurement data of the surface microgeometry. The spacing of the asperities is calculated using the chosen asperity shape and the characteristics of the surface microgeometry obtained from the measurements (Demkin, 1970).

In addition to the approximate description of the surface microgeometry (its roughness), approximate methods of solution of Eqs. (2.1), (2.3) and (2.4) are used to analyse the multiple contact problem. The first investigations into the mechanics of discrete contact did not account for the interaction between contact spots, that is, the stress-strain state of bodies in the vicinity of one contact spot was determined by the load applied to this contact, neglecting the deformation caused by the loads applied to the remaining asperities. Under this assumption the operator A in Eq. (2.1) depends only on the function $p_i(x, y)$, if $(x; y) \in \omega_i$. This assumption gives good agreement between theory and experiment for low contact density, i.e. for low ratio of the real contact area to the nominal one. However, under certain conditions, there are discrepancies between experimental results and predictions. For example, investigating the contact area of elastomers, Bartenev and Lavrentiev (1972) revealed the effect of saturation, that is, the real contact area A_r is always smaller than the nominal contact area A_a, however great a compression load is used. Based on the experimental data, they obtained the following relation

$$\lambda = 1 - \exp\left(-\frac{\beta p}{E}\right), \tag{2.5}$$

where $\lambda = \dfrac{A_r}{A_a}$ is the relative contact area, β is the parameter of roughness, p is a contact pressure, and E is the elasticity modulus of the elastomers. It follows from Eq. (2.5) that $\lambda < 1$ for a finite value of p.

However, if we use the simple theory neglecting the interaction between asperities, we may obtain $\lambda = 1$. For example, it follows from the Hertz solution that for waviness modelled by cylinders of radius R with axes parallel to the half-space surface and spaced at the distance l from each other, $\lambda = 1$ if the load P applied to unit of length of one cylinder is $P = \dfrac{\pi E^* l^2}{16R}$, where

$$E^* = \left(\frac{1-\nu_1^2}{E_1} + \frac{1-\nu_2^2}{E_2}\right)^{-1},$$

E_1, ν_1 and E_2, ν_2 are the moduli of elasticity of the cylinders and the half-space, respectively.

In contact mechanics of rough surfaces, the method of calculation of contact characteristics developed by Greenwood and Williamson (1966) is widely used. They considered a model of a rough surface consisting of a system of spherical asperities of equal radii; the height of an asperity was a random function with some probability distribution. The deformation of each asperity obeyed the Hertz equation. The additional displacement of the surface because of the average (nominal) pressure distribution within the nominal contact area was also taken into account in this model.

For surfaces with regular microgeometry (for example, wavy surfaces) the methods of solution of periodic contact problems can be used to analyze Eqs. (2.2), (2.3) and (2.4). The 2-D periodic contact problem for elastic bodies in the absence of friction was investigated by Westergaard (1939) and Staierman (1949). Kuznetsov and Gorokhovsky (1978a, 1978b, 1980) obtained the solution of a 2-D periodic contact problem with friction force, and analysed the stress-strain state of the surface layer for different parameters characterizing the surface shape. Johnson, Greenwood and Higginson (1985) developed a method of analysis of a multiple contact problem for an elastic body, the surface of which in two mutually perpendicular directions was described by two sinusoidal functions; the counter body had a smooth surface.

We will start the investigation of a multiple contact problem from the analysis of a 3-D periodic contact problem for a system of asperities of regular shape.

2.2 Periodic contact problem

2.2.1 One-level model

We consider a system of identical axisymmetric elastic indenters $\left(z = f(r)\right)$ of the same height (one-level model), interacting with an elastic half-space (Fig. 2.1). The axes of the indenters are perpendicular to the half-space surface $z = 0$ and intersect this surface at points which are distributed uniformly over the plane $z = 0$. As an example of such a system we can consider indenters located at the sites of a quadratic or hexagonal lattice.

Let us fix an arbitrary indenter and locate the origin O of a polar system of coordinates (r, θ) in the plane $z = 0$ at the point of intersection of the axis of this indenter with the plane $z = 0$ (see Fig. 2.1(a)). The tops of the indenters have the coordinates (r_i, θ_{ij}) ($i = 1, 2, \ldots$; $j = 1, 2, \ldots, m_i$, where m_i is the number of indenters located at the circumference of the radius r_i, $r_i < r_{i+1}$).

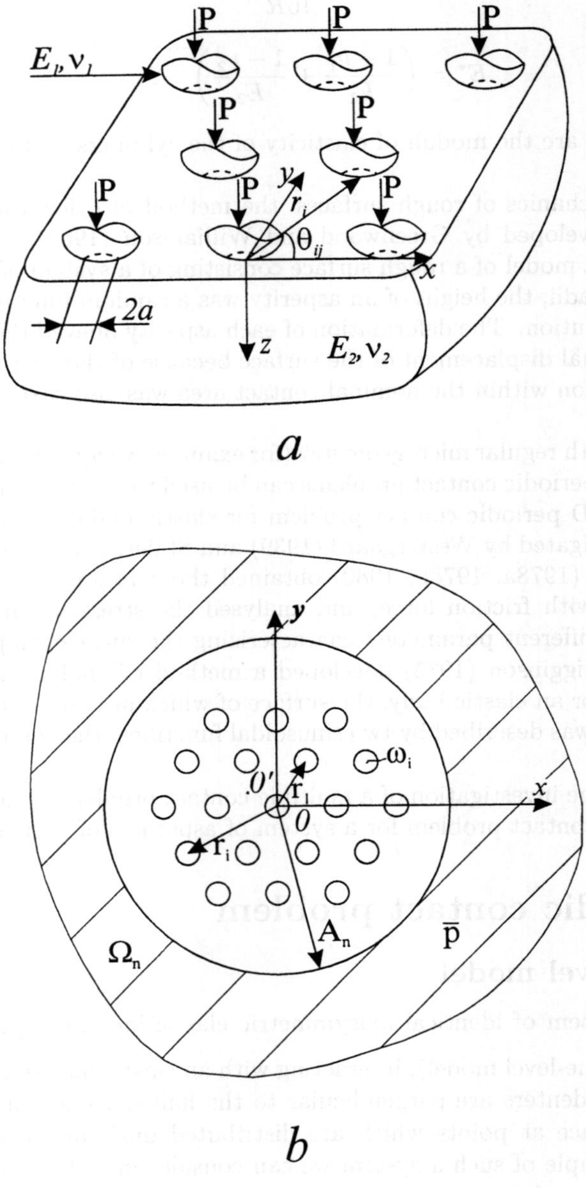

Figure 2.1: Scheme of contact of a periodic system of indenters and an elastic half-space (a) and representation of the contact region based on the principle of localization (b) (the nominal pressure $\bar{p}$ is applied to the shaded region).

Due to the periodicity of the problem, each contact occurs under the same conditions. We assume that contact spots are circles of radius a, and that only normal pressure $p(r, \theta)$ acts at each contact spot ($r \le a$) (the tangential stress is negligibly small). To determine the pressure $p(r, \theta)$ acting at an arbitrary contact spot with a center O, we use the solution of a contact problem for an axisymmetric indenter $\left(z = f(r)\right)$ and an elastic half-space subjected to the pressure $q(r, \theta)$, distributed outside the contact region (Galin, 1953). The contact pressure $p(r, \theta)$ ($r \le a$) is determined by the formula

$$p(r,\theta) = G(r) + \frac{c(\theta)}{\sqrt{a^2 - r^2}} - \frac{1}{\sqrt{a^2 - r^2}} \int_a^{+\infty} \int_0^{2\pi} q(r', \theta') H_2(r, \theta, r', \theta') r' dr' d\theta', \tag{2.6}$$

where

$$G(r) = \frac{E^*}{4\pi^2} \int_0^a \Delta f(r') H_1(r, r')\, dr', \tag{2.7}$$

$$H_1(r, r') = \int_0^{2\pi} \frac{2r'}{\sqrt{r^2 - 2rr'\cos\theta' + r'^2}} \arctan \frac{\sqrt{a^2 - r^2}\sqrt{a^2 - r'^2}}{a\sqrt{r^2 - 2rr'\cos\theta' + r'^2}} d\theta', \tag{2.8}$$

$$H_2(r, \theta, r', \theta') = \frac{\sqrt{r'^2 - a^2}}{\pi^2 \left[r^2 + r'^2 - 2rr'\cos(\theta - \theta')\right]}, \tag{2.9}$$

$$E^* = \left(\frac{1 - \nu_1^2}{E_1} + \frac{1 - \nu_2^2}{E_2}\right)^{-1}. \tag{2.10}$$

Here E_1, ν_1 and E_2, ν_2 are the moduli of elasticity of the indenters and the half-space, respectively. The function $c(\theta)$ depends on a shape of the indenter $f(r)$. For example, if the indenter is smooth (the function $f'(r)$ is continuous at $r = a$), then the contact pressure is zero at $r = a$, i.e. $p(a, \theta) = 0$, and the function $c(\theta)$ has the form

$$c(\theta) = \int_a^{+\infty} \int_0^{2\pi} q(r', \theta') H_2(a, \theta, r', \theta') r' dr' d\theta'. \tag{2.11}$$

The first term in Eq. (2.6) means the pressure that occurs under a single axisymmetric indenter of the shape function $f(r)$ penetrating into an elastic half-space, the last two terms are the additional contact pressure occurring due to the pressure $q(r, t)$ distributed outside the contact region.

For the periodic contact problem the function $q(r, \theta)$ coincides with the pressure $p(r, \theta)$ at each contact spot located at (r_i, θ_{ij}) ($r_i > a$), and is zero outside contact spots. So we obtain the following integral equation from Eq. (2.6), on the assumption that $f'(r)$ is a continuous function ($p(a,\theta)=0$):

$$p(r,\theta)-\int_0^a\int_0^{2\pi} K(r,\theta,r',\theta')\,p(r',\theta')r'dr'd\theta' = G(r), \tag{2.12}$$

where

$$K(r,\theta,r',\theta')=\sum_{i=1}^{\infty} K_i(r,\theta,r',\theta'), \tag{2.13}$$

$$K_i(r,\theta,r',\theta')=\frac{1}{\pi^2\sqrt{a^2-r^2}}\sum_{j=1}^{m_i}\left[K_{ij}(a,\theta,r',\theta')-K_{ij}(r,\theta,r',\theta')\right], \tag{2.14}$$

$$K_{ij}(r,\theta,r',\theta')=$$

$$\frac{\sqrt{r_i^2+r'^2+2r_ir'\cos(\theta_{ij}-\theta')-a^2}}{(r\cos\theta-r'\cos\theta'-r_i\cos\theta_{ij})^2+(r\sin\theta-r'\sin\theta'-r_i\sin\theta_{ij})^2}. \tag{2.15}$$

It is worth noting that similar reasoning can be used to obtain the integral equation for the system of punches with a given contact region (for example, cylindrical punches with a flat base); the equation will have the same structure as Eq. (2.12).

The kernel $K(r,\theta,r',\theta')$ of Eq. (2.12) is represented as a series (2.13). A general term (2.14) of this series can be transformed to the form:

$$K_i(r,\theta,r',\theta')=$$

$$\frac{1}{\pi^2\sqrt{a^2-r^2}}\sum_{j=1}^{m_i}\left\{\frac{2(a-r)\cos(\theta_{ij}-\theta)}{r_i^2}+\right. \tag{2.16}$$

$$\left.\frac{(a-r)\left[-a-r-6r'\cos(\theta_{ij}-\theta')\cos(\theta_{ij}-\theta)+2r'\cos(\theta'-\theta)\right]}{r_i^3}+O\left(\frac{1}{r_i^4}\right)\right\}.$$

We assume that for the periodic system of indenters under consideration, each contact spot with center $(r_i;\theta_{ij})$ has a partner with center at the point $(r_i;\pi+\theta_{ij})$. So the sum on the first line of Eq. (2.16) is zero. Hence, the general term of the series (2.13) has order $O\left(\frac{1}{r_i^2}\right)$, since $m_i\sim r_i$, and the series converges.

2.2.2 Principle of localization

In parallel with Eq. (2.12) we consider the following equation

$$p(r,\theta)-\int_0^a\int_0^{2\pi}\sum_{i=1}^{n} K_i(r,\theta,r',\theta')\,p(r',\theta')r'dr'd\theta' = \tag{2.17}$$

$$G(r)+\frac{2\bar{N}P}{\pi}\arctan\frac{\sqrt{a^2-r^2}}{\sqrt{A_n^2-a^2}},$$

where P is a load applied to each contact spot. This load satisfies the equilibrium equation

$$P = \int_0^a \int_0^{2\pi} p(r,\varphi) r \, dr d\varphi. \tag{2.18}$$

To obtain Eq. (2.17) we substitute integration over region Ω_n ($\Omega_n : r \geq A_n, 0 \leq \theta \leq 2\pi$) for summation over $i > n$ in Eq. (2.13), taking into account that the centers of contact spots are distributed uniformly over the plane $z = 0$ and their number per unit area is characterized by the value $\bar{N}$. Actually, the following transformation demonstrates the derivation Eq. (2.17)

$$J_n = \sum_{i=n+1}^{\infty} K_i(r,\theta,r',\theta') \approx$$

$$\bar{N} \int_{A_n}^{+\infty} \int_0^{2\pi} \frac{\sqrt{x^2 + r'^2 + 2xr'\cos(\phi - \theta') - a^2}}{\pi^2\sqrt{a^2 - r^2}} \times$$

$$\left[\frac{1}{(a\cos\theta - r'\cos\theta' - x\cos\phi)^2 + (a\sin\theta - r'\sin\theta' - x\sin\phi)^2} - \right.$$

$$\left. \frac{1}{(r\cos\theta - r'\cos\theta' - x\cos\phi)^2 + (r\sin\theta - r'\sin\theta' - x\sin\phi)^2} \right] x \, dx d\phi.$$

Changing the variables $y\cos\varphi = x\cos\phi + r'\cos\theta'$, $y\sin\varphi = x\sin\phi + r'\sin\theta'$ and taking into account that $r' \leq a \ll A_n$, we finally obtain

$$J_n \approx \frac{\bar{N}}{\pi^2\sqrt{a^2 - r^2}} \times$$

$$\int_{A_n}^{+\infty} \int_0^{2\pi} \sqrt{y^2 - a^2} \left[\frac{1}{a^2 + y^2 - 2ay\cos\varphi} - \frac{1}{r^2 + y^2 - 2ry\cos\varphi} \right] y \, dy d\varphi =$$

$$\frac{2\bar{N}}{\pi} \arctan \frac{\sqrt{a^2 - r^2}}{\sqrt{A_n^2 - a^2}},$$

where A_n is the radius of a circle in which there are $\sum_{i=1}^{n} m_i + 1$ central indenters. It is apparent that

$$A_n^2 = \frac{1}{\pi\bar{N}} \left(\sum_{i=1}^{n} m_i + 1 \right). \tag{2.19}$$

We note that the solution of Eq. (2.17) tends to the solution of Eq. (2.12) if $n \to \infty$.

Let us analyze the structure of Eq. (2.17). The integral term on the left side of Eq. (2.17) governs the influence of the real pressure distribution at the neighboring contact spots ($r_i < A_n$), on the pressure at the fixed contact spot with center $(0,0)$ (local effect). The effect of the pressure distribution at the remaining contact spots which have centers (r_i, θ_{ij}), $r_i > A_n$, is taken into account by the second term in the right side of Eq. (2.17). This term describes the additional pressure $p_a(r)$ which arises within a contact spot ($r < a$) from the nominal pressure $\bar{p} = P\bar{N}$ in the region Ω_n ($r > A_n$). Indeed, from Eqs. (2.6) and (2.11) it follows that the additional pressure $p_a(r)$ within the contact spot ($r \leq a$) arising from the pressure $q(r,\theta) = \bar{p}$ distributed uniformly in the region Ω_n has the form

$$p_a(r) = \frac{\bar{p}}{\pi^2\sqrt{a^2 - r^2}} \times$$

$$\int_{A_n}^{+\infty}\int_0^{2\pi} \sqrt{r'^2 - a^2}\left[\frac{1}{a^2 + r'^2 - 2ar'cos\theta} - \frac{1}{r^2 + r'^2 - 2rr'cos\theta}\right] r'dr'd\theta =$$

$$\frac{2\bar{p}}{\pi} \arctan \frac{\sqrt{a^2 - r^2}}{\sqrt{A_n^2 - a^2}}.$$

Thus, the effect of the real contact pressure distribution over the contact spots ω_i far away from the contact spot under consideration ($\omega_i \in \Omega_n$) can be taken into account to sufficient accuracy by the nominal pressure $\bar{p}$ distributed over the region Ω_n (Fig. 2.1(b)).

This conclusion stated for the periodic contact problem is a particular case of a general contention which we call *a principle of localization*: in conditions of multiple contact, the stress-strain state near one contact spot can be calculated to sufficient accuracy by taking into account the real contact conditions (real pressure, shape of bodies, etc.) at this contact spot and at the nearby contact spots (in the local vicinity of the fixed contact), and the averaged (nominal) pressure over the remaining part of the region of interaction (nominal contact region). This principle will be supported by results of investigation of some particular problems considered in this chapter.

Eqs. (2.17) and (2.18) are used to determine the contact pressure $p(r,\theta)$ and the radius a of each spot. The stress distribution in the subsurface region ($z > 0$) arising from the real contact pressure distribution at the surface $z = 0$ can then be found by superposition, using the potentials of Boussinesq (1885) or the particular solution of the axisymmetric problem given by Timoshenko and Goodier (1951).

To simplify the procedure, we can use the principle of localization for determination of internal stresses, substituting the real contact pressure at distant contact spots by the nominal contact pressure. We give here the analytical expressions for the additional stresses which occur on the axis of symmetry of any fixed contact spot from the action of the nominal pressure $\bar{p}$ within the region $\Omega_n(r > A_n)$.

$$\sigma_z = -\frac{\bar{p}z^3}{(A_n^2 + z^2)^{3/2}},$$

$$\sigma_r = \sigma_\theta = \frac{\bar{p}z}{\sqrt{A_n^2 + z^2}}\left[\frac{z^2}{2\,(A_n^2 + z^2)} - (1+\nu)\right], \tag{2.20}$$

$$\tau_{rz} = \tau_{\theta z} = \tau_{r\theta} = 0.$$

2.2.3 System of indenters of various heights

The method described above is used to determine the real pressure distribution in contact interaction between a periodic system of elastic indenters of the various heights, and an elastic half-space. We assume that the shape of an indenter is described by a continuously differentiable function $z = f_m(r) + h_m$, where h_m is a height of indenters of a given level m ($m = 1, 2, \ldots, k$), k is the number of levels. An example of positions of indenters of each level for $k = 3$ for a hexagonal lattice is shown in Fig. 2.2(a). We assume also that the contact spot of the m-th level is a circle of radius a_m.

Let us fix any indenter of the m-th level and place the origin of the polar system of coordinates at the center of its contact spot (Fig. 2.2(b)). Using the principle of localization, we take into account the real pressure $p_j(r,\theta)$ ($j = 1, 2, \ldots, k$) at the contact spots which are inside the region $\bar{\Omega}_m$ which is a circle of radius A_m ($\bar{\Omega}_m : r \le A_m$):

$$A_m^2 = \frac{1}{\pi}\left(\sum_{j=1}^{k}\frac{k_{jm}}{\bar{N}_j} + \frac{1}{\bar{N}_m}\right),$$

where k_{jm} is the number of indenters of the j-th level inside the region $\bar{\Omega}_m$, $\bar{N}_j$ is the density of indenters of the j-th level, which is the number of indenters at the j-th level for the unit area. It must be noted that the number of indenters of the m-th level ($j = m$) inside the region $\bar{\Omega}_m$ is $k_{mm} + 1$. Replacing the real contact pressure at the removed contact spots ($r_i > A_m$) by the nominal pressure $\bar{p}$ acting within the region ($r > A_m$)

$$\bar{p} = \sum_{j=1}^{k}\bar{N}_j\int_0^{a_j}\int_0^{2\pi} p_j(r',\theta')r'dr'd\theta',$$

we obtain the following relationship similar to Eq. (2.17)

$$p_m(r,\theta) - \sum_{j=1}^{k}\int_0^{a_j}\int_0^{2\pi} K_n\left(a_m, r, \theta, r', \theta'\right)p_j(r',\theta')r'dr'd\theta' = \\ G_m(r) + \frac{2\bar{p}}{\pi}\arctan\frac{\sqrt{a_m^2 - r^2}}{\sqrt{A_m^2 - a_m^2}}. \tag{2.21}$$

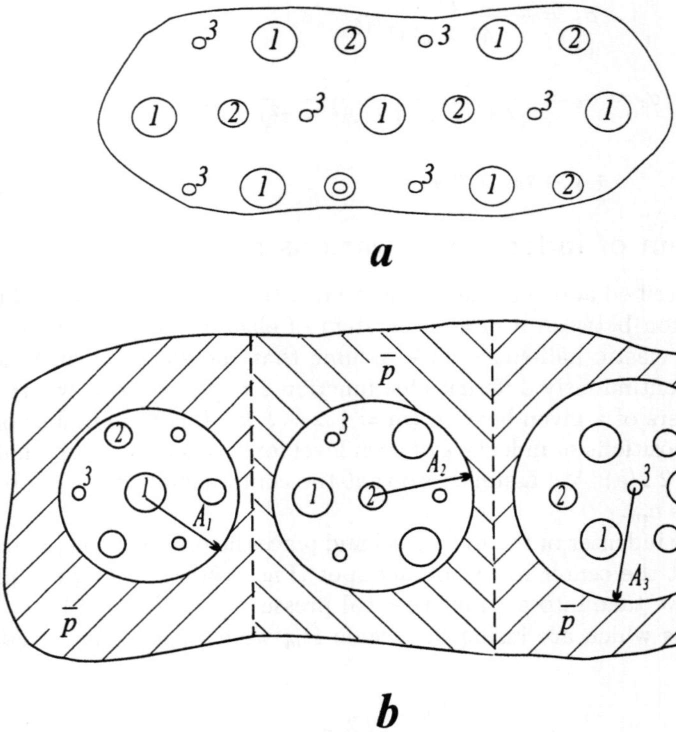

Figure 2.2: The location of indenters of each level in the model ($k = 3$) (a) and scheme of calculations based on Eqs. (2.21)–(2.23) for $n = 1$ (b).

The kernel of Eq. (2.21) has the form

$$K_n(a_m, r, \theta, r', \theta') = \sum_{i=1}^{n} K_i(a_m, r, \theta, r', \theta'),$$

where functions $K_i(a_m, r, \theta, r', \theta')$ are determined by Eqs. (2.14) and (2.15), in which we must put $a = a_m$. The function $G_m(r)$ is determined by Eq. (2.7), where $a = a_m$ and $f(r) = f_m(r)$.

Repeating the same procedure for indenters of each level (see Fig. 2.2(b)), we obtain the system of k integral equations (2.21) ($m = 1, 2, \ldots, k$) for determination of the pressure $p_m(r, \theta)$ within the contact spot ($r \leq a_m$) of each level.

Usually the radius of a contact spot a_m is unknown. If an origin of a polar system of coordinates is placed in the center O_m of the m-th level contact spot, we can write

$$h_m = \frac{1}{\pi E^*}\left[\int_0^{a_m}\int_0^{2\pi} p_m(r,\theta)drd\theta + 2\pi\bar{p}(A_\infty - A_m) + \right.$$
$$\left.\sum_{j=1}^{k}\sum_{i=1}^{k_{jm}}\int_0^{a_j}\int_0^{2\pi}\frac{p_j(r,\theta)rdrd\theta}{\sqrt{r^2 - 2rr_{ij}^{(m)}\cos\left(\theta-\theta_{ij}^{(m)}\right) + r_{ij}^{(m)2}}}\right], \tag{2.22}$$

where $r_{ij}^{(m)}, \theta_{ij}^{(m)}$ are the coordinates with respect to the system $(O_m r\theta)$ of the centers of contact spots located within the region Ω_m $(a_m < r_{ij}^{(m)} < A_m,\ 0 < \theta_{ij}^{(m)} < 2\pi)$, A_∞ is a constant which can be excluded from the system of Eqs. (2.22) by consideration of differences of heights $h_1 - h_m$, where h_1 is the largest height. The system of equations is completed if we add the equilibrium condition

$$\bar{p}\pi A_m^2 = \sum_{j=1}^{k} k_{jm}\int_0^{a_j}\int_0^{2\pi} p_j(r,\theta)rdrd\theta + \int_0^{a_m}\int_0^{2\pi} p_m(r,\theta)rdrd\theta. \tag{2.23}$$

It should be remarked that for given height distribution h_m all indenters enter into contact only if the nominal pressure reaches the definite value $\bar{p}^*$. For $\bar{p} < \bar{p}^*$ there are less than k levels of indenters in contact.

2.2.4 Stress field analysis

We use the relationships obtained in § 2.2.1–2.2.3 to analyze a real contact pressure distribution and the internal stresses in a periodic contact problem for a system of indenters and the elastic half-space. Particular emphasis will be placed upon the influence of the geometric parameter which describes the density of indenter location, on the stress-strain state. This will allow us to determine the range of parameter variations in which it is possible to use the simplified theories which neglect the interaction between contact spots (the integral term in Eq. (2.12)) or the local effect of the influence of the real pressure distribution at the neighboring contact spots on the pressure at the fixed spot (the integral term in Eqs. (2.17)).

Numerical results are presented here for a system of spherical indenters, $\left(f(r) = \dfrac{r^2}{2R},\ R \text{ is a radius of curvature}\right)$, located on a hexagonal lattice with a constant pitch l. Fig. 2.2(a) shows the location of indenters of different levels at the plane $z = 0$ for a three-level model ($k = 3$). We introduce the following dimensionless parameters and functions

$$\rho = \frac{r}{R},\quad A_n^1 = \frac{A_n}{R},\quad a^1 = \frac{a}{R},\quad l^1 = \frac{l}{R},$$
$$p^1(\rho,\theta) = \frac{\pi p(\rho R,\theta)}{2E^*},\quad P^1 = \frac{\pi P}{2E^* R^2}. \tag{2.24}$$

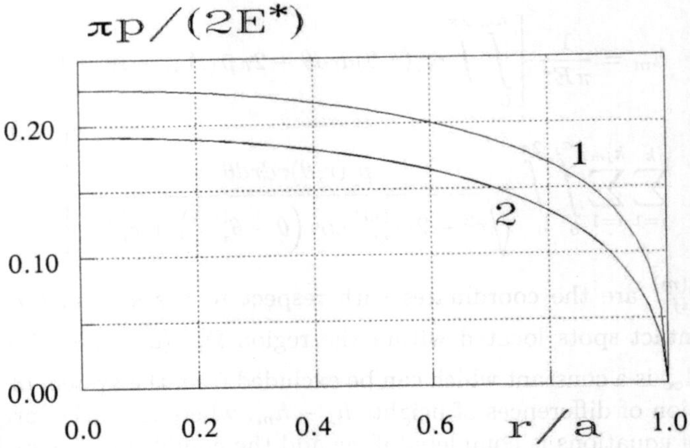

Figure 2.3: Pressure distribution within a contact spot, calculated from Eq. (2.17) for $n = 0$ (curve 1), $n = 1$ and $n = 2$ (curve 2) and $a/R = 0.1$, $l/R = 0.2$ (one-level model).

The systems of Eqs. (2.17) and (2.18) for the one-level model and of Eqs. (2.21)-(2.23) for the three-level model are solved by iteration. The density $\bar{N}_j$ of arrangement of indenters in the three-level model under consideration is determined by the formula

$$\bar{N}_j = \frac{2}{3l^2\sqrt{3}}, \quad (j = 1, 2, 3). \tag{2.25}$$

For the one-level model $\bar{N} = 3\bar{N}_j = \dfrac{2}{l^2\sqrt{3}}$.

For determination of the radius A_n of the circle ($r \le A_n$) where the real pressure distribution within a nearby contact spots is taken into account (local effect) and the corresponding value of n which gives an appropriate accuracy of the solution of Eq. (2.17), we calculated the contact pressure $p^1(\rho, \theta)$ from Eqs. (2.17) and (2.18) for $n = 0$, $n = 1$, $n = 2$ and so on. For $n = 0$, the integral term on the left of Eq. (2.17) is zero, so that the effect of the remaining contact spots surrounding the fixed one (with the center at the origin of coordinate system O) is taken into account by a nominal pressure distributed outside the circle of radius A_0 (the second term in the right side of Eq. (2.17)), where A_0 is determined by Eq. (2.19). For $n = 1$ we take into account the real pressure within 6 contact spots located at the distance l from the fixed one, for $n = 2$ they are 12 contact spots, six located at the distance l and the another six at a distance $l\sqrt{3}$, and so on. Fig. 2.3 illustrates the results calculated for $a^1 = 0.1$ and $l^1 = 0.2$, i.e. $\dfrac{a}{l} = 0.5$, this case corresponds to the limiting value of contact density. The results show that the contact pressure calculated for $n = 1$ and $n = 2$ differ from one another

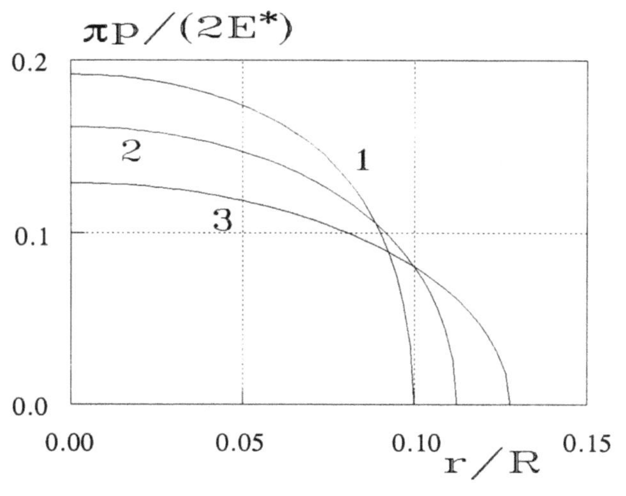

Figure 2.4: Pressure distribution under an indenter acted on by the force $P^1 = 0.0044$ for the one-level model characterized by the various distances between indenters: $l/R = 0.2$ (curve 1), $l/R = 0.25$ (curve 2), $l/R = 1$ (curve 3).

less than 0.1%. If contact density decreases $\left(\frac{a}{l} \text{ decreases}\right)$ this difference also decreases. Based on this estimation, we will take $n = 1$ in subsequent analysis.

We first analyze the effect of interaction between contact spots and pressure distribution. Fig. 2.4 illustrates the contact pressure under some indenter of the one-level system for different values of the parameter l^1 characterizing the distance between indenters. In all cases, the normal load $P^1 = 0.0044$ is applied to each indenter. The results show that the radius of the contact spot decreases and the maximum contact pressure increases if the distance l between indenters decreases; the contact density characterized by the parameter $\frac{a}{l}$ also increases $\left(\frac{a}{l} = 0.128\right.$ (curve 3), $\frac{a}{l} = 0.45$ (curve 2), $\frac{a}{l} = 0.5$ (curve 1)$\left.\right)$. The curve 3 practically coincides with the contact pressure distribution calculated from Hertz theory which neglects the influence of contact spots surrounding the fixed one. So, for small values of parameter $\frac{a}{l}$, it is possible to neglect the interaction between contact spots for determination of the contact pressure.

The dependencies of the radius of a contact spot on the dimensionless nominal pressure $\bar{p}^1 = \frac{\bar{p}\pi}{2E^*}$ calculated for different values of parameter l^1 and a one-level model are shown in Fig. 2.5 (curves 1, 2, 3). The results of calculation based on the Hertz theory are added for comparison (curves 1′, 2′, 3′). The results show that under a constant nominal pressure $\bar{p}$ the radius of each contact spot and, hence

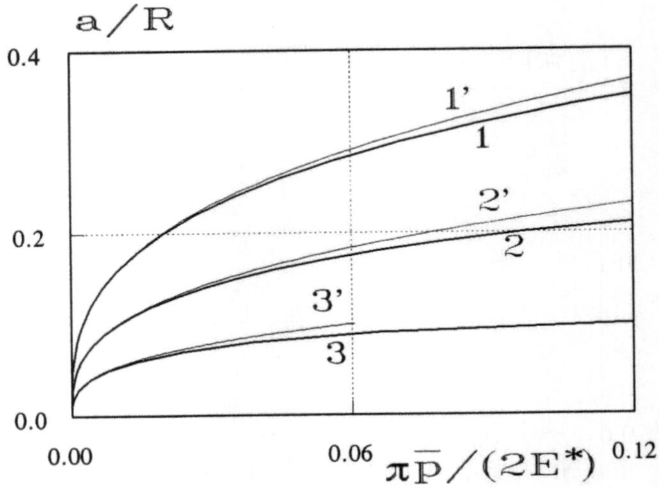

Figure 2.5: Dependence of the radius of a contact spot on the nominal pressure for $l = 1$ (curves 1, 1′), $l = 0.5$ (curves 2, 2′), $l = 0.2$ (curves 3, 3′), calculated from Eq. (2.17) (1, 2, 3) and from Hertz theory (1′, 2′, 3′).

the real contact area, decreases if the relative distance $\frac{l}{R}$ between contact spots decreases. The comparison of these results with the curves calculated from Hertz theory makes it possible to conclude that for $\frac{a}{l} < 0.25$ the discrepancy between the results predicted from the multiple contact theory and Hertz theory does not exceed 2.5%. For higher nominal pressure and, hence higher contact density, the discrepancy becomes serious. Thus, for $l = 0.5$ (curves 2, 2′) and $\frac{a}{l} = 0.44$ the calculation of the real contact area from Hertz theory gives an error of about 15%.

Investigation of contact characteristics in the three-level model is a subject of particular interest because this model is closer to the real contact situation than is the one-level model. The multiple contact model developed in this section takes into account the influence of the density of contact spots on the displacement of the surface between contact spots, and so the load, which must be applied to bring a new level of indenters into contact, depends not only on the height difference of the indenters, but also on the contact density. The calculations were made for a model with fixed height distribution: $\dfrac{h_1 - h_2}{R} = 0.014$ and $\dfrac{h_1 - h_3}{R} = 0.037$. Fig. 2.6 illustrates the pressure distribution within the contact spots for each level if $P^1 = 0.059$ where P^1 is the load applied to 3 indenters ($P^1 = P_1^1 + P_2^1 + P_3^1$). The curves 1, 2, 3 and the curves 1′, 2′, 3′ correspond to the solutions of the periodic contact problem and to the Hertz problem, respectively. The results show that the smaller the height of the indenter, the greater is the difference between the contact pressure calculated from the multiple contact and Hertz theory.

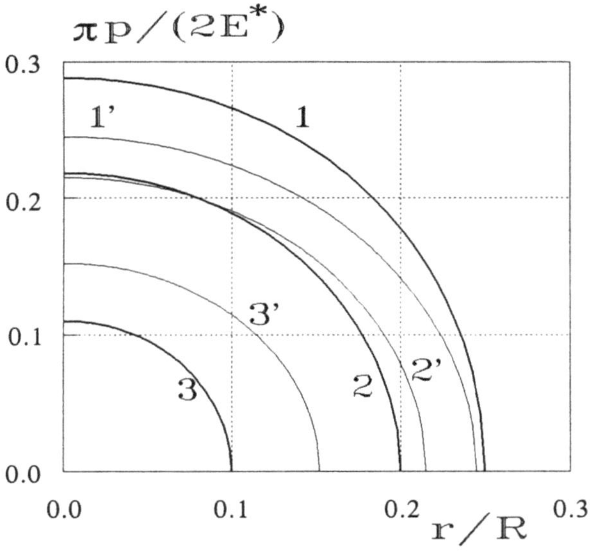

Figure 2.6: Pressure distribution at the contact spots of indenters with the heights h_1 (curves 1, 1′), h_2 (curves 2, 2′) and h_3 (curves 3, 3′) for the three-level model ($(h_1 - h_2)/R = 0.014$, $(h_1 - h_3)/R = 0.037$, $P^1 = 0.059$) calculated from Eqs. (2.21) - (2.23) (1, 2, 3) and from Hertz theory (1′, 2′, 3′).

We also investigated the internal stresses for the one-level periodic problem and compared them with the uniform stress field arising from the uniform loading by the nominal pressure $\bar{p}_n$. It follows from the analysis that for periodic loading by the system of indenters, there is a nonuniform stress field in the subsurface layer, the thickness of which is comparable with the distance l between indenters. The stress field features depend essentially on the contact density parameter $\frac{a}{l}$. Fig. 2.7 illustrates the principal shear stress $\frac{\tau_1}{\bar{p}}$ along the z-axis which coincides with the axis of symmetry of the indenter (curves 1, 2) and along the axis $O'z$ (curves 1′, 2′) equally spaced from the centers of the contact spots (see Fig. 2.1). The results are calculated for the same nominal pressure $\bar{p}^1 = 0.12$, and the different distances $\frac{l}{R}$ between the indenters: $\frac{l}{R} = 1$, $\left(\frac{a}{R} = 0.35\right)$ (curves 1, 1′) and $\frac{l}{R} = 0.5$ $\left(\frac{a}{R} = 0.21\right)$ (curves 2, 2′). The maximum value of the principal shear stress is related to the nominal pressure; the maximum difference of the principal shear stress at the fixed depth decreases as the parameter $\frac{a}{l}$ increases. The maximum value of the principal shear stress occurs at the point $r = 0$, $\frac{z}{a} = 0.43$

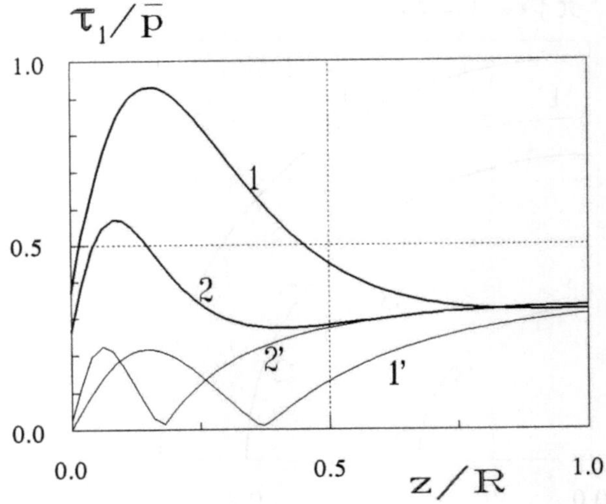

Figure 2.7: The principal shear stress $\tau_1/\bar{p}$ along the axes Oz (curves 1, 2) and $O'z$ (curves $1'$, $2'$) for $l/R = 1$ (1, 1'), $l/R = 0.5$ (2, 2'), $\bar{p}^1 = 0.12$.

for $\frac{a}{l} = 0.35$ (curve 1) and at the point $r = 0$, $\frac{z}{a} = 0.38$ for $\frac{a}{l} = 0.42$ (curve 2). At infinity the principal shear stresses depend only on the nominal stress $\bar{p}$. The results show that internal stresses differ noticeably from ones calculated from the Hertz model if the parameter $\frac{a}{l}$ varies between the limits $0.25 < \frac{a}{l} \leq 0.5$.

Fig. 2.8 illustrates contours of the function $\frac{\tau_1}{\bar{p}}$ at the plane $\frac{z}{R} = 0.08$, which is parallel to the plane Oxy. The principal shear stresses are close to the maximum values at the point $x = 0$, $y = 0$ of this plane. Contours are presented within the region $\left(-\frac{l^1}{2} < x < l^1,\ -\frac{l^1\sqrt{3}}{4} < y < \frac{l^1\sqrt{3}}{2}\right)$ for $a^1 = 0.2$ and $l^1 = 1$ (Fig. 2.8(a)) and $l^1 = 0.44$ (Fig. 2.8(b)). The results show that the principal shear stress at the fixed depth varies only slightly if the contact density parameter is close to 0.5. Similar conclusions follow for all the components of the stress tensor.

Thus, as a result of the nonuniform pressure distribution at the surface of the half-space (discrete contact), there is a nonuniform stress field dependent on the contact density parameter in the subsurface layer. The increase of stresses in some points of the layer may cause plastic flow or crack formation. The results obtained here coincide with the conclusions which follow from the analysis of the periodic contact problem for the sinusoidal punch and an elastic half-plane (2-D contact problem) in Kuznetsov and Gorokhovsky (1978a, 1978b).

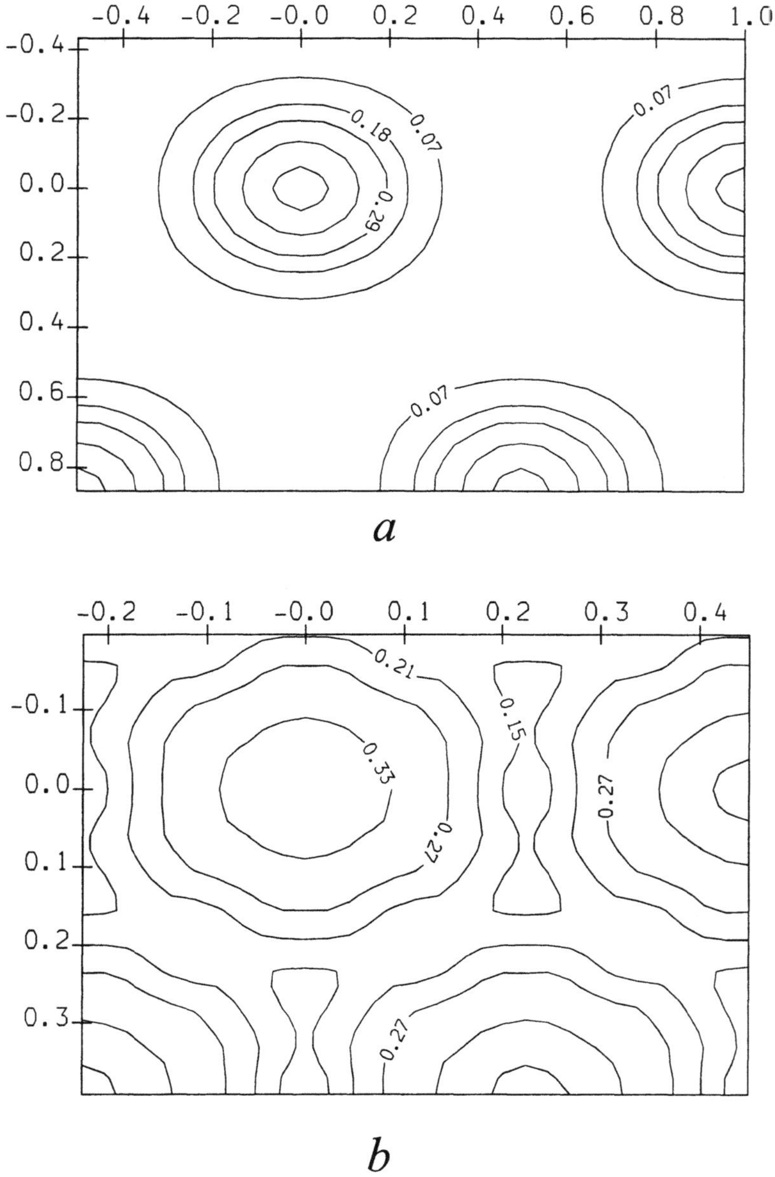

Figure 2.8: Contours of the function $\tau_1/\bar{p}$ at the plane $z/R = 0.08$ for $l^1 = 1$ (a) and $l^1 = 0.44$ (b); $a^1 = 0.2$.

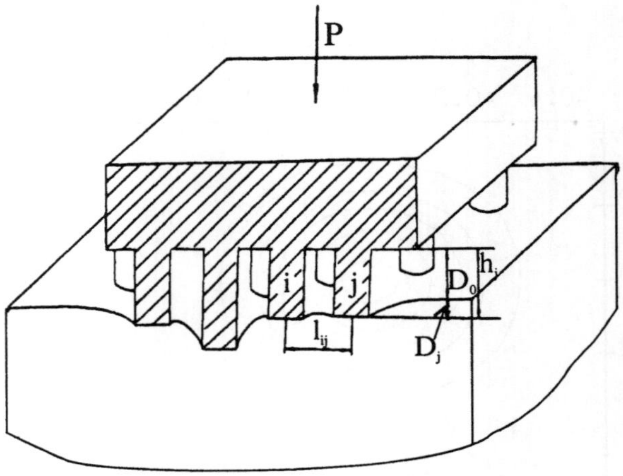

Figure 2.9: Scheme of contact of a system of punches and an elastic half-space.

2.3 Problem with a bounded nominal contact region

A distinctive feature of periodic contact problems is the uniform distribution of the nominal pressure on the half-space surface. The nominal pressure is the ratio of the load to the area, for one cell. Within one period, the load distribution between contact spots depends only on the difference of heights of indenters and variations in contact density.

For a finite number of indenters interacting with an elastic half-space, the nominal contact region is bounded. A nonuniform load distribution between indenters which are rigidly bonded, arises not only from the differences in indenter height and their arrangement density, but also from the different locations of the indenters within the nominal contact region. The load distribution for such a system of indenters is nonuniform even though all indenters have the same height and they are arranged uniformly within a bounded nominal contact region.

In what follows we will investigate the contact problem for a finite number of punches and an elastic half-space, and analyze the dependence of the contact characteristics (load distribution, real contact area, etc.) on the spatial arrangement of the punches.

2.3.1 Problem formulation

We consider the contact interaction of a system of punches with an elastic half-space (Fig. 2.9). The system of punches is characterized by:

- the total number N;

- the shape of the contact surface of an individual punch $f_j(r)$ (it is assumed that each punch is a body of revolution with its axis perpendicular to the undeformed surface of the half-space, and r is the polar radius for the coordinate system related to the axis of the punch);

- the distance l_{ij} between the axes of symmetry of the i-th and j-th punches;

- the heights of punches h_j.

The region of contact of the system of punches with the elastic half-space is a set of subregions ω_i $(i = 1, 2, \ldots, N)$. The remaining boundary of the half-space is stress free.

We introduce the coordinate system $Oxyz$. The Oz-axis is chosen to coincide with the axis of revolution of an arbitrary fixed i-th punch and the Oxy plane coincides with the undeformed half-space surface. For convenience, the directions of the axes Ox and Oy are chosen to coincide where possible with axes of symmetry of the system of punches.

Let us formulate the boundary conditions for the i-th punch and replace the action of the other punches on the boundary of the elastic half-space by the corresponding pressure, distributed over the aggregate region $\bigcup_{\substack{j=1\\ j\neq i}}^{N} \omega_j$. The elastic displacement of the half-space surface in the z-axis direction within the region ω_i caused by the pressure $p_j(x, y)$, $(x, y) \in \omega_j$, $(j = 1, 2, \ldots, N,\ i \neq j)$ is calculated from Boussinesq's solution

$$u_z^i(x, y) = \frac{1-\nu^2}{\pi E} \sum_{\substack{j=1\\ j\neq i}}^{N} \iint_{\omega_j} \frac{p_j(x', y')\, dx'dy'}{\sqrt{(X_j + x' - x)^2 + (Y_j + y' - y)^2}}.$$

Generally speaking, the pressure $p_j(x, y)$ is not known in advance. To simplify the problem, we approximate $u_z^i(x, y)$ by the following function

$$u_z^{i1}(x, y) = \frac{1-\nu^2}{\pi E} \sum_{\substack{j=1\\ j\neq i}}^{N} \frac{P_j}{\sqrt{(X_j - x)^2 + (Y_j - y)^2}}, \tag{2.26}$$

where P_j is the concentrated force, $P_j = \iint_{\omega_j} p_j(x, y)\, dxdy$, which is applied at the center of the subregion with coordinates (X_j, Y_j). The high accuracy of this approximation follows from the estimation made for the particular case of the axially symmetric function $p_j(x', y') = p(r)$, $(r \leq a)$

$$\int_0^{2\pi}\int_0^{a} \frac{p(r) r\, dr d\theta}{\sqrt{r^2 + l^2 - 2rl\cos\theta}} - \frac{P}{l} =$$

$$\int_0^a p(r)\left[\frac{4}{r+l}K\left(\frac{2\sqrt{rl}}{r+l}\right) - \frac{2\pi}{l}\right] r\, dr = 4\int_0^a p(r)\frac{r}{l}\left[K\left(\frac{r}{l}\right) - \frac{\pi}{2}\right] dr =$$

$$\frac{\pi}{2}\int_0^a p(r)\left[\left(\frac{r}{l}\right)^3 + O\left(\left(\frac{r}{l}\right)^5\right)\right] dr \leq \frac{P}{4}\left(\frac{a^2}{l^3} + O\left(\frac{a^4}{l^5}\right)\right),$$

where $l = \sqrt{(X_j - x)^2 + (Y_j - y)^2}$, $P = 2\pi \int_0^a p(r) r\, dr$, $K(x)$ is the elliptic integral of the first kind. The following relations have been used to obtain this estimation

$$\int_0^{\pi} \frac{d\theta}{\sqrt{r^2 + l^2 - 2rl\cos\theta}} = \frac{2}{r+l}K\left(\frac{2\sqrt{rl}}{r+l}\right), \tag{2.27}$$

$$K\left(\frac{2\sqrt{x}}{1+x}\right) = (1+x)K(x). \tag{2.28}$$

The superposition principle, which is valid for the linear theory of elasticity, makes it possible to present the displacements of the boundary of the elastic half-space along the axis Oz under the i-th punch, as the sum of the displacement $u_z^{i1}(x, y)$ and the elastic displacement $u_z^{i2}(x, y)$ due to the pressure $p_i(x, y)$ distributed over the i-th punch base within the subregion ω_i.

As a result, the pressure $p_i(x, y)$ can be determined from the solution of the problem of the elasticity theory for the half-space with the mixed boundary conditions

$$\begin{aligned} u_z^{i1}(x,y) + u_z^{i2}(x,y) &= D_i - f_i\left(\sqrt{x^2+y^2}\right), \\ \tau_{zx} = \tau_{zy} &= 0, \quad (x,y) \in \omega_i, \\ \sigma_z = \tau_{zx} = \tau_{zy} &= 0, \quad (x,y) \notin \omega_i, \end{aligned} \tag{2.29}$$

where D_i is the displacement of the punch along the z-axis.

For further consideration it is necessary to determine the relation between the loads P_i, acting upon the punches, and the depths of penetration of punches D_i. We use Betti's theorem to obtain this relation. We assume that the contact region ω_i of an axially symmetric punch with the curved surface of the elastic half-space is close to a circular one of radius a_i. For an axisymmetric punch with a flat base of radius a_i, penetrating into the half-space to a depth D^*, the pressure $p^*(r)$ $\left(r = \sqrt{x^2 + y^2}\right)$ is determined by the formula (see, for example, Galin, 1953 or

Gladwell, 1980)

$$p^*(r) = \frac{ED^*}{\pi(1-\nu^2)\sqrt{a_i^2 - r^2}}.$$

From Betti's theorem it follows that

$$\int_0^{2\pi}\int_0^{a_i} p_i(r,\theta) D^* r\,dr d\theta = \int_0^{2\pi}\int_0^{a_i} p^*(r) u_z^{i2}(r,\theta) r\,dr d\theta \quad \left(\theta = \arctan\frac{y}{x}\right)$$

or

$$P_i = \frac{E}{\pi(1-\nu^2)} \int_0^{2\pi}\int_0^{a_i} \frac{u_z^{i2}(r,\theta) r\,dr d\theta}{\sqrt{a_i^2 - r^2}}. \tag{2.30}$$

Substituting Eqs. (2.26) and (2.29) in the right-hand side of Eq. (2.30) we calculate the integrals using Eqs. (2.27), (2.28) and the following relations

$$\int_0^{x^2} \frac{dx'}{\sqrt{(x^2 - x')(1 - x'\sin^2\phi)}} = \frac{1}{\sin\phi} \ln\frac{1 + x\sin\phi}{1 - x\sin\phi}, \quad (|x| < 1),$$

$$\int_0^{\pi/2} \frac{1}{\sin\phi} \ln\frac{1 + x\sin\phi}{1 - x\sin\phi} d\phi = \pi \arcsin x.$$

Then we get from Eq. (2.30)

$$P_i = \frac{2E}{1-\nu^2} \int_0^{a_i} (D_i - f(r)) \frac{r\,dr}{\sqrt{a_i^2 - r^2}} - \frac{2}{\pi} \sum_{\substack{j=1 \\ j\neq i}}^{N} P_j \arcsin\frac{a_i}{l_{ij}}, \tag{2.31}$$

where $l_{ij} = \sqrt{X_j^2 + Y_j^2}$.

Considering the relations (2.31) for each punch of the system in combination with the contact condition

$$D_i = h_i - D_0, \tag{2.32}$$

(D_0 is the approach of bodies under the load P (Fig. 2.9)), we get $2N$ equations for determining the values of D_i and P_i ($i = 1, 2, \ldots, N$).

If the approach of the bodies D_0 is unknown, and the load P is given, then in order to determine D_0 one should add to Eqs. (2.31) and (2.32) the equilibrium condition

$$\sum_{i=1}^{N} P_i = P. \tag{2.33}$$

When we study the contact interaction of a system of smooth axially symmetric punches with the elastic half-space, the radius of each contact spot a_i is the unknown value. We can find this value from the condition

$$p(a_i, \theta) = 0, \quad 0 < \theta \leq 2\pi.$$

It follows from this relation and the equilibrium equation

$$P_i(a_i) = \int_0^{a_i}\int_0^{2\pi} p(r,\theta) r \, dr d\theta$$

that $\dfrac{\partial P_i}{\partial a_i} = 0$. Differentiation of Eq. (2.31) with respect to a_i gives

$$D_i = -a_i \int_0^{a_i} \frac{f'(r)\, dr}{\sqrt{a_i^2 - r^2}} + \frac{1-\nu^2}{\pi E} \sum_{\substack{j=1 \\ j\neq i}}^{N} \frac{P_j}{\sqrt{l_{ij}^2 - a_i{}^2}}. \tag{2.34}$$

Eqs. (2.34) in conjunction with Eqs. (2.31) and (2.32) give the complete system of equations to determine the values of D_i, a_i and P_i for a system of punches, the shapes of which are described by a continuously differentiable function.

2.3.2 A system of cylindrical punches

We consider a system of cylindrical punches with flat bases of radii a_i ($f(r) = 0$) penetrating into the elastic half-space, and assume that the contact is complete, that is, it occurs within the subregion ω_i, ($r \leq a_i$). Then we obtain from Eqs. (2.31) the following relationship for the i-th punch penetration ($i = 1, 2, \ldots, N$)

$$D_i = \frac{P_i\,(1-\nu^2)}{2a_i E} + \frac{1-\nu^2}{\pi a_i E} \sum_{\substack{j=1 \\ j\neq i}}^{N} P_j \arcsin\frac{a_i}{l_{ij}}. \tag{2.35}$$

It follows from Eq. (2.35) that the penetration of the punch depends only on the total load applied to the punches located at the distant l_{ij} from the fixed one (circumference of radius l_{ij}).

Eqs. (2.35) in combination with the contact condition (2.32) and the equilibrium equation (2.33) are used to calculate the load distribution P_i between punches. Then the pressure at the i-th contact spot can be approximately determined from the formula (2.6), by substitution of the concentrated loads $P_j = \iint_{\omega_j} p_j(x,y)\, dxdy$, applied to the centers of the contact spots ω_j, ($j \neq i$) for the real pressure $p_j(x,y)$

$$p_i(x,y) = \frac{ED_i}{\pi(1-\nu^2)\sqrt{a^2 - x^2 - y^2}} -$$

$$\frac{1}{\pi^2\sqrt{a^2 - x^2 - y^2}} \sum_{\substack{j=1 \\ j\neq i}}^{N} \frac{P_j\sqrt{X_j^2 + Y_j^2 - a_i^2}}{(x - X_j)^2 + (y - Y_j)^2}.$$

For definiteness we consider a system of N cylindrical punches which are rigidly bonded and acted on by the force P directed along the z-axis. Each punch has a flat base of radius a. We introduce the following notation

$$\Theta_j = \frac{1-\nu^2}{2aED_0}P_j, \quad \delta_j = \frac{h_j}{D_0} - 1, \quad b_{ij} = \begin{cases} 1, & j = i, \\ \dfrac{2}{\pi}\arcsin\dfrac{a}{l_{ij}}, & j \neq i. \end{cases} \tag{2.36}$$

In this case Eqs. (2.32), (2.33) and (2.35) take the form

$$B\Theta = \delta, \tag{2.37}$$

where B is a square nonsingular matrix with elements b_{ij}, δ is a column vector with elements $\delta_i > 0$, Θ is a column vector with elements θ_i. We assume that the column vector δ provides the conditions $\Theta_j > 0$ $(j = 1, 2, \ldots, N)$ which occur if all punches are in contact with the elastic half-space.

In view of nonsingularity of matrix B it follows from Eq. (2.37) that

$$\Theta = B^{-1}\delta. \tag{2.38}$$

Adding up N equations in the system (2.38) and taking into account the equilibrium equation (2.33), we obtain

$$\frac{1-\nu^2}{2aED_0}P = \sum_{i=1}^{N}\sum_{j=1}^{N}\tilde{b}_{ij}\delta_i, \tag{2.39}$$

where $\tilde{b}_{ij}$ are the elements of the inverse matrix B^{-1}. Eq. (2.39) makes it possible to determine the relation between the load P applied to the system of punches, and its penetration D for different spatial arrangement of punches (their height distribution h_j and location within the nominal region Ω).

The system of equations (2.37) and the relationship (2.39) have been used for calculation of loads acting on the punches, and for determination of the relation between the total load and the depth of penetration for a system of N cylindrical punches of radius a that are embedded in a rigid plate. The traces of the axes of the cylinders form a hexagonal lattice with a constant pitch l, and the flat faces of the cylinders are at the same level $h_j = h$ for all $j = 1, 2, \ldots, N$. The punches are located symmetrically relative to the central punch, so the nominal region is close to a circle. The density of the contact is determined by the parameter $\dfrac{a}{l}$. The scheme of the punch arrangement is presented in Fig. 2.13.

Fig. 2.10 illustrates the loads acting on the punches located at the various distances $\dfrac{l_{1j}}{l}$ from the central punch, for different values of the parameter $\dfrac{a}{l}$ and $N = 91$. The results show that for high density $\left(\dfrac{a}{l} = 0.5\text{, dark-coloured rectangles}\right)$ the punches in the outlying districts are acted on by a load rough-

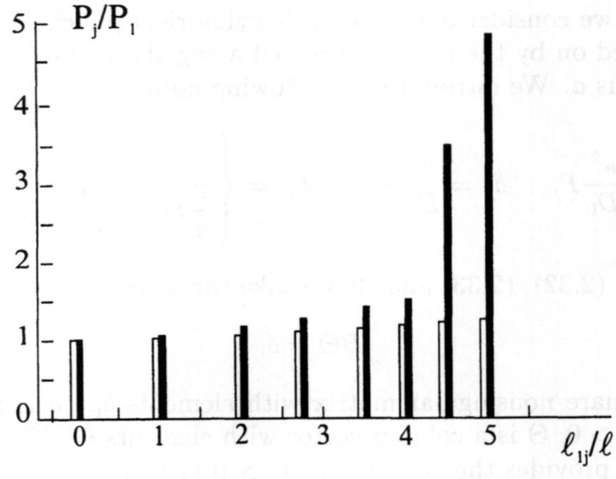

Figure 2.10: Load distribution between the cylindrical punches located at the distance l_{1j} from the central punch. The model parameters are $N = 91$, $a/l = 0.5$ (dark-coloured rectangles), $a/l = 0.2$ (light-coloured rectangles). A schematic diagram of an arrangement of the punches is shown in Fig. 2.13.

ly 5 times greater than the load acting on the central punch; for lower density $\left(\frac{a}{l} = 0.2\right.$, light-coloured rectangles$\left.\right)$ this ratio is equal to 1.14.

It follows from Eq. (2.39), that for the system of punches under consideration the relation between the total load P and the depth of penetration D $(D = h - D_0)$ has the form

$$P = j_0 \beta D, \tag{2.40}$$

where $j_0 = \dfrac{2aE}{1-\nu^2}$ is the ratio of the load acting on an isolated cylindrical punch of radius a, to its penetration (contact stiffness of an isolated cylindrical punch), $\beta = \sum_{i=1}^{N} \sum_{j=1}^{N} \tilde{b}_{ij}$. The value β can be approximated by the function (Goryacheva and Dobychin, 1988)

$$\beta = kN^{\alpha}, \quad 0.5 \le \alpha \le 1, \tag{2.41}$$

where the coefficient k and the power α depend on the parameter $\frac{a}{l}$. For $\frac{a}{l} = 0.5$, i.e. the punches are arranged with a maximum possible density, $\alpha = 0.5$. We can reason as follows. If we arrange the punches with the maximum possible density, the whole system of punches can be regarded as a single punch having radius r_N; obviously, in that case $\pi r_N^2 \approx \pi a^2 N$ or $r_N \sim \sqrt{N}$. Since the stiffness of an isolated punch is proportional to its radius, the stiffness of the whole system must be proportional to $\sqrt{N}$. On the other hand, if the punches are thinly scattered

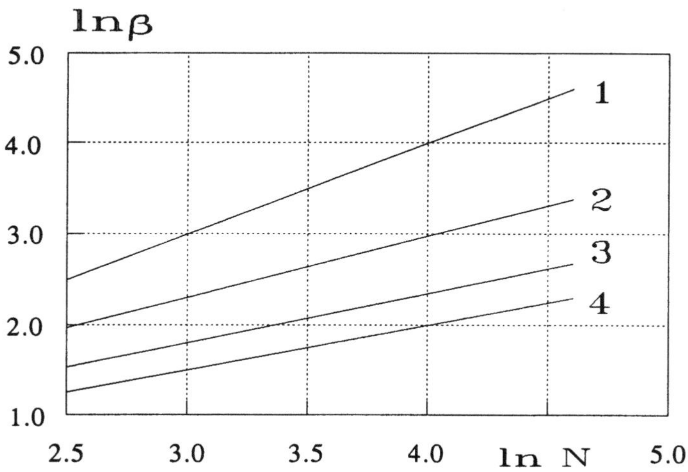

Figure 2.11: The dependence of β upon N for the various values of parameter a/l: $a/l = 0$ (curve 1), $a/l = 0.125$ (curve 2), $a/l = 0.3$ (curve 3), $a/l = 0.5$ (curve 4).

$\left(\frac{a}{l} \to 0\right)$, their mutual influence is practically negligible, $P_j = \frac{P}{N}$ and, as follows from Eq. (2.40), $\beta = N$. The variations of β with N, calculated from Eq. (2.40) for different values of the parameter $\frac{a}{l}$ are presented in Fig. 2.11. The estimated values in the system of coordinates $\ln\beta - \ln N$ cluster near the straight lines, which testifies to the appropriateness of the approximation function (2.41).

Thus, when the interaction of contact spots is neglected(the second term in Eq. (2.35) becomes zero and, hence, $P = j_0 ND$) the contact stiffness $\frac{P}{D}$ of the system of punches is overestimated, the error grows with the number of punches and the density of contact.

The approach described above has been used to analyze the relation between the load and the depth of penetration for different shapes of nominal region in which the punches are arranged (ellipses with different eccentricity are considered). For the models under consideration, the number N of punches and the contact density $\frac{a}{l}$ were all the same. The results of calculation showed that as the eccentricity of the nominal region increases, the contact stiffness of the model increases moderately, the contact stiffness difference for an elongated contour and circular one is small (Goryacheva and Dobychin, 1988). It is interesting to note that the same result was obtained by Galin (1953) for an isolated punch with a flat base of an elliptic shape.

For calculations of the depth of penetration and the real area of contact of bodies with surface microgeometry, of great interest is the case when the tops of

Layer number	1	2	3	4	5	6	7	8
Number of punches	1	6	6	6	12	6	6	12
$\dfrac{h}{h_{\max}}$ (1)	0.9	0.7	0.5	0.3	1.0	0.6	0.8	0.4
$\dfrac{h}{h_{\max}}$ (2)	0.9	0.7	0.5	0.3	0.4	0.6	0.8	1.0

Table 2.1: The parameters of the model with different spatial arrangement of punches (the layer number is counted from the center to the periphery).

the punches are distributed in height rather than lying at the same level. Numerical calculations were carried out for a system of 55 flat-ended cylindrical punches which were located at sites of a hexagonal lattice (see Fig. 2.13). Different variants of the spatial arrangement of punches were considered. Two of them are presented in the Table 2.1. The punches of the j-th layer are located at the same distance l_{1j} from the central punch of the system. For the models under consideration, the number of punches that are intersected by the plane located at an arbitrary distance from the faces of the highest punches (with the height $h_{\max}$), was the same in all variants (the layers of the model with given heights changed positions, but the number of punches in the layers was the same), i.e. the models were characterized by the same height distribution function. The results of the calculations have been described in details in the monograph by Goryacheva and Dobychin (1988). Fig. 2.12 shows the dependence of the real area of contact $\dfrac{A_r}{A_r^*}$ ($A_r^* = 55\pi a^2$) upon load $\dfrac{P}{P^*}$ (P^* is the smallest load necessary for the complete contact of all the punches of the system in case of $\dfrac{a}{l} = 0$) for $\dfrac{a}{l} = 0$ (solid line) and $\dfrac{a}{l} = 0.45$ (broken line). It must be noted that the dependence is a piecewise-constant function for the model under consideration. The broken line represents an averaged curve which reflects the ratio of the contact area and the load for the different variants of punch positions (the variants 1 and 2, presented in the Table 2.1 are indicated by triangles and squares, respectively).

The calculations showed that as the parameter $\dfrac{a}{l}$ increases, the load which is necessary for the complete contact of all punches of the system also increases.

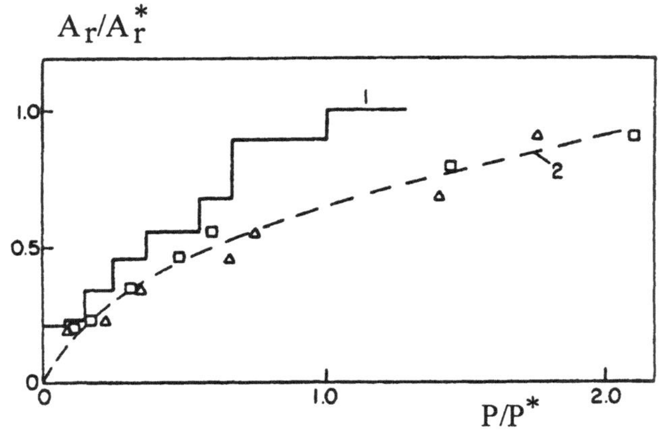

Figure 2.12: Real area of contact as a function of load (cylindrical punches distributed in height): $a/l = 0$ (curve 1), $a/l = 0.45$ (curve 2).

This can be explained by the interaction between the individual contact spots in the contact problem for the system of punches and the elastic half-space.

In order to evaluate the contribution of the simplifying assumptions made in the present model, experiments were made to study the dependence of the load upon the depth of penetration for a system of cylindrical punches with flat bases in contact with an elastic half-space.

The test sample was a steel plate with pressed-in steel cylinders of diameter $2a = 3$ mm. When viewed from the top, the traces of the axes of the cylinders form a hexagonal lattice with a constant pitch l, and the flat faces of the cylinders are all at the same level. Two samples with $\frac{a}{l} = 0.25$ and with $\frac{a}{l} = 0.125$ were tested. The number of punches in each model was $N = 55$. A block of rubber was used as the elastic body. Its elastic constant had been estimated in advance: $\frac{E}{1-\nu^2} = 21.2$ MPa. Fig. 2.13 shows the results of experiments for these two samples. The theoretical dependencies obtained from Eq. (2.40) are given for comparison.

Thus, in full accord with the theory, the relation between the depth of penetration and the load is linear. The theoretical angular coefficients of these dependencies, which are equal to 1.37 and 0.86 N/m, respectively, are sufficiently close to the experimental values (1.44 and 0.93, respectively). A slight difference between the theoretical and experimental data can be accounted for by the influence of the tangential stresses on the contact surfaces, which are not taken into account in the statement of the problem, but are not excluded by the experimental conditions. There will also be an error arising from the simplifying assumptions of the model, by which the real pressure distribution at neighboring punches is replaced by

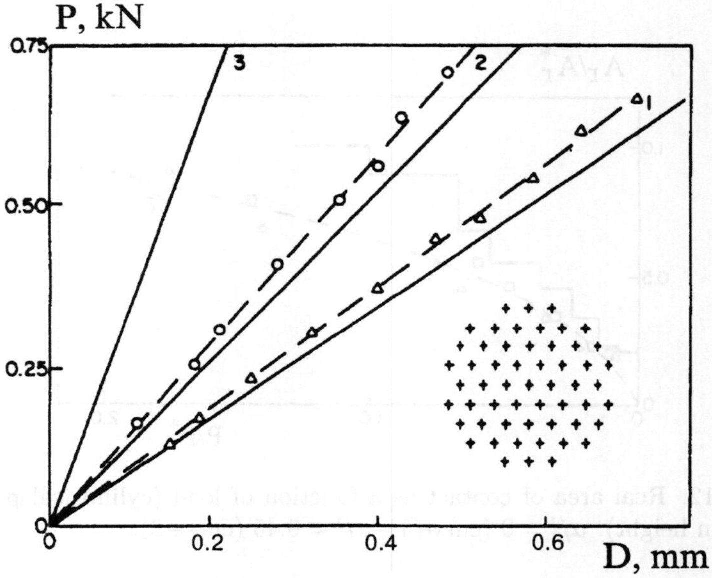

Figure 2.13: Relation between the normal load and the depth of penetration for $a/l = 1/4$ (1), $a/l = 1/8$ (2), $a/l = 0$ (3); (solid line – theory, broken line – experiment). In the lower right-hand corner a schematic diagram of an arrangement of cylindrical punches on the test sample is shown.

concentrated forces.

The present model has been used to predict experimental results obtained by Kendall and Tabor (1971). The theoretical and experimental results are in good agreement (Goryacheva and Dobychin, 1980).

2.3.3 A system of spherical punches

For punches with a spherical contact surface of radius R, $f(r) = \dfrac{r^2}{2R}$, and the given spatial arrangement Eqs. (2.31)-(2.33) take the form

$$h_i - D_0 = \frac{a_i^2}{R} + \frac{1-\nu^2}{\pi E} \sum_{\substack{j=1 \\ j \neq i}}^{N} \frac{P_j}{\sqrt{l_{ij}^2 - a_i^2}},$$

$$P_i = \frac{4Ea_i^3}{3R(1-\nu^2)} + \frac{2}{\pi} \sum_{\substack{j=1 \\ j \neq i}}^{N} P_j \left[\frac{a_i}{\sqrt{l_{ij}^2 - a_i^2}} - \arcsin \frac{a_i}{l_{ij}} \right], \tag{2.42}$$

$$P = \sum_{i=1}^{N} P_i, \quad i = 1, 2, \ldots, N.$$

The system (2.42) determines the distribution of forces P_i among N punches, which are loaded with the total force P and interact with the elastic half-space, the radii a_i of the contact subregions ω_i, the total real area of contact $A_r = \pi \sum_{i=1}^{N} a_i^2$ and the dependence of the approach upon load $D_0(P)$.

It follows from the second group of Eqs. (2.42) that the radius of the i-th contact spot can be determined with accuracy of order $\left(\frac{a_i}{l_{ij}}\right)^3$ by the Hertz formula

$$a_i = \sqrt[3]{\frac{3RP_i\left(1-\nu^2\right)}{4E}}.$$

Then the real area of contact can be approximated by the formula

$$A_r = \pi \sum_{i=1}^{N} \sqrt[3]{\left[\frac{3RP_i\left(1-\nu^2\right)}{4E}\right]^2},$$

where P_i is determined from the first group of Eqs. (2.42).

Fig. 2.14 shows plots of the relative area of contact $\frac{A_r}{A_a}$ (A_a is the nominal area of contact) versus the pressure $\bar{p} = \frac{P}{A_a}$ calculated from Eqs. (2.42) (curve 1) for the system of $N = 52$ spherical punches of radius R, located at the same height and distributed at the sites of square lattice (l is the lattice pitch) with $\frac{l}{R} = 0.5$. Curve 2 is calculated using the Hertz theory and neglecting the redistribution of the loads applied to each contact spot due to the interaction between contact spots. From $\frac{A_r}{A_a} = 0.3$ there is a noticeable error in the calculation of the real area of contact from the theory which ignores interaction.

Fig. 2.15 shows the dependence of the depth of penetration D upon the load P for the system of spherical asperities. The higher is the contact density $\left(\text{i.e. the smaller is the parameter } \frac{l}{R}\right)$, the smaller is the load required to achieve the given depth of penetration. Analogous results were obtained theoretically and experimentally when studying the interaction of a system of cylindrical punches, located at the same level, with an elastic half-space (Fig. 2.13).

From the results of the analysis we conclude that the calculation methods which do not take into account the interaction of the contact spots give overestimated values for the contact stiffness $\frac{dP}{dD}$ and the real area of contact A_r; the error increases with the number of contacts and their density.

The geometrical imperfections of a surface, in particular its waviness and distortion, which are caused by inaccurate conjunctions and deviations from the ideal system of external loads, lead to the localization of contact spots within the so-called contour regions. The nominal region can include a few or many contour

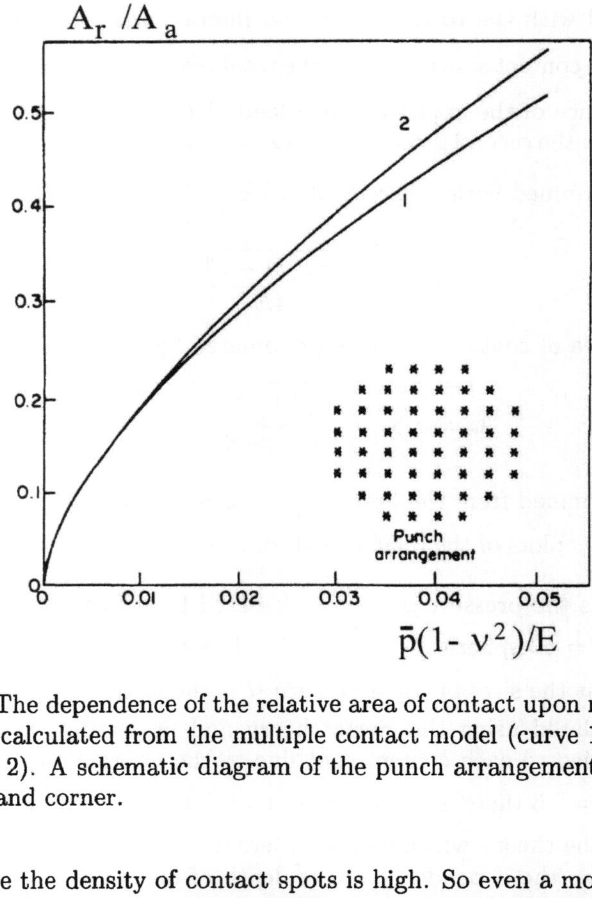

Figure 2.14: The dependence of the relative area of contact upon nominal pressure at $l/R = 0.5$ calculated from the multiple contact model (curve 1) and the Hertz model (curve 2). A schematic diagram of the punch arrangement is shown in the lower right-hand corner.

regions, where the density of contact spots is high. So even a moderate load provides a high relative contact area within the contour regions, and the error of calculation based on the simplified theory can be large.

It is worth noting that the investigation of the multiple contact problem based on the approach described in this section and in § 2.2 necessitates the knowledge of the additional parameter characterizing the density of the arrangement of contact spots. This parameter can be determined, in particular, from modelling of rough surfaces based on the theory of random functions (Sviridenok, Chijik and Petrokovets, 1990).

2.4 The additional displacement function

2.4.1 The function definition

We again direct our attention to Eq. (2.2), which determines the displacement $u_z(x, y)$ of the half-space surface loaded by the pressure $p_i(x, y)$ within contact

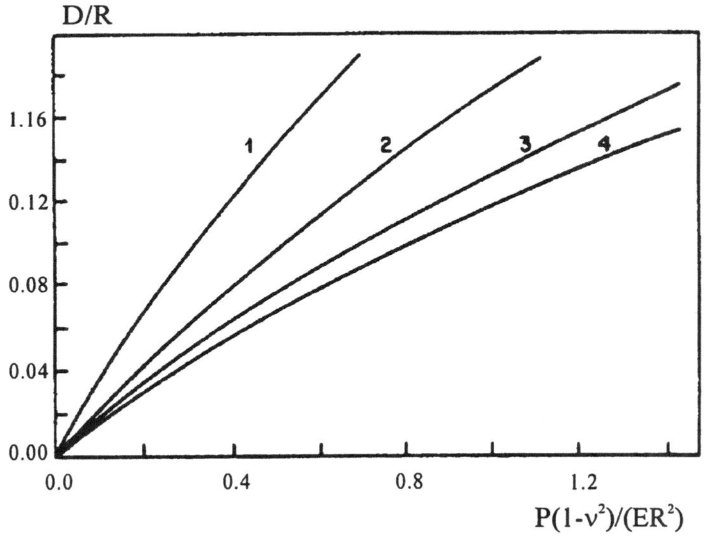

Figure 2.15: The dependence of the depth of penetration upon the load for the various values of the parameter l/R: $l/R = 0.5$ (curve 1), $l/R = 1$ (curve 2), $l/R = 1.5$ (curve 3), $l/R = 2$ (curve 4).

spots ω_i, and substitute the nominal pressure $p(x,y)$ within the region $\Omega \setminus \Omega_0$ (Ω_0 is the circle with the center (x,y)) for the real pressure distributed within the contact spots $\omega_i \in \Omega \setminus \Omega_0$, i.e.

$$\frac{\pi E}{1-\nu^2} u_z(x,y) = \sum_{i=1}^{n} \iint\limits_{\omega_i} \frac{p_i(x',y')\,dx'dy'}{\sqrt{(x'-x)^2+(y'-y)^2}} + \iint\limits_{\Omega\setminus\Omega_0} \frac{p(x',y')\,dx'dy'}{\sqrt{(x'-x)^2+(y'-y)^2}}, \tag{2.43}$$

where $\omega_i \in \Omega_0$ $(i = 1, 2, \ldots, n)$.

The principle of localization formulated in § 2.2.2 shows that this substitution can be carried out with a high degree of accuracy. The radius R_0 of the region Ω_0 can be determined from the following limiting estimate. We assume that there are N concentrated normal forces P_i $(i = 1, 2, \ldots, N)$ within the annular domain ($\Omega_{R_0} : R_0 < r < R_1$), and the nominal pressure is uniformly distributed within this region, i.e. $p(x,y) = \bar{p}$ (see Fig. 2.16). This simulates the limiting case of a discrete contact. We determine the difference Δu_z of displacements at the center (x,y) of the annulus Ω_{R_0} which arises from the concentrated forces on the one hand, and from the nominal pressure on the other hand, which are distributed

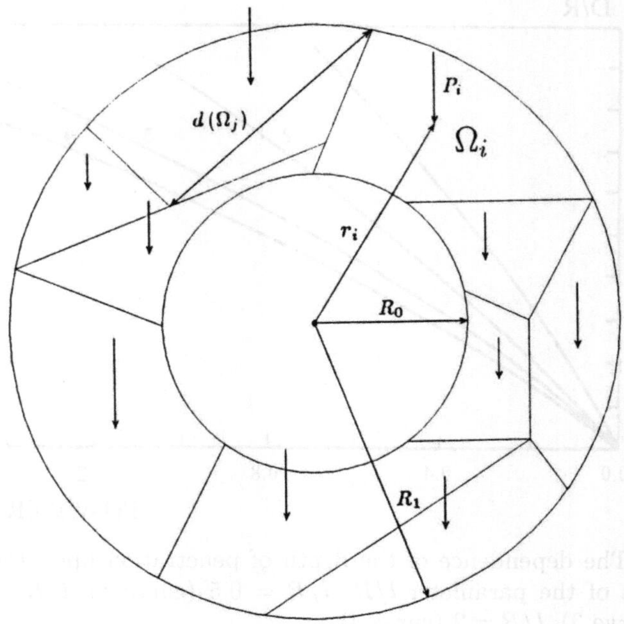

Figure 2.16: Scheme of an arrangement of the concentrated normal forces inside the nominal region Ω_{R_0}.

within the region Ω_{R_0}, that is

$$\frac{\pi E}{1-\nu^2}\Delta u_z = \sum_{i=1}^{N}\frac{P_i}{r_i} - \iint\limits_{\Omega_{R_0}} \frac{\bar{p} r\, dr\, d\varphi}{r}, \tag{2.44}$$

where r_i is the distance from the point (x, y) to the point where the concentrated force P_i is applied. We divide the region Ω_{R_0} into N subregions Ω_i so that only one force is within each subregion and the condition $\bar{p} = \dfrac{P_i}{A_{\Omega_i}}$ is satisfied (A_{Ω_i} is the area of Ω_i). Then we obtain on the basis of the law of the mean

$$\iint\limits_{\Omega_i} \frac{\bar{p} r\, dr\, d\varphi}{r} = \frac{P_i}{\tilde{r}_i}, \tag{2.45}$$

where $\tilde{r}_i$ is the distance from the point (x, y) to some point inside the subregion Ω_i. Then it follows from Eqs. (2.44) and (2.45), and conditions $r_i \geq R_0$, $\tilde{r}_i \geq R_0$ that

$$\frac{\pi E}{1-\nu^2}\Delta u_z = \sum_{i=1}^{N}\left(\frac{1}{r_i} - \frac{1}{\tilde{r}_i}\right) P_i \leq \frac{\max\limits_i d(\Omega_i)}{R_0^2} \max\limits_i P_i,$$

where $d(\Omega_i)$ is the characteristic linear size of the region Ω_i.

If concentrated forces with the same value $P_i = P$ are uniformly distributed over the region Ω_{R_0}, this estimate takes a simple form

$$\frac{\pi E}{1-\nu^2}\Delta u_z \leq \frac{2P}{R_0^2}\sqrt{\frac{R_1^2 - R_0^2}{N}}.$$

We write the contact condition (2.3) in the following form

$$u_z(x,y) = D - f(x,y) + h(x,y), \quad x, y \in \omega_0, \tag{2.46}$$

where the function $f(x,y)$ describes the macroshape of the indenter, and the function $h(x,y)$ describes the shape of an asperity within the contact spot ω_0.

From Eqs. (2.43) and (2.46), and substituting the integral over $\Omega \setminus \Omega_0$ in Eq. (2.43) by the difference of integrals over regions Ω and Ω_0, we can derive the following integral equation

$$D - f(x,y) = \beta(x,y) + \frac{1-\nu^2}{\pi E}\iint\limits_{\Omega}\frac{p(x',y')\,dx'dy'}{\sqrt{(x'-x)^2+(y'-y)^2}}, \tag{2.47}$$

where

$$\beta(x,y) = -h(x,y) +$$

$$\frac{1-\nu^2}{\pi E}\left[\sum_{i=1}^{n}\iint\limits_{\omega_i}\frac{p_i(x',y')\,dx'dy'}{\sqrt{(x'-x)^2+(y'-y)^2}} - \iint\limits_{\Omega_0}\frac{p(x',y')\,dx'dy'}{\sqrt{(x'-x)^2+(y'-y)^2}}\right]. \tag{2.48}$$

The function $\beta(x,y)$ depends only on the parameters of loading and microgeometry in the vicinity of the point (x,y) (within the region Ω_0).

It should be noted that there are two length scales in the problem: the macroscale connected with the nominal contact area and the macroshape of the indenter, and the microscale related to the size and distance between the contact spots. In what follows, we assume that all functions related to the macroscale, i.e. $p(x,y)$, $f(x,y)$, $\beta(x,y)$, etc., change negligibly little for distances of the order of the distance between neighbouring contact spots.

We will demonstrate below that under this assumption the function $\beta(x,y)$ (we call it *the additional displacement*) can be presented as a function $C(p)$ of the nominal pressure $p(x,y)$, and determine the form of this function for some particular models of surface microgeometry.

2.4.2 Some particular cases

We consider Eq. (2.48) at the point $(x_0, y_0) \in \omega_0$, where the top of an asperity with height h_0 is located. Taking into account the assumption concerning two geometry scales for the problem under consideration, we suppose that the nominal pressure within the region Ω_0 which is a circle of radius R_0 with a center at the point (x_0, y_0) is uniform and equal to $p(x_0, y_0)$. Inside the region Ω_0 we consider also the real pressure distribution at the contact spots $\omega_i \in \Omega_0$ $(i = 1, 2, \ldots, n)$ (local effect). Then Eq. (2.48) can be reduced to the form

$$\beta(x_0, y_0) = -h_0 +$$
$$\frac{1-\nu^2}{\pi E}\left[\sum_{i=0}^{n} \iint_{\omega_i} \frac{p_i(x', y')\, dx'dy'}{\sqrt{(x'-x_0)^2 + (y'-y_0)^2}} - 2\pi R_0 p(x_0, y_0)\right]. \tag{2.49}$$

So the value $\beta(x_0, y_0)$ characterizes the additional displacement of the region Ω_0 (which is acted by the nominal pressure $p(x_0, y_0)$) arising from the penetration of asperities into the elastic half-space inside this region. Since $A_{\Omega_0} \ll A_\Omega$, we can neglect the curvature of the surface at the point (x_0, y_0) when determining the value of $\beta(x_0, y_0)$. This suggests that it might be convenient to use the solution of the periodic contact problem for determination of $\beta(x_0, y_0)$. In this case the periodic contact problem must be considered for the system of indenters which models the real surface geometry in the region Ω_0 and which is loaded by the nominal pressure $p(x_0, y_0)$. It was shown in § 2.2 that for the given nominal pressure p and the known spatial arrangement of indenters we can uniquely determine the real contact pressure $p_i(x.y)$ from the systems of equations (2.17) or 2.21 – 2.23 and, hence, the value $\beta(x_0, y_0)$ from Eq. (2.49). So the dependence of the additional displacement upon the nominal pressure $C(p)$ can be constructed at each point (x_0, y_0) based on Eq. (2.49).

We note that to sufficient accuracy the function $C(p)$ can be written in analytical form for some surfaces with a regular microgeometry. Using the law of the mean, we reduce Eq. (2.49) to the following form

$$\beta(x_0, y_0) = -h_0 + \phi(P_0) + \frac{1-\nu^2}{\pi E}\left[\sum_{i=1}^{n} \frac{P_i}{l_{0i}} - 2\pi R_0 p(x_0, y_0)\right], \tag{2.50}$$

where

$$\phi(P_0) = \frac{1-\nu^2}{\pi E} \iint_{\omega_0} \frac{p_0(x', y')\, dx'dy'}{\sqrt{(x'-x_0)^2 + (y'-y_0)^2}},$$

$$P_i = \iint_{\omega_i} p_i(x', y')\, dx'dy'$$

and l_{0i} is the distance from the point (x_0, y_0) to some internal point of the contact spot $\omega_i \in \Omega_0$ $(i = 1, 2, \ldots, n)$.

As an example, we consider a surface for which the microgeometry can be simulated by asperities of the same height located at the sites of a hexagonal lattice with constant pitch l. In § 2.2 it was shown that to sufficient accuracy we can take $n = 6$ in Eq. (2.49). Then we obtain from Eq. (2.19)

$$R_0 = \sqrt{\frac{7}{\pi \bar{N}}},$$

where $\bar{N}$ is a number of asperities per unit area. For the hexagonal lattice we have $\bar{N} = \frac{2}{l^2\sqrt{3}}$. Since all asperities within the region Ω_0 are undergoing the same conditions, they are loaded uniformly and so the load P applied to one asperity is obtained from the equilibrium condition

$$P = \frac{1}{7}p(x_0, y_0)\pi R_0^2 = \frac{p(x_0, y_0)}{\bar{N}}.$$

For a cylindrical asperity with a flat base of radius a, the function $\phi(P)$ in Eq. (2.50) has the form

$$\phi(P) = \frac{(1-\nu^2)\,P}{2aE}.$$

Substituting the relations obtained above in Eq. (2.50) on the assumption that $l_{0i} \approx l$, gives the following form for the additional displacement function:

$$\frac{1}{l}C[p(x,y)] = \frac{1}{l}\beta(x,y) =$$
$$\frac{1-\nu^2}{E}p(x,y)\left(\frac{\sqrt{3}}{4}\cdot\frac{l}{a} + \frac{3\sqrt{3}}{\pi}\cdot\frac{l}{a}\arcsin\frac{a}{l} - \sqrt{\frac{14\sqrt{3}}{\pi}}\right). \tag{2.51}$$

The height of asperities h is not present in Eq. (2.51) because this value can be taken into account in the right side of Eq. (2.47) for models with asperities of the same height.

For elastic asperities of spherical shape, i.e. $f(r) = \frac{r^2}{2R}$, located at the sites of a hexagonal lattice with a pitch l, the function $C(p)$ can be reduced in a similar way based on the results of § 2.2. The final expression has the form

$$\frac{C[p(x,y)]}{R} = \frac{\beta(x,y)}{R} = \frac{p(x,y)}{E^*}\cdot\frac{l}{R}\left(\frac{3\sqrt{3}}{\pi} - \sqrt{\frac{14\sqrt{3}}{\pi}}\right) +$$
$$\frac{3}{4}\cdot\left(\frac{l}{R}\right)^{\frac{4}{3}}\left(\frac{p(x,y)}{E^*}\right)^{\frac{2}{3}}\left(1 + \frac{p(x,y)}{E^*}\cdot\frac{R}{l}\cdot\frac{3\sqrt{3}}{2\pi}\right)^{\frac{1}{3}}. \tag{2.52}$$

2.4.3 Properties of the function

The equation of the type (2.47) was first introduced by Staierman (1949) for determination of the nominal pressure and nominal contact area for the contact of rough bodies. He proposed that for contact interaction of bodies with a surface microstructure, it is necessary to take into account the additional compliance (analogous to soft interlayer) caused by asperity deformation. As a rule, it is taken to be a linear or power additional displacement function in Eq. (2.47)

$$C[p(x,y)] = Bp^{\kappa}(x,y), \quad \kappa \le 1. \tag{2.53}$$

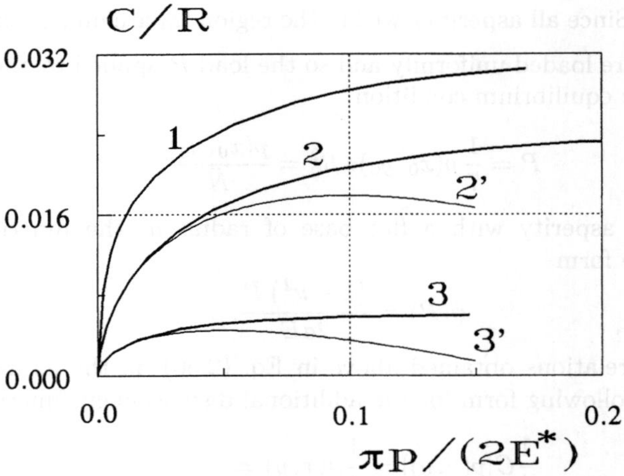

Figure 2.17: The additional displacement function for the three-level model (1) and the one-level model with $l/R = 0.6$ (2, 2′) and $l/R = 0.3$ (3, 3′), calculated from Eqs. (2.17) and (2.21)–(2.23) (curves 1, 2, 3) and from Eq. (2.52) (curves 2′, 3′).

The coefficient B and the exponent κ are usually obtained experimentally. However the experimental determination of the function $C[p(x,y)]$ for rough bodies is a complicated and laborious problem.

The method developed above makes it possible to calculate the additional displacement function for different kinds of surface microgeometry. It is based on the model representation of the microgeometry of rough surfaces. Fig. 2.17 illustrates the functions $C(p)$ calculated for the three-level system of spherical indenters of radius R located at the sites of a hexagonal lattice with pitch $\frac{l}{R} = 0.6$ and characterized by the following relative difference in the heights of the levels: $\frac{h_0 - h_1}{R} = 0.01$, $\frac{h_0 - h_2}{R} = 0.015$ (curve 1); and for the one-level system of spherical indenters located at the sites of a hexagonal lattice with the pitch $\frac{l}{R} = 0.6$ (curve 2) and $\frac{l}{R} = 0.3$ (curve 3). The calculations were based on Eq. (2.49), where the functions $p_i(x,y)$ were obtained from the integral equations (2.17) and (2.21) – (2.23) for the one-level model and three-level one, respectively. The results indicate that the rate of change of the function $C(p)$ decreases as the nominal pressure $\bar{p}$ increases. If the real contact area is close to saturation, i.e. $\frac{A_r}{A_a} \approx 1$, the additional displacement function is close to a constant value, i.e. $\frac{dC}{dp} = 0$.

The curves 2 and 3 in Fig. 2.17 calculated for the one-level model illustrate this conclusion. The results calculated from Eq. (2.52) for the corresponding models are also presented in Fig. 2.17 (curves 2′ and 3′). The coincidence of the curves 2, 2′ and 3, 3′ for relatively small values of the nominal pressure shows that it is possible to use the approximate analytical relationship (2.52) for calculation of the additional displacement, if $\frac{a}{l} < 0.2$. The discrepancy between the results for the higher values of the parameter $\frac{a}{l}$ is explained by the essential effect of the real pressure distribution at the contact spots nearest to the chosen one. This effect in Eq. (2.52) is taken into account approximately by the corresponding values of the concentrated forces applied to these contact spots.

Thus, as the nominal pressure increases, the additional compliance $\frac{dC}{dp}$ caused by the existence of a surface microgeometry, is progressively reduced and tends to zero in going from the discrete to continuous contact.

We note that the power function (2.53) does not describe this process, so it can be used only for low values of the nominal pressure, for which continuous contact does not occur.

2.5 Calculation of contact characteristics

2.5.1 The problem of continuous contact

We consider the contact of two elastic bodies with the macroshape described by the function $z = f(x, y)$ and take into account parameters of their surface microgeometry. There are two scales of size in the problem: the characteristic dimension R_a of the nominal contact region Ω, and the characteristic distance l_a between contact spots. The relation between R_a and l_a can vary in the contact interaction. For small loads it is conceivable that $R_a \sim l_a$, i.e. there are a finite number of asperities in the contact. In this case the method described in § 2.3 can be used for the determination of the contact characteristics (the nominal and real contact area, the load distribution between contact spots, the real pressure distribution, etc.).

If $l_a \ll R_a$ there are many asperities within the nominal contact region. In this case the nominal (averaged) pressure can be determined from the integral equation (2.47) in which $C(p)$ is the additional displacement function. The method for its determination is described in § 2.4. Eq. (2.47) completely determines the nominal pressure $p(x, y)$ if the nominal contact region Ω and the penetration D are prescribed. If the nominal contact region is not known in advance, the problem is reduced to the determination of the nominal contact pressure $p(x, y)$ and the region Ω with its boundary $\partial\Omega$ from the system of equations

$$\begin{aligned} &u_z(x,y) = C(p(x,y)) + \iint\limits_{\Omega} K(x,y,x',y')p(x',y')\,dx'dy', \\ &p(x,y) > 0, \quad \text{if} \quad u_z(x,y) = D - f(x,y), \quad (x,y) \in \Omega, \\ &p(x,y) = 0, \quad \text{if} \quad (x,y) \in \partial\Omega, \\ &p(x,y) = 0, \quad \text{if} \quad u_z(x,y) > D - f(x,y), \quad (x,y) \notin \Omega. \end{aligned} \tag{2.54}$$

The equilibrium equation

$$P = \iint\limits_{\Omega} p(x',y')\,dx'dy' \tag{2.55}$$

is added to this system to obtain the unknown value D if the load P applied to the indenter is known in advance. Eq. (2.47) or the system of equations (2.54) have been analyzed in Staierman (1949), Popov and Savchuk (1971), Aleksandrov and Kudish (1979), Goryacheva (1979b), Galanov (1984), etc. for different types of the function $C(p)$ and different kernels $K(x,y,x',y')$ of the integral operator which are typical for contact problems. In what follows we will describe the method of investigation of these equations for plane and axisymmetric contact problems.

2.5.2 Plane contact problem

We consider the contact of a strip punch or a long elastic cylinder, with an elastic layer of thickness h ($|x| < \infty,\ 0 < z < h$), lying on a rigid foundation (Fig. 2.18). This problem can be analyzed in a 2-D formulation. The indenter macroshape is given by the equation $z = f(x)$. The load P is applied to the indenter in the z-axis direction. The tangential stress within the contact region is supposed to be negligibly small. We investigate two types of contact conditions at the boundary between the strip (layer) and foundation ($z = h$):

1. The strip lies on the rigid foundation without friction; then

$$\tau_{xz}(x,h) = 0, \quad u_z(x,h) = 0, \quad |x| < +\infty.$$

2. The strip is bonded with the foundation; then

$$u_x(x,h) = u_z(x,h) = 0.$$

The boundary conditions at the surface $z = 0$ are

$$\begin{aligned} &\tau_{xz}(x,0) = 0, \quad \sigma_z(x,0) = 0, \qquad a < |x| < \infty, \\ &\tau_{xz}(x,0) = 0, \quad u_z(x,0) = D - f(x), \qquad |x| < a. \end{aligned} \tag{2.56}$$

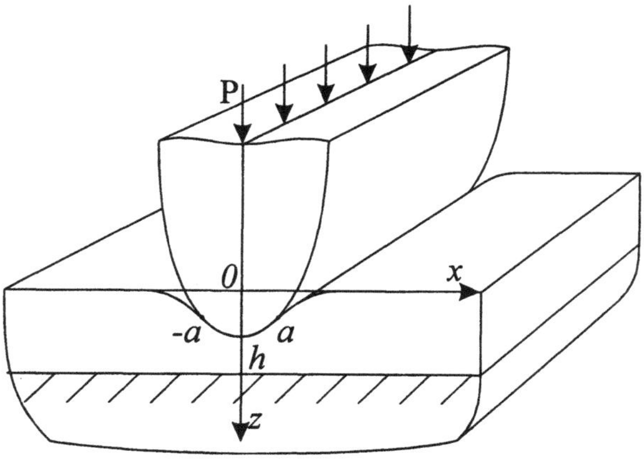

Figure 2.18: Scheme of the contact of a rough punch and an elastic layer lying on a rigid foundation.

The main integral equation (2.54) taking into account the additional displacement $C(p)$ caused by the surface roughness of the contacting bodies takes the following form for the problem under consideration

$$D - f(x) = C(p(x)) + \frac{2(1-\nu^2)}{\pi E} \int_{-a}^{a} k\left(\frac{x'-x}{h}\right) p(x')\, dx'. \tag{2.57}$$

It has been shown in Vorovich, Aleksandrov and Babeshko (1974), that the kernel of the integral operator in Eq. (2.57) has the form

$$k(t) = \int_{0}^{+\infty} \frac{L(u)}{u} \cos ut\, du. \tag{2.58}$$

The form of the function $L(u)$ depends on the boundary conditions at the plane $z = h$.

In case 1

$$L(u) = \frac{\cosh 2u - 1}{\sinh 2u + 2u}. \tag{2.59}$$

In case 2

$$L(u) = \frac{2\kappa_* \sinh 2u - 4u}{2\kappa_* \cosh 2u + 4u^2 + 1 + \kappa_*^2}, \qquad \kappa_* = 3 - 4\nu. \tag{2.60}$$

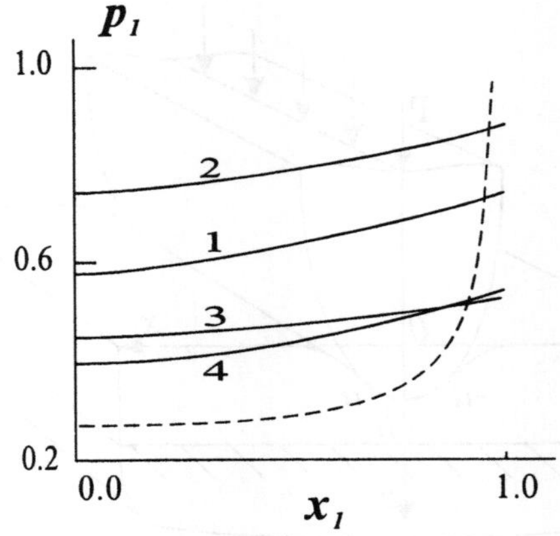

Figure 2.19: The pressure in the contact of a punch with flat rough base and a thick elastic strip, for various roughness parameters (solid lines) and for a smooth surface (broken lines).

We rewrite Eq. (2.57) and the equilibrium condition (2.55) in dimensionless form. To do this, we introduce the following notation:

$$x_1 = \frac{x}{a}, \qquad \delta = \frac{D}{a}, \quad \lambda = \frac{a}{h},$$
$$f_1(x_1) = \frac{f(ax_1)}{a}, \qquad C_1(p_1(x_1)) = \frac{C(p(ax_1))}{a}, \tag{2.61}$$
$$p_1(x_1) = \frac{2\,(1-\nu^2)}{\pi E} p(ax_1), \qquad P_1 = \frac{2\,(1-\nu^2)\,P}{\pi E a}.$$

Then we obtain

$$\int_{-1}^{1} k(\lambda(t-x_1))p_1(t)\,dt + C_1(p_1(x_1)) = \delta - f_1(x_1), \tag{2.62}$$

$$\int_{-1}^{1} p_1(x)\,dx = P_1. \tag{2.63}$$

If the contact half-width a is known in advance, Eqs. (2.62) and (2.63) form the complete system of equations for the determination of the dimensionless contact pressure $p_1(x_1)$ and the penetration δ of the indenter.

Eq. (2.62) is an equation of Hammerstein type. It can be reduced to canonical form. To do this, we introduce the function

$$\psi(x_1) = C_1(p_1(x_1)) + f_1(x_1) - \delta. \tag{2.64}$$

Then

$$\int_{-1}^{1} k(\lambda(t - x_1))C_1^{-1}(\psi(t) - f_1(t) + \delta)\,dt + \psi(x_1) = 0, \qquad (2.65)$$

where $C_1^{-1}(x)$ is the inverse function to $C_1(x)$.

To solve this equation, we can use iteration. We take $\psi_0(x_1) = 0$ as the initial solution and then calculate the subsequent values from the recurrence relation

$$\psi_{n+1}(x_1) = -\int_{-1}^{1} k(\lambda(t - x_1))C_1^{-1}(\psi_n(t) - f_1(t) + \delta)\,dt.$$

The convergence of the method can be proved for some particular forms of the function $C(p)$. It was indicated in § 2.4 that $C(p)$ can be approximated by the power function (2.53), valid for relatively low values of the nominal pressure p and, hence, for the case in which the real contact area is much less than the nominal contact area. For the function $C(p) = Bp^{\kappa}$ (2.53), successive approximations $\psi_n(x_1)$ converge to the unique solution of the equation (2.65), if the parameters of the problem satisfy to the following inequality (Goryacheva, 1979b)

$$\left(\frac{1}{\kappa B_1^{1/\kappa}}\right)^2 \int_{-1}^{1} [\delta - f_1(x)]^{2/\kappa - 2} \left[\int_{-1}^{1} k^2(\lambda(t - x))\,dt\right] dx < 1, \qquad (2.66)$$

where

$$B_1 = \frac{B}{2a}\left[\frac{\pi E}{2\,(1 - \nu^2)}\right]^{\kappa}. \qquad (2.67)$$

For the other values of parameters, the Newton–Kantorovich method (Kantorovich and Krylov, 1952) can be used to solve the problem. Then the dimensionless pressure can be found from the formula (2.64). If the penetration D of the indenter is not known in advance, we use also Eq. (2.63) to solve the problem.

The study makes it apparent that for the function $C(p)$ (2.53), the contact pressure does not tend to infinity at the ends of the contact region. To prove this fact, we anticipate that pressure has an integrable singularity of the type $(1 - x_1)^{-\theta}$ $(0 < \theta < 1)$ at the point $x_1 = 1$. We take into account also that the kernel of the integral operator has a singularity of the type $\ln(1 - x_1)$. Then we conclude that the left side of Eq. (2.62) has a singularity of the form $(1 - x_1)^{-k\theta}$, whereas there is no singularity at the right side of the equation. This contradiction proves the proposition mentioned above. Thus, the consideration of the additional displacement caused by the asperity penetration leads to the disappearance of the singularity of the contact pressure at the ends of the contact zone which occurs for the problem formulation neglecting the surface microgeometry, for bodies whose macroshape $f(x)$ provides a discontinuity of the derivative of the surface displacement $u_z'(x)$ at the ends of the contact region (for example, $f'(x) = 0$ for $x < a$).

For linear contact of elastic cylindrical bodies with rough surfaces we use the additional condition that the contact pressure is equal to zero at the ends of the contact region, i.e. $p_1(-1) = p_1(1) = 0$, and also the relation $C(0) = 0$. Then the integral equation (2.62) for the nominal contact pressure determination can be reduced to the form

$$\int_{-1}^{1} (k(\lambda(t - x_1)) - k(\lambda(t - 1)))\, p_1(t)\, dt + C_1(p_1(x_1)) = f_1(1) - f_1(x_1), \quad (2.68)$$

where

$$p_1(x_1) = \frac{2}{\pi E^*} p\,(a x_1)\,. \quad (2.69)$$

This is also a Hammerstein type integral equation which can be solved by iteration or the Newton-Kantorovich method.

The solution of Eq. (2.68) with the function $C(p)$ of the form (2.53), where $0 < \kappa < 1$, has zero derivative at the ends of the contact region, i.e. $p_1'(-1) = p_1'(1) = 0$. This can be proved as follows. Upon differentiating Eq. (2.68) with respect to x_1 and setting $x_1 = -1$ (the case $x_1 = 1$ can be analyzed in a similar manner), we obtain

$$\int_{-1}^{1} (-\lambda k'(\lambda(t - x_1)))\, p_1(t)\, dt + B_1 \kappa p_1^{\kappa-1}(-1) p_1'(-1) = -f_1'(-1), \quad (2.70)$$

where B_1 is determined from Eq. (2.67).

Since the function $p_1(x_1)$ is continuously differentiable, $p_1(-1) = p_1(1) = 0$, and the kernel $k(t)$ (2.58) is presented as (see Vorovich, Aleksandrov and Babeshko, 1974)

$$k(t) = -\ln|t| + F(t),$$

where $F(t)$ is an analytical function, the integral term on the left side of Eq. (2.70) is bounded. The second term in the left side of this equation has to be also bounded, as the value $f_1'(-1)$ is bounded on the right side of Eq. (2.70). This holds for $0 < \kappa < 1$, only if $p_1'(-1) = 0$.

As an example, we consider the problem of frictionless contact between a thick rough layer and a punch with the flat base, $f(x) = 0$. For the nominal pressure determination, we use Eq. (2.62) in which $f_1(x_1) = 0$, and the kernel $k(t)$ has the form $k(t) = -\ln|t| + a_0$; $a_0 = -0.352$ for case 1, and $a_0 = -0.527$ for case 2 ($\nu = 0.3$) (Vorovich et al., 1974). This asymptotic representation of the kernel holds for the comparatively thick layer $\left(\lambda \leq \frac{1}{2}\right)$. The function $C(p)$ is used in the form of Eq. (2.53).

The problem is attacked by solving Eq. (2.65) by iteration. Then we obtain the nominal contact pressure as

$$p_1(x_1) = B_1^{-\frac{1}{\kappa}} \left[\psi(x_1) + \delta\right]^{\frac{1}{\kappa}},$$

where $\psi(x_1)$ is the limit of the function sequence $\{\psi_n(x_1)\}$ determined by

$$\psi_{n+1}(x_1) = B_1^{-\frac{1}{\kappa}} \int_{-1}^{1} (\ln|t - x_1| + c_0) [\psi_n(t) + \delta]^{\frac{1}{\kappa}} dt,$$

$$c_0 = \ln\left(\frac{a}{h}\right) - a_0.$$

This limit exists if the condition (2.66) holds, which has the following form in this case

$$\frac{\delta^{\frac{2}{\kappa} - 2} \left(2c_0^2 - 3c_0 + 3.5\right)}{\kappa^2 B_1^{\frac{2}{\kappa}}} < 1.$$

For the numerical calculation, the following values of parameters are used: $\kappa = 0.4$, $c_0 = -3.352$. Fig. 2.19 illustrates the pressure distribution for different values of the dimensionless load P_1 and the roughness parameter B_1. The curves 1 and 2 are drawn for $B_1 = 1$ and $P_1^{(1)} = 0.6 \cdot 10^{-2}$ (curve 1) and $P_1^{(2)} = 0.75 \cdot 10^{-2}$ (curve 2). Penetration for the cases $P_1^{(1)}$ and $P_1^{(2)}$ are $\delta^{(1)} = 0.15$, $\delta^{(2)} = 0.17$. The results indicate that for the same roughness parameter, the pressure increases especially at the periphery of the contact region, as the load increases. For fixed load $0.41 \cdot 10^{-2}$, the penetration and the pressure distribution depend on the roughness parameters B_1 and κ. For the case $B_1 = 0.75$ ($\kappa = 0.4$), the penetration is $\delta = 0.1$; for $B_1 = 0.35$ (the smoother surface) the penetration is smaller, $\delta = 0.06$. The graphs of pressure distribution for the cases are shown in Fig. 2.19 by the curves 3 and 4, respectively; the pressure distribution for the smooth punch is shown by the broken line. The calculation showed the fast convergence of the iteration method. For an accuracy of 10^{-5}, it is sufficient to take 15–20 iterations.

2.5.3 Axisymmetric contact problem

We consider the contact of an axisymmetric punch or elastic indenter with the macroshape described by the function $z = f(r)$ ($f(0) = 0$), and the elastic half-space ($z < 0$). The contact region Ω is a circle of the radius a. Using the Boussinesq's solution (see Galin, 1976b, Gladwell, 1980, etc.), we write the integral term in Eq. (2.54) which indicates the elastic displacements $u_z^{(n)}$ of the half-space surface caused by the nominal pressure $p(r)$ distributed within the circle of the radius a, in the following form

$$u_z^{(n)} = \frac{1}{\pi E^*} \int_0^a \int_0^{2\pi} \frac{p(r')r'dr'd\theta}{\sqrt{r^2 + r'^2 - 2rr'\cos\theta}} = \frac{1}{\pi E^*} \int_0^a H(r, r')p(r')r'dr',$$

where

$$H(r, r') = \frac{4}{r + r'} K\left(\frac{2\sqrt{rr'}}{r + r'}\right),$$

and $K(t)$ is the complete elliptic integral of the first kind.

To write the integral equation in dimensionless form, we introduce the notation

$$\rho = \frac{r}{a}, \quad \delta = \frac{D}{a}, \quad f_1(\rho) = \frac{f(\rho a)}{a},$$

$$p_1(\rho) = \frac{p(\rho a)}{\pi E^*}, \quad P_1 = \frac{P}{\pi E^* a^2},$$

$$C_1(p_1(\rho)) = \frac{C(p(\rho a))}{a}.$$

If we consider the contact of a rough punch and an elastic half-space, and the radius a of the contact region is fixed due to the special punch shape (for example, if the punch has a flat base), the integral equation for the determination of the nominal pressure has the form

$$\int_0^1 H(\rho, \rho')p(\rho')\rho' d\rho' + C_1(p_1(\rho)) = \delta - f_1(\rho). \tag{2.71}$$

If the radius of the contact is not known in advance ($f_1(\rho)$ is a smooth function), we use the additional conditions $p_1(1) = 0$ and $C(0) = 0$, and obtain the following integral equation

$$\int_0^1 [H(\rho, \rho') - H(1, \rho')]p(\rho')\rho' d\rho' + C_1(p_1(\rho)) = f_1(1) - f_1(\rho). \tag{2.72}$$

Since the elliptic integral $K(t)$ for $t \approx 1$ has a logarithmic singularity of the same kind as the principal part of the kernel analyzed in § 2.5.2, Eqs. (2.71) and (2.72) can be analyzed in the same way as in § 2.5.2 for the given function $C(p)$. The conclusions of § 2.5.2 concerning the properties of the function $p_1(\rho)$ at the boundary of the contact region for the function $C(p)$ of the form (2.53) are valid also for axisymmetric contact problems, i.e. the value $p(a)$ is always bounded above and $p(a) = p'(a) = 0$ if $f'(\rho)$ is continuous at $\rho = a$.

We note that for a linear additional displacement function, i.e. $C = Bp$, Eq. (2.54) is a Fredholm integral equation of the second kind, which can be solved by standard methods (for example, reduction to the linear algebraic equations). The dependence of the penetration of a punch with flat base upon the load is linear in this case. The results of calculations show that the contact stiffness $\frac{P}{D}$ decreases as the roughness coefficient B increases.

2.5.4 Characteristics of the discrete contact

The nominal pressure obtained from Eq. (2.54) or its particular forms (Eqs. (2.57) and (2.71)) can be used to determine the characteristics of a discrete contact

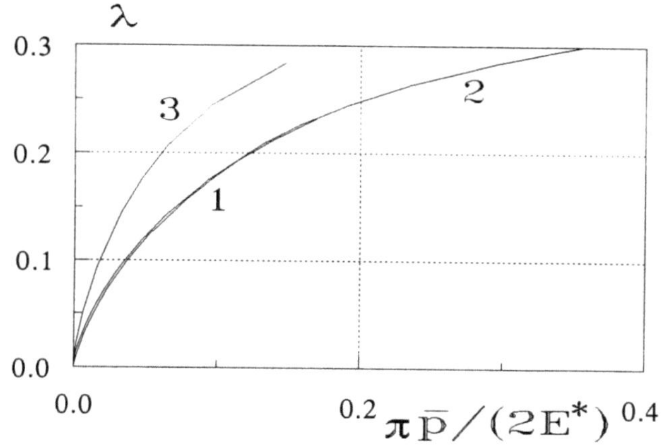

Figure 2.20: The dependence of the relative real area of contact on the nominal pressure for various models of the surface microgeometry.

which are needed for the study of friction and wear in the contact interaction (see Chapters 3, 5), or for calculation of the contact electric and heat conductivity, leak-proofness of seals, etc.

We describe the method of calculation of the discrete contact characteristics on the example of the calculation of the real area of contact A_r. For the given parameters characterizing the surface microgeometry of the contacting bodies, we can obtain the additional displacement $C(p)$ and the relative area of contact $\lambda(p)$ as functions of the nominal contact pressure p from the solution of the multiple contact problem. For example, for microgeometry modelled by a uniformly distributed system of asperities of different or the same height, these functions can be determined from the periodic contact problem for the system of asperities and the elastic half-space using the methods of §§ 2.2 and 2.4. The functions $C(p)$ for some given values of the microgeometry parameters are shown in Fig. 2.17. Fig. 2.20 illustrates the variation of the relative real area of contact $\lambda = \dfrac{4\pi\left(a_1^2 + a_2^2 + a_3^2\right)}{l^2\sqrt{3}}$ with the dimensionless nominal contact pressure $\bar{p}^1 = \dfrac{\pi p}{2E^*}$ calculated for the one-level ($a_1 = a_2 = a_3$) and the three-level models of asperity arrangement for the same parameters of surface microgeometry as in Fig. 2.17.

The function $C(p)$ calculated for the given parameters of the surface microgeometry is then used to determine the nominal contact pressure $p(x, y)$ and the nominal contact region Ω from Eqs. (2.54) and (2.55) if we know the macroshapes of contacting bodies and the load applied to them. Thus, for the given parameters which describe the surface macroshape and microgeometry, the real area of contact

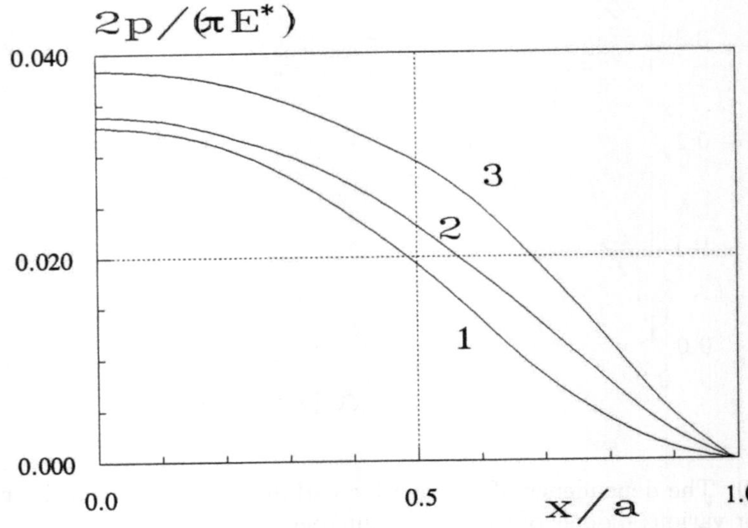

Figure 2.21: Nominal pressure distribution for the contact of a rough cylinder and a thick elastic layer for various microgeometry parameters.

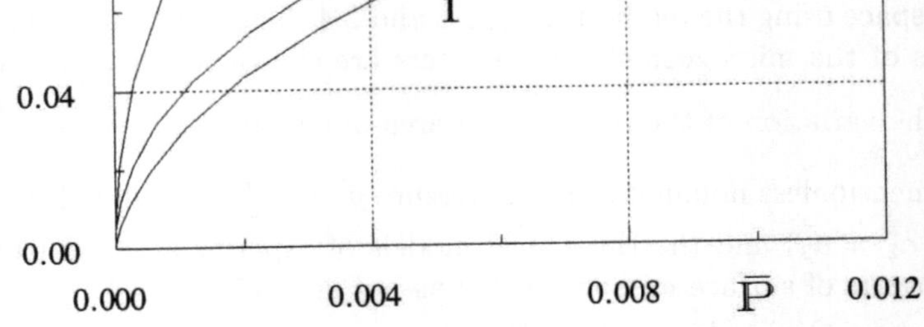

Figure 2.22: The variation of the relative real contact area with the load applied to the cylinder for the various microgeometry parameters.

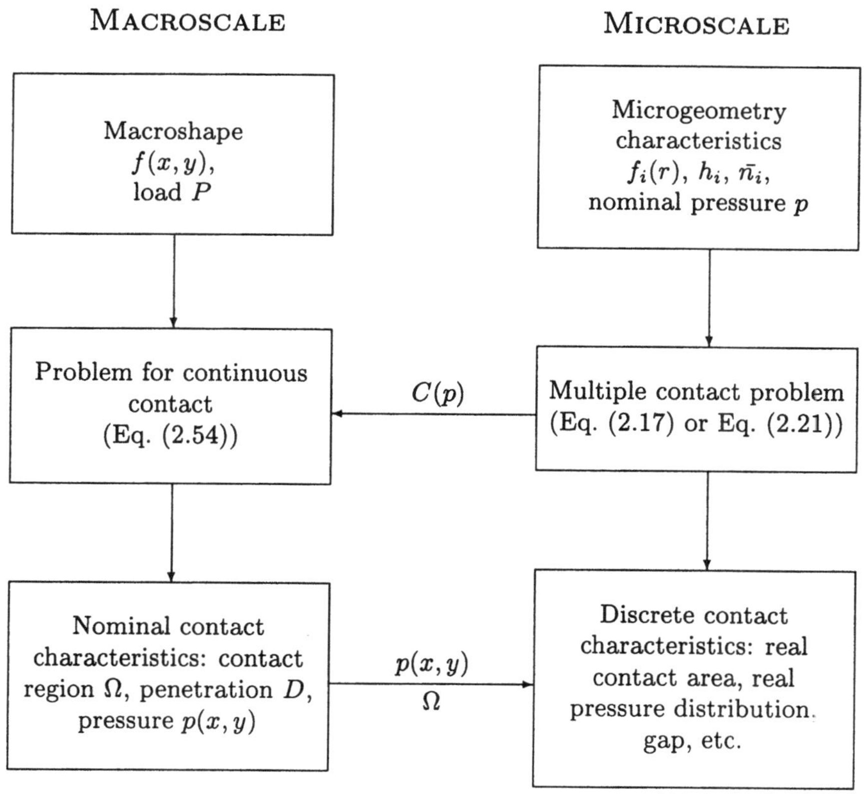

Figure 2.23: Scheme of the analysis of the contact characteristics, taking into account micro- and macro- geometry of the bodies in contact.

is determined from the formula

$$A_r = \iint\limits_{\Omega} \lambda\left(p(x,y)\right)\, dxdy. \tag{2.73}$$

By way of example, let us consider the 2-D contact problem for an elastic cylinder whose macroshape is described by the function $f(x) = \dfrac{x^2}{2R_0}$ (R_0 is the radius of the cylinder), and an elastic thick layer bonded with a rigid foundation, for the various parameters characterizing their surface microgeometry. We investigate the microgeometry modelled by the one-level or three-level systems of spherical indenters uniformly distributed over the surface of the contacting body. The functions $C(p)$ and $\lambda(p)$ for these kinds of microgeometry with given parameters of the density of asperity arrangement are shown in Fig. 2.17 and in Fig. 2.20, respectively.

Using the function $C(p)$, we determine the nominal pressure $p(x)$ and the contact half-width $\frac{a}{R_0}$ from Eqs. (2.63) and (2.68) for the given value of the dimensionless load $\bar{P}$ $\left(\bar{P} = \frac{2(1-\nu^2)P}{\pi E R_0}\right)$ applied to the cylinder. Fig. 2.21 illustrates the nominal pressure distribution within the nominal contact region for $\bar{P} = 3.2 \cdot 10^{-3}$ and the functions $C(p)$ presented in Fig. 2.17. The number of curves in Fig. 2.17 and Fig. 2.21 correspond to the particular model of the surface microgeometry. The half-widths of the nominal contacts for the models under consideration are $\frac{a}{R_0} = 0.09$ (curve 1), $\frac{a}{R_0} = 0.08$ (curve 2), $\frac{a}{R_0} = 0.065$ (curve 3).

Then the relative real area of contact $\frac{A_r}{A_a}$ where A_r is determined by Eq. (2.73) and A_a is the width of the nominal contact region ($A_a = 2a$) is

$$\frac{A_r}{A_a} = \int_{-1}^{1} \lambda(p(x))\, dx.$$

Fig. 2.22 illustrates the variation of the relative area of contact $\frac{A_r}{A_a}$ with the dimensionless load $\bar{P}$ for the various parameters describing the surface microgeometry (the curves with the same number in Fig. 2.17, Fig. 2.21 and Fig. 2.22 correspond to the same parameters of the surface microgeometry).

In a similar way it is possible to calculate the gap between the contacting bodies arising from their surface microgeometry, the number of asperities in contact, etc.

The estimation of the real contact pressure and its maximum values in contact of rough bodies is of interest in studies of internal stresses in the thin subsurface layers and the surface fracture (the wear) of bodies in contact interaction (see Chapter 5). If the microgeometry of the contacting bodies has a homogeneous structure along the surface, the maximum value of the real pressure occurs at the contact spots where the nominal pressure reaches its peak. This can be calculated from the multiple contact problem solution for the given maximum value of the nominal pressure.

Fig. 2.23 illustrates the general stages in calculation of the characteristics of the nominal and the real contact described above by the example of the determination of the relative real area of contact.

Chapter 3

Friction in Sliding/Rolling Contact

3.1 Mechanism of friction

The causes of friction have been explored for many years. According to the modern conception of tribology there are two main causes of energy dissipation which give rise to a resistance in sliding contact.

The first one is associated with the work done in making and breaking adhesion bonds formed in the points of contact of sliding surfaces. The force necessary to shear these bonds is termed the adhesive (molecular) component of the friction force. The mechanism for the formation of adhesion bonds depends on the properties of the contacting bodies and on the friction conditions. For sliding contact of metal surfaces, it is realized as the rupture of the welded *bridges* between the contacting surfaces. For sliding contact of rubbers and rubber-like polymers, the energy dissipation takes place in the process of thermal jumping of the molecular chains from one equilibrium state to another. The adhesive component of the friction force depends on the surface properties of both contacting bodies. An interesting approach to modelling of the adhesive interaction in sliding contact was developed in papers by Godet (1984), Alekseev and Dobychin (1994), where the motion of the substance of the *third-body* was investigated. The third-body is a thin layer at the interface between the contacting bodies. Its properties depend on the mechanical properties of the surface layers of the contacting bodies, the boundary film etc. However, up to now, there is no theoretical model for calculating the adhesive component of the friction force.

The adhesive friction is taken into account in the formulation of contact problems by some relationship between the stresses in the contact zone. The law of friction established experimentally by Coulomb (1785) is usually used to describe the relation between the normal p and tangential τ stresses in the contact zone:

$$\tau = \tau_0 + \mu p \tag{3.1}$$

Here τ_0 and μ are parameters of the friction law. It has been found that the value τ_0 is very small for polymers and boundary lubrication (see Kragelsky, Dobychin and Kombalov, 1982). Eq. (3.1) is used in the formulation of contact problems for elastic bodies in sliding contact (§ 3.2 and § 3.3).

The second cause of the energy dissipation is the cyclic deformation of the bodies in sliding contact. The resistive force connected with this process is termed the mechanical component of friction. It depends on the mechanical properties of the bodies in sliding contact, the geometry of their surfaces, the applied forces etc. Unlike the adhesive component, the mechanical component of friction force depends in the main on the deformation of the bodies in contact, and thus can be studied by the methods of contact mechanics.

Since there is no energy dissipation in the deformation of elastic bodies, the mechanical component of the friction force is equal to zero for elastic bodies. For example, in sliding contact of elastic cylinders the contact pressure is distributed symmetrically within the contact zone (which is also symmetrically placed with respect to the symmetry axis of the cylinder) for the case $\tau = 0$ and so there is no resistance to the relative motion. To study the mechanical component of the friction force, imperfect elasticity of contacting bodies must be taken into account. This is the reason for considering contact problems for viscoelastic bodies in this Chapter.

In tribology, the adhesive and mechanical components of friction force are usually considered as independent. However there are some experimental results which argue against this statement (see Moore, 1975). It has been established that the relation between the components of the friction force depends on friction conditions, mechanical properties of contacting bodies etc. The investigation of the sliding contact of viscoelastic bodies (§ 3.4) makes it possible to analyze the dependence between the mechanical and adhesive components of the friction force.

Both causes of energy dissipation also occur in rolling contact. It has been shown theoretically and experimentally that the resistance to rolling is caused by the following:

1. Friction due to the relative slip of the surfaces within the contact area arising from the differences of the curvature of the contacting surfaces, and their different mechanical properties. Reynolds (1875) was the first to establish this fact. It was also supported by experimental results of Heathcote (1921), Konvisarov and Pokrovskaia (1955), Pinegin and Orlov (1961) etc.

2. Imperfect elasticity of the contacting bodies (Tabor, 1952, Flom and Bueche, 1959, Flom, 1962, etc.).

3. The adhesive forces in the contact (Tomlinson, 1929).

The question is what is the contribution of each process to rolling resistance for different operating conditions? To answer this question the rolling contact of viscoelastic bodies is considered, taking into account the partial slip in the contact zone (§ 3.5).

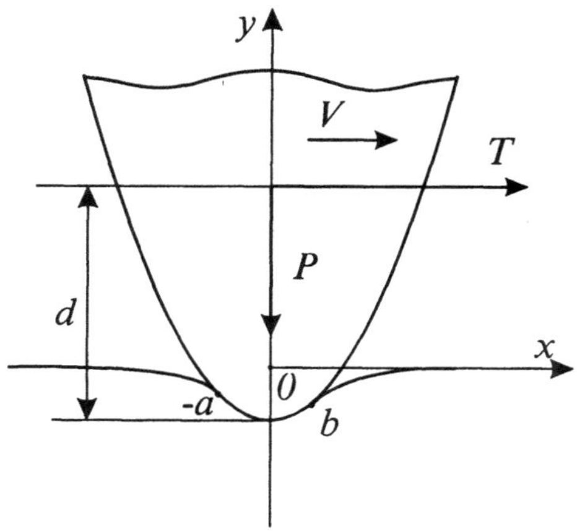

Figure 3.1: Sliding contact of a cylindrical punch and an elastic half-space.

As has been mentioned in Chapter 2, the roughness is usually modelled by a system of asperities described by some simple shape and a specific spatial distribution. The first stage of investigation of the contact of rough bodies is the consideration of the contact of two asperities. The methods of contact mechanics can be applied to this problem. So some of the results obtained in this Chapter can be used to describe the resistance to the relative motion of isolated asperities and rough surfaces.

3.2 Two-dimensional sliding contact of elastic bodies

3.2.1 Problem formulation

We consider a sliding contact of a rigid cylinder and an elastic half-space (Fig. 3.1). The shape of the rigid body is described by the function $y = f(x)$. External forces also are independent of the z-coordinate. This problem is considered as a two-dimensional (plane) problem for a punch and an elastic half-plane. The two-term friction law (3.1) is assumed to hold within the contact zone (-a,b):

$$\tau_{xy}(x) = (\tau_0 + \mu p(x)) \operatorname{sgn} V, \tag{3.2}$$

where $p(x) = -\sigma_y(x)$ and $\tau_{xy}(x)$ are the normal pressure and tangential stress at the surface of the elastic half-plane ($y = 0$), and V is the velocity of the cylinder.

Applied tangential T and normal P forces cause the body to be in the limiting equilibrium state, or to move with a constant velocity. This motion occurs so slowly that dynamic effects may be neglected.

In the moving coordinate system connected with the rigid cylinder, the following boundary conditions hold ($y = 0$)

$$\begin{aligned} \sigma_y = 0, \quad \tau_{xy} = 0, \quad (-\infty < x < -a, \quad b < x < +\infty), \\ v = f(x) - D, \quad \tau_{xy} = (\tau_0 + \mu p)\,\mathrm{sgn}\, V, \quad (-a \le x \le b), \end{aligned} \tag{3.3}$$

where v is the normal displacement of the half-plane surface, D is the approach of the contacting bodies.

The relationship between stresses and the normal displacement gradient at the boundary $y = 0$ of the lower half-plane has the form (Galin, 1980)

$$\frac{\pi E}{2(1-\nu^2)} \cdot \frac{\partial v}{\partial x} = \int_{-\infty}^{+\infty} \sigma_y \frac{dt}{t-x} - \frac{1-2\nu}{2-2\nu} \pi \tau_{xy}. \tag{3.4}$$

Using Galin's method (Galin, 1980), we introduce a function $w_1(z)$ of a complex variable in the lower half-plane $y \le 0$

$$w_1(z) = U_1 - iV_1 = \int_{-\infty}^{+\infty} \sigma_y \frac{dt}{t-z}. \tag{3.5}$$

Using (3.3), (3.4) and the limiting values of the Cauchy integral (3.5) as $z \to x - i0$, we can derive the following boundary conditions for the function $w_1(z)$

$$\begin{aligned} V_1 = 0, \quad (-\infty < x < -a, \quad b < x < +\infty), \\ U_1 + \mu\vartheta V_1 \,\mathrm{sgn}\, V = \pi F(x), \quad (-a \le x \le b), \end{aligned} \tag{3.6}$$

where

$$\begin{aligned} F(x) = \vartheta \tau_0 \,\mathrm{sgn}\, V + \frac{f'(x)}{\pi K}, \\ K = \frac{2(1-\nu^2)}{\pi E}, \quad \vartheta = \frac{1-2\nu}{2(1-\nu)}. \end{aligned} \tag{3.7}$$

So the problem is reduced to the determination of the analytic function $w_1(z)$ (3.5) based on the relationships (3.6) between its real and imaginary parts U_1, V_1 at the boundary of the region of its definition. This is a particular case of the Riemann–Hilbert problem.

The solution of this problem that satisfies the condition $w_1(z) \sim \frac{P}{z}$ as $z \to \infty$ and has the integrated singularities at the boundary is the following function

$$w_1(z) = \frac{\varrho}{X(z)} \int_{-a}^{b} F(t) X^+(t) \frac{dt}{t-z} + \frac{P}{X(z)}, \tag{3.8}$$

where

$$\begin{aligned} X(z) &= (z+a)^{1/2+\eta}(z-b)^{1/2-\eta}, \\ X^+(t) &= (t+a)^{1/2+\eta}(b-t)^{1/2-\eta}, \end{aligned}$$

$$\varrho = \frac{1}{\sqrt{1+\mu^2\vartheta^2}}, \quad \eta = \frac{1}{\pi}\arctan(\mu\vartheta)\,\mathrm{sgn}\,V, \quad |\eta| < \frac{1}{2}. \tag{3.9}$$

Using the function (3.8), we can determine the stress-strain state of the elastic half-plane. For example, Eq. (3.5) implies that the normal stress at the x-axis $\sigma_y(x,o)$ is the imaginary part of the function (3.8) as $z \to x - i0$. The limiting value of the Cauchy integral

$$\Phi(z) = \frac{1}{2\pi i}\int_{-\infty}^{+\infty} \varphi(t)\frac{dt}{t-z}$$

as $z \to x - i0$ can be determined by the Plemelj (1908) formula (see also Muskhelishvili, 1949)

$$\Phi^-(x) = \frac{1}{2\pi i}\int_{-\infty}^{+\infty} \varphi(t)\frac{dt}{t-x} - \frac{1}{2}\varphi(x).$$

The limiting value of the function $\dfrac{1}{X(z)}$ as $z \to x - i0$ is determined by the formula

$$\left[\frac{1}{(z+a)^{1/2+\eta}(z-b)^{1/2-\eta}}\right]_{z=x-i0} =$$

$$= \begin{cases} -\dfrac{1}{(-a-x)^{1/2+\eta}(b-x)^{1/2-\eta}}, & (-\infty < x < -a), \\ \dfrac{\sin\pi\eta + i\cos\pi\eta}{(x+a)^{1/2+\eta}(b-x)^{1/2-\eta}}, & (-a \le x \le b), \\ \dfrac{1}{(x+a)^{1/2+\eta}(x-b)^{1/2-\eta}}, & (b < x < +\infty). \end{cases}$$

So the contact pressure $p(x) = -\sigma_y(x,0) = -\dfrac{1}{\pi}V_1(x,0)$, where $V_1(x,0)$ is the imaginary part of the function $w_1(z)$ as $z \to x - i0$, is given by

$$p(x) = -F(x)\varrho\sin\pi\eta +$$

$$\frac{\varrho}{\pi}\cdot\frac{\cos\pi\eta}{X^+(x)}\int_{-a}^{b} F(t)X^+(t)\frac{dt}{t-x} + \frac{P\cos\pi\eta}{\pi X^+(x)}, \quad x \in (-a,b). \tag{3.10}$$

3.2.2 Contact problem for a cylinder

We consider the particular case of a sliding contact of a rigid cylinder and an elastic half-space. For this case $f(x) = \dfrac{x^2}{2R}$ and the function $F(x)$ (3.7) becomes

$$F(x) = \tau_0\vartheta\,\mathrm{sgn}\,V + \frac{x}{\pi K R}. \tag{3.11}$$

Substituting (3.11) in (3.10) and using the following relationships (Gradshteyn and Ryzhik, 1963)

$$\int_{-a}^{b} (a+t)^{-1/2+\eta}(b-t)^{-1/2-\eta}\frac{dt}{t-x} =$$

$$\pi \tan \pi\eta (a+x)^{-1/2+\eta}(b-x)^{-1/2-\eta}, \quad (-a < x < b),$$

$$\int_{-a}^{b} (a+x)^{\mu-1}(b-x)^{\nu-1}\,dx = (a+b)^{\mu+\nu-1}B(\mu,\nu), \quad (\mu > 0,\ \nu > 0),$$

we obtain the expression for the contact pressure

$$p(x) = \frac{\varrho L(x)}{(x+a)^{-1/2+\eta}(b-x)^{-1/2-\eta}},$$

$$L(x) = \frac{P}{\pi} + \frac{(a+b)^2\left(\frac{1}{4}-\eta^2\right)}{2\pi KR} + \tau_0\vartheta(a+b)\left(\frac{1}{2}-\eta\right)\operatorname{sgn} V \tag{3.12}$$

$$+\frac{(a+b)x}{\pi KR}\left(\frac{1}{2}-\eta\right) - \tau_0\vartheta(x+a)\operatorname{sgn} V + \frac{(a+x)x}{\pi KR}.$$

The contact pressure (3.12) has to be bounded at the ends of the contact zone. Equation (3.12) shows that if it is bounded there, it must in fact be zero there, i.e. $p(-a) = p(b) = 0$ and

$$l^2 = (a+b)^2 = \frac{2RPK}{\frac{1}{4}-\eta^2}, \tag{3.13}$$

$$\frac{a-b}{2} = l\eta + \tau_0\vartheta KR\pi \operatorname{sgn} V. \tag{3.14}$$

So that

$$p(x) = \frac{\varrho}{\pi KR}(x+a)^{1/2-\eta}(b-x)^{1/2+\eta}. \tag{3.15}$$

The relationships (3.13), (3.14) and (3.15) determine the contact width, the shift of the contact zone and the contact pressure, respectively. Equations (3.13) and (3.15) coincide with the ones obtained by Galin (1953), where the contact problem in the analogous formulation with Amontons'(1699) law of friction $\tau_{xy} = \mu\sigma_y$ was considered.

The results indicate that the magnitude τ_0 in the law (3.2) influences only the contact displacement (3.14).

It follows from Eq. (3.15) that the contact pressure is an unsymmetrical function. It provides the moment M

$$M = \int_{-a}^{b} p(x)x\,dx = -P\left[\frac{4}{3}l\eta + \pi\vartheta K R\tau_0 \operatorname{sgn} V\right], \tag{3.16}$$

where

$$P = \int_{-a}^{b} p(x)\,dx.$$

If there is no active moment applied to the cylinder, the moment M is equal to the moment of the tangential force T

$$T = \int_{-a}^{b} \tau_{xy} dx = (\tau_0 l + \mu P) \operatorname{sgn} V. \tag{3.17}$$

In this case, it follows from the equilibrium conditions that the force T must be applied at the point $(0, d)$ (Fig. 3.1): $d = \left|\frac{M}{T}\right|$.

Note that in most cases $\mu\vartheta \ll 1$, so that we may approximate Eq. (3.9) by

$$|\eta| \approx \frac{\mu}{\pi}\vartheta \ll 1.$$

Based on this estimation, it follows from Eqs. (3.13), (3.14) and (3.15) that the friction coefficient μ has no essential influence on the contact pressure, the shift or the width of contact zone.

The analysis of subsurface stresses revealed that the effect of the parameter τ_0 on the stress-strain state in an elastic body is similar to a friction coefficient μ: it moves the point where the maximum principal shear stress $(\tau_1)_{\max}$ takes place closer to the surface, and it increases the magnitude of $(\tau_1)_{\max}$ (Fig. 3.2).

Eqs. (3.13) – (3.16) can be used to determine contact characteristics (contact width and displacement, contact pressure etc.) for sliding contact of two elastic bodies with radii of curvature R_1 and R_2. We replace the parameters K, ϑ, R and η (see Eqs. (3.7) and (3.9)) by the parameters K^*, ϑ^*, R^*, η^*. For plane stress

$$K^* = \frac{2}{\pi}\left(\frac{1}{E_1} + \frac{1}{E_2}\right), \quad \vartheta^* = \frac{1}{K\pi}\left(\frac{1-\nu_2}{E_2} - \frac{1-\nu_1}{E_1}\right), \tag{3.18}$$

and for plane strain

$$\begin{gathered} K^* = \frac{2}{\pi}\left(\frac{1-\nu_1^2}{E_1} + \frac{1-\nu_2^2}{E_2}\right), \\ \vartheta^* = \frac{1}{K\pi}\left[\frac{(1+\nu_2)(1-2\nu_2)}{E_2} - \frac{(1+\nu_1)(1-2\nu_1)}{E_1}\right], \end{gathered} \tag{3.19}$$

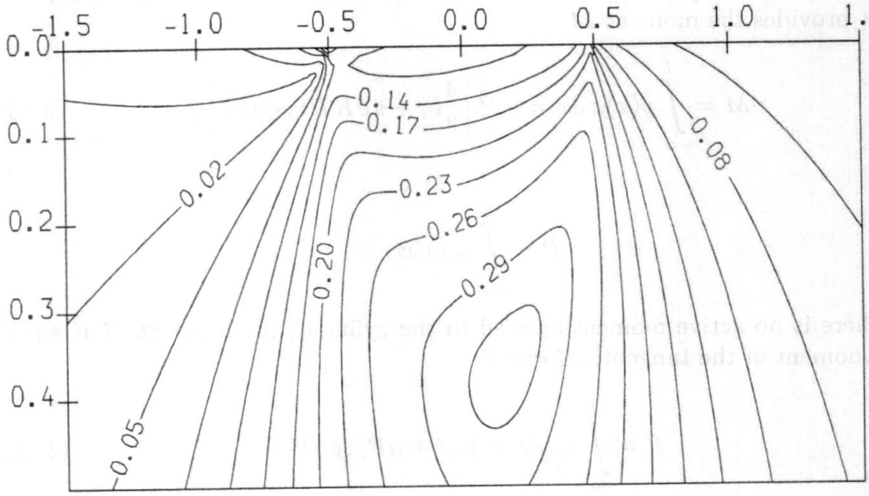

Figure 3.2: Contours of the principal shear stress beneath a sliding contact ($\mu = 0,\ \tau_0/p_0 = 0.1$).

and

$$\frac{1}{R^*} = \frac{1}{R_1} + \frac{1}{R_2}, \quad \eta^* = \frac{1}{\pi}\arctan(\mu\vartheta^*)\,\mathrm{sgn}\,V.$$

Provided that $l \ll R_i$, $(i = 1, 2)$ we can consider the cylinders as half-planes. So we use Eq. (3.4) to determine the gradient of normal displacement for both cylinders, taking into account the relationship: $\tau_{xy}^{(1)} = -\tau_{xy}^{(2)}$.

3.2.3 Contact problem for a flat punch

We consider sliding contact of a punch with a flat base (Fig. 3.3). Under the applied forces, the punch has the inclination γ. So the equation for the punch shape is $f(x) = -\gamma x - D$.

The function $F(x)$ (3.7) has the following form

$$F(x) = \tau_0\vartheta\,\mathrm{sgn}\,V - \frac{\gamma}{\pi K} \quad (-a \le x \le b). \tag{3.20}$$

We introduce the dimensionless parameter

$$\kappa = \frac{b}{P}\left(\tau_0\vartheta\,\mathrm{sgn}\,V - \frac{\gamma}{\pi K}\right). \tag{3.21}$$

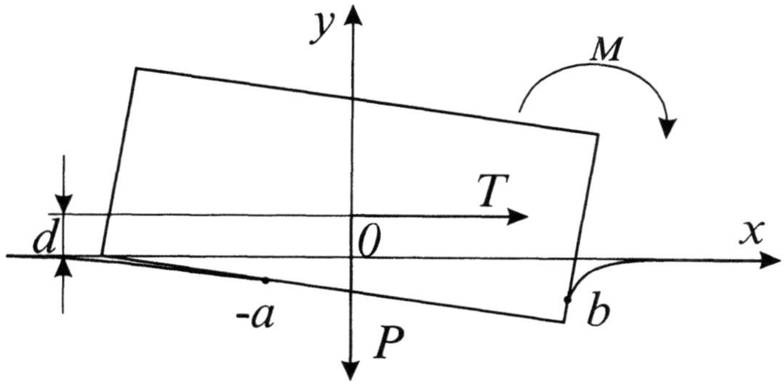

Figure 3.3: Sliding contact of a flat punch and an elastic half-plane.

Substituting Eq. (3.20) in Eq. (3.10) and transforming this equation, we have

$$p(x) = \frac{P\varrho}{\pi b} \cdot \frac{b + \pi\kappa\left[b - x - \left(\frac{1}{2} + \eta\right)(a+b)\right]}{(x+a)^{1/2+\eta}(b-x)^{1/2-\eta}}. \tag{3.22}$$

Eq. (3.22) shows that the contact pressure near the ends of contact zone ($x \to +0$) can be represented as

$$p(-a+x) = \frac{A_1}{x^{1/2+\eta}} + O\left(x^{1/2-\eta}\right),$$
$$A_1 = \frac{P\varrho}{\pi b} \cdot \frac{\left[b + \pi\kappa(a+b)\left(\frac{1}{2} - \eta\right)\right]}{(a+b)^{1/2-\eta}}, \tag{3.23}$$

$$p(b-x) = \frac{A_2}{x^{1/2-\eta}} + O\left(x^{1/2+\eta}\right),$$
$$A_2 = \frac{P\varrho}{\pi b} \cdot \frac{\left[b - \pi\kappa(a+b)\left(\frac{1}{2} + \eta\right)\right]}{(a+b)^{1/2+\eta}}. \tag{3.24}$$

We consider the case of a complete contact of a flat punch and an elastic half-plane. Setting $a = b$ in Eq. (3.22) we have

$$p(x) = \frac{P\varrho}{\pi b} \cdot \frac{[b - \pi\kappa(x + 2b\eta)]}{(x+b)^{1/2+\eta}(b-x)^{1/2-\eta}}. \tag{3.25}$$

The contact pressure is a nonnegative function, $p(x) \geq 0$ ($-b \leq x \leq b$), and hence

$$\kappa_1 \leq \kappa \leq \kappa_2, \tag{3.26}$$

where

$$\kappa_1 = -\frac{1}{\pi(1-2\eta)}, \quad \kappa_2 = \frac{1}{\pi(1+2\eta)}. \tag{3.27}$$

The contact pressure $p(x)$ given by Eq. (3.25) tends to infinity at the edges of the punch ($x = \pm b$), if $\kappa \in (\kappa_1,\ \kappa_2)$. If $\kappa = \kappa_1$ or $\kappa = \kappa_2$, the contact pressure is zero at the left end or at the right end of the contact zone, respectively.

If the parameter $\kappa \notin [\kappa_1,\ \kappa_2]$, there is only partial contact. If $\kappa \leq \kappa_1 < 0$ the separation of the punch base from the half-plane appears at the left-hand end of the contact zone at the point $x = -a$. The contact width is found according to Eq. (3.23)

$$a + b = -\frac{b}{\pi\kappa\left(\frac{1}{2} - \eta\right)}. \tag{3.28}$$

Using Eqs. (3.22) and (3.28), we obtain the contact pressure

$$p(x) = \frac{P\varrho}{\pi b} \cdot \frac{\left[b + \pi\kappa(b - x)\left(\frac{1}{2} - \eta\right)\right]}{\left(\frac{1}{2} - \eta\right)(x + a)^{1/2+\eta}(b - x)^{1/2-\eta}}. \tag{3.29}$$

If $\kappa > \kappa_2 > 0$ the contact pressure is zero at the right-hand end of the contact zone at the point $x = b$, where $|b| < a$ (a is the half-width of the punch in this case). Using (3.22) and (3.24), we find the equation for the contact pressure

$$p(x) = \frac{P\kappa\varrho}{b}\left(\frac{b - x}{x + a}\right)^{1/2+\eta}.$$

It follows from Eqs. (3.21) and (3.24) that the coordinate $x = b$ is determined by the formula

$$b = -a + \frac{P}{\pi\left(\tau_0\vartheta \operatorname{sgn} V - \frac{\gamma}{\pi K}\right)\left(\frac{1}{2} + \eta\right)}. \tag{3.30}$$

The contact pressure distributions for different values of the parameter κ are shown in Fig. 3.4. The curves 1 – 4 correspond to the cases of complete contact and pressure approaching to infinity at the ends of contact zone ($\kappa = 0$), complete contact when $p(-b) = 0$ ($\kappa = \kappa_1$, see Eqs. (3.25) and (3.27)), and partial contact ($\kappa = -0.5$ and $\kappa = -0.75$), respectively. For the calculations we used $|\mu\vartheta| = 0.057$ ($\mu = 0.2$, $\nu = 0.3$). Note that for frictionless contact ($\mu = 0$, $\tau_0 = 0$) the results obtained in this part coincide with those obtained by Galin (1953).

The parameter κ depends on the inclination γ (see Eq. (3.21)). For definiteness, let us consider the punch moving in the x-axis direction ($V > 0$). The parameter γ can be found using the equilibrium conditions for the punch. The normal load P, the tangential force T, and the active moment M are applied to the punch (see Fig. 3.3). The contact pressure $p(x)$ and the tangential stress $\tau_{xy}(x)$ form the resistance forces which satisfy the following equilibrium conditions:

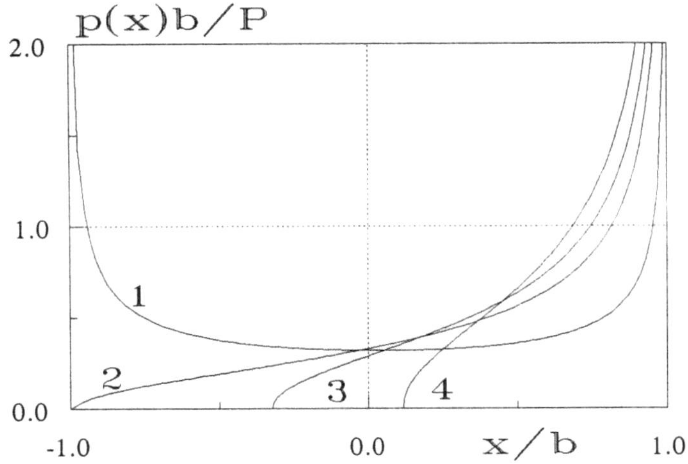

Figure 3.4: Contact pressure under a flat inclined punch sliding on an elastic half-plane ($\mu\vartheta = 0.057$); $\kappa = 0$ (curve 1); $\kappa = \kappa_1 = -0.33$ (curve 2); $\kappa = -0.5$ (curve 3); $\kappa = -0.75$ (curve 4).

$$P = \int_{-a}^{b} p(x)dx,$$
$$T = \int_{-a}^{b} \tau_{xy}(x)dx = \tau_0(a+b) + \mu P, \tag{3.31}$$

$$\int_{-a}^{b} (b-x)p(x)dx - Pb + Td - M = 0, \tag{3.32}$$

where $(0, d)$ are the coordinates of the point where the force T is applied, and M is the active moment relative to the point $x = b$.

Using Eqs. (3.22) and (3.31), we can transform Eq. (3.32) to the following relation

$$-P(a+b)\left(\frac{1}{2} - \eta\right) + Pa + \frac{\pi P\kappa(a+b)^2}{2b}\left(\frac{1}{4} - \eta^2\right) + \tau_0(a+b)d + \mu Pd - M = 0. \tag{3.33}$$

Eqs. (3.20) and (3.33) are used to determine the inclination γ, which depends on both quantities d and M.

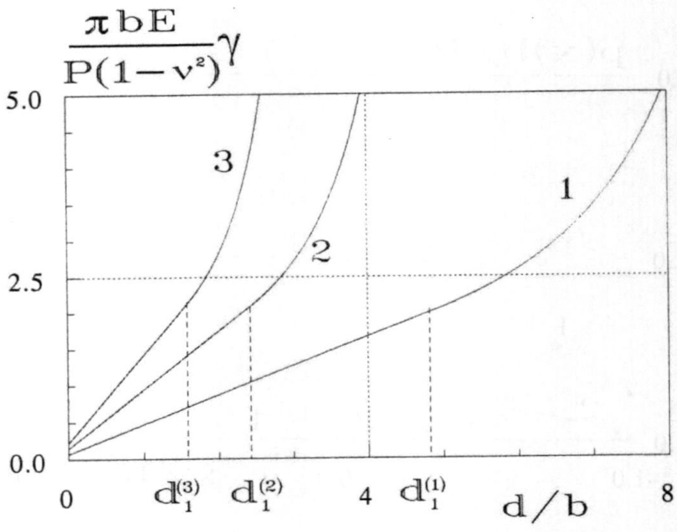

Figure 3.5: The effect of the position of the point of application of the tangential force T on the inclination of a punch ($\nu = 0.3$, $\tau_0 = 0$); $\mu = 0.1$ (curve 1), $\mu = 0.2$ (curve 2), $\mu = 0.3$ (curve 3); $d_1^{(i)}$, ($i = 1, 2, 3$) indicates the transition point from complete to partial contact.

Let us consider the particular case $M = 0$ and analyze the dependence of the inclination γ on the distance d. Using Eqs. (3.21), (3.26) and (3.33) we conclude that the complete contact occurs for $d \in (0, d_1)$, where

$$d_1 = \frac{Pb\left(\frac{1}{2} - \eta\right)}{2b\tau_0 + \mu P}. \tag{3.34}$$

The inclination γ for this case is

$$\gamma = \frac{2bP\eta + (2b\tau_0 + \mu P)d}{2b^2\left(\frac{1}{4} - \eta^2\right)}K + \tau_0 \pi \vartheta K. \tag{3.35}$$

If $d \in (d_1, d_2)$, the partial contact occurs with the separation point $x = -a$, where $|a| < b$; d_2 is determined by the condition $-a = b$, i.e. there is point contact. It follows from Eq. (3.33), that $d_2 = \dfrac{b}{\mu}$. The inclination γ of the punch for the case

$d_1 \le d < d_2$ is determined from Eqs. (3.21), (3.28) and (3.33)

$$\gamma = \frac{P\left(\dfrac{1}{2}+\eta\right) + 2\tau_0 d}{2\left(\dfrac{1}{2}-\eta\right)(b-\mu d)} K + \tau_0 \pi \vartheta K. \tag{3.36}$$

It follows from Eq. (3.36) that $\gamma \to +\infty$ (the punch is overturned) as $d \to d_2 - 0$. Fig. 3.5 illustrates the dependence of the inclination γ on the distance $d \in [0,\, d_2)$ for different magnitudes of the coefficient μ and $\tau_0 = 0$. The Eqs. (3.35) and (3.36) have been used to plot the curves.

The results of this analysis can be used in the design of devices for tribological tests. If two specimens with flat surfaces come into contact, the hinge is used to provide their complete contact. The results show that the hinge must be fixed at a distance $d \in (0, d_1)$ from the specimen base. The limiting distance d_1 essentially depends on the friction coefficient μ. If $\tau_0 = 0$, we obtain from Eq. (3.34)

$$\frac{d_1}{b} \approx \frac{1}{2\mu} - \frac{1-2\nu}{2\pi(1-\nu)}.$$

3.3 Three-dimensional sliding contact of elastic bodies

We investigate three dimensional contact problems under the assumption that friction forces are parallel to the motion direction. This case holds if the punch slides along the boundary of an elastic half-space with anisotropic friction. The friction depends in magnitude and direction on direction of sliding. The description of the anisotropic friction has been made by Vantorin (1962) and Zmitrovicz (1990). This friction occurs, for example, in sliding of monocrystals, which have properties in different directions which depend on the orientation of the crystal. Seal (1957) investigated friction between two diamond samples, and showed that the friction coefficient changes from 0.07 to 0.21, depending on the mutual orientation of the samples. A similar phenomenon was observed by Tabor and Wynne-Williams (1961) in experiments on polymers, where polymeric chains at the surface have special orientations.

For arbitrary surfaces, the assumption that friction forces are parallel to the motion direction is satisfied approximately.

3.3.1 The friction law has the form $\tau_{xz} = \mu p$

We consider the contact of a punch sliding along the surface of an elastic half-space. We assume the problem to be quasistatic, which imposes a definite restriction on the sliding velocity, and we introduce a coordinate system (x, y, z) connected with

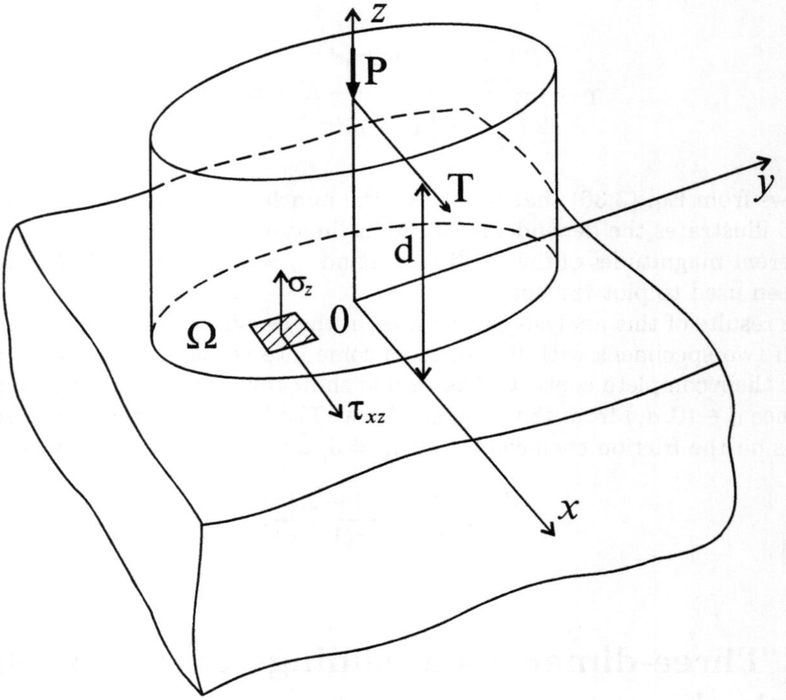

Figure 3.6: Sliding contact of a punch and an elastic half-space.

the moving punch (Fig. 3.6). The tangential stresses within the contact region Ω are assumed to be directed along the x-axis, and $\tau_{xz} = \mu p(x,y)$, where $p(x,y) = -\sigma_z(x,y,0)$ is the contact pressure ($p(x,y) \geq 0$). The boundary conditions have the form

$$\begin{aligned} &w = f(x,y) - D, \quad \tau_{xz} = -\mu\sigma_z, \quad \tau_{yz} = 0, \quad x,y \in \Omega, \\ &\sigma_z = \tau_{xz} = \tau_{yz} = 0, \qquad\qquad\qquad\qquad x,y \notin \Omega. \end{aligned} \tag{3.37}$$

Here $f(x,y)$ is the shape of the punch, and D is its displacement along the z-axis.

The displacement w of the half-space boundary in the direction of the z-axis can be represented as the superposition of the displacements caused by the normal pressure $p(x,y)$ and the tangential stress τ_{xz} within the contact zone. The solution of the problem for the elastic half-space loaded by a concentrated force at the origin with components T_x, T_z along the x- and z-axis, gives the vertical displacement

w on the plane $z = 0$ as

$$w = \frac{1-\nu^2}{\pi E} \cdot \frac{T_z}{R} + \frac{(1+\nu)(1-2\nu)}{2\pi E} \cdot \frac{xT_x}{R^2},$$
$$\left(R = \sqrt{x^2+y^2}\right). \tag{3.38}$$

Integrating (3.38) over the contact area Ω and taking into account conditions (3.37), we obtain the following integral equation to determine the contact pressure $p(x, y)$

$$\iint\limits_{\Omega} \left[\frac{1}{\sqrt{(x-x')^2+(y-y')^2}} + \frac{\mu\vartheta(x-x')}{(x-x')^2+(y-y')^2}\right] p(x', y')dx'dy'$$
$$= \frac{\pi E}{1-\nu^2}\left[D - f(x,y)\right], \tag{3.39}$$
$$\vartheta = \frac{1-2\nu}{2-2\nu}.$$

The coefficient ϑ is equal to zero when $\nu = 0.5$, i.e. the elastic body is incompressible; in this case, friction forces do not affect the magnitude of the normal pressure. For real bodies, Poisson's ratio ν satisfies the inequality $0 < \nu < 0.5$, hence the coefficient ϑ varies between the limits $0.5 > \vartheta > 0$; for example, $\vartheta = 0.286$ for $\nu = 0.3$. Moreover, it should be remembered that the magnitude of the friction coefficient μ is also small. For dry friction of steel on steel, $\mu = 0.2$. In the case $\nu = 0.3$, $\mu\vartheta \approx 0.057$. For lubricated surfaces, the coefficient $\mu\vartheta$ takes a still smaller value.

We investigate Eq. (3.39), assuming the parameter $\mu\vartheta = \varepsilon$ to be small, and use the notation $p_0(x, y)$ for the solution of the integral equation (3.39) in the case $\mu\vartheta = 0$. We represent the function $p(x, y)$ in the form of the series

$$p(x, y) = p_0(x, y) + \varepsilon p_1(x, y) + \cdots + \varepsilon^n p_n(x, y) + \cdots. \tag{3.40}$$

Substituting the series (3.40) into the integral equation (3.39), we obtain a recurrent system of equations for the unknown functions $p_n(x, y)$

$$A[p_n(x, y)] = B[p_{n-1}(x, y)], \quad n = 1, 2, \ldots. \tag{3.41}$$

Here the following notations are introduced for operators

$$A[\omega] = \iint\limits_{\Omega} \frac{\omega(x', y')dx'dy'}{\sqrt{(x-x')^2+(y-y')^2}},$$
$$B[\omega] = -\iint\limits_{\Omega} \frac{\omega(x', y')(x-x')dx'dy'}{(x-x')^2+(y-y')^2}. \tag{3.42}$$

The convergence of the series (3.40) was proved (Galin and Goryacheva, 1983) for the case of a bounded function ω.

As an illustration, let us consider sliding contact of an axisymmetric punch of circular planform, $f(r) = \dfrac{r^2}{2R}$, ($r \leq a$, a is the radius of the contact region Ω, R is the radius of curvature of the punch surface). We introduce the polar coordinates (r, θ), i.e.

$$x = r\cos\theta, \quad y = r\sin\theta.$$

As is known (see, for example, Galin (1953) or Johnson (1987)), in this case the function $p_0(x, y) = p_0(r)$ is

$$p_0(r) = \frac{4}{R\pi^2 K}\sqrt{a^2 - r^2}.$$

where K is determined in Eq. (3.7).

To find the next term $p_1(r, \theta)$ in the series (3.40), first we find $B[p_0(r)]$, which is the result of integration

$$B\left[p_0(r)\right] = b(r)\cos\theta,$$

$$b(r) = \frac{8}{3R\pi K r}\left[\left(a^2 - r^2\right)^{3/2} - a^3\right].$$

Then we solve the equation

$$A[p_1(r, \theta)] = b(r)\cos\theta. \tag{3.43}$$

We will seek the solution of the equation (3.43) in the form

$$p_1(r, \theta) = q(r)\cos\theta.$$

Changing to polar coordinates in Eq.(3.42) we obtain

$$A[p_1(r, \theta)] = \int_0^a\int_0^{2\pi} \frac{q(r')\cos(\theta')r'\,dr'\,d\theta'}{\sqrt{r^2 + r'^2 - 2rr'\cos(\theta - \theta')}}.$$

Using tables of Gradshteyn and Ryzhik (1963, 3.674), we calculate the integral

$$\int_0^{2\pi} \frac{\cos(\theta')\,d\theta'}{\sqrt{r^2 + r'^2 - 2rr'\cos(\theta - \theta')}} = Q(r, r')\cos\theta,$$

where

$$Q(r, r') = \begin{cases} \dfrac{4}{r'}\left[K\left(\dfrac{r'}{r}\right) - E\left(\dfrac{r'}{r}\right)\right], & r' < r, \\[2ex] \dfrac{4}{r}\left[K\left(\dfrac{r}{r'}\right) - E\left(\dfrac{r}{r'}\right)\right], & r' > r. \end{cases}$$

$K(x)$ and $E(x)$ are the complete elliptic integrals of the first and second kinds, respectively. So Eq.(3.43) reduces to the equation for determining the function $q(r)$

$$\int_0^r \left[K\left(\frac{r'}{r}\right) - E\left(\frac{r'}{r}\right)\right] q(r')dr' + \frac{1}{r}\int_r^a \left[K\left(\frac{r}{r'}\right) - E\left(\frac{r}{r'}\right)\right] r'q(r')dr' = \frac{1}{4}b(r).$$

The other terms in the series (3.40) have the form (Galin and Goryacheva, 1983)

$$p_n(r,\theta) = \sum_{k=1}^{n} q_{nk}(r)\cos k\theta.$$

So in the case of sliding contact with friction, the contact pressure has the form $p(r,\theta) = p_0(r) + \varepsilon q(r)\cos\theta + O\left(\varepsilon^2\right)$ which indicates, in particular, that the contact pressure is distributed nonsymmetrically, so that there is an additional moment M_y with respect to the y-axis:

$$M_y = \int_0^a\int_0^{2\pi} p(r,\theta)r^2\cos\theta dr\, d\theta = \varepsilon\pi\int_0^a q(r)r^2 dr + O\left(\varepsilon^2\right).$$

It follows from the equilibrium condition that the force T directed along the x-axis that causes the punch motion, should be applied at a distance $d = \left|\dfrac{M_y}{\mu P}\right|$ from the base. When this is not satisfied, the punch has an inclined base, which implies a change of the boundary conditions (3.37).

The contact problem for the punch with the flat circular base was investigated in the paper of Galin and Goryacheva (1983). It has been shown that the contact pressure can be presented in the form

$$p(r,\theta) = \frac{\psi(r,\theta)}{(a-r)^{1/2+\eta}},$$

where $\eta = \dfrac{1}{\pi}\arctan(\varepsilon\cos\theta)$, and $\psi(r,\theta)$ is a bounded and continuous function. To obtain this function, we again use the method of series-expansion with respect to the small parameter ε.

For the flat punch, the function $w(r,\theta)$ in (3.37) has the form $w(r,\theta) = \gamma r\cos\theta - D$. The unknown coefficient γ governing the inclination of the punch can be found from the equilibrium condition for the moments acting on the punch (see § 3.2).

3.3.2 The friction law has the form $\tau_{xz} = \tau_0 + \mu p$

Consider the sliding contact of the punch and an elastic half-space, and assume that tangential stresses within the contact region are directed along the x-axis and satisfy the friction law (3.1). Based on Eq. (3.38), we obtain the following integral equation for the contact pressure $p(x,y)$

$$\iint\limits_{\Omega}\left[\frac{1}{\sqrt{(x-x')^2+(y-y')^2}}+\frac{\mu\vartheta(x-x')}{(x-x')^2+(y-y')^2}\right]p(x',y')\,dx'dy'$$
$$+\vartheta\tau_0\iint\limits_{\Omega}\frac{x-x'}{(x-x')^2+(y-y')^2}\,dx'dy'=\frac{2}{K}\left[D-f(x,y)\right]. \tag{3.44}$$

The second integral in the left-hand part of Eq. (3.44) can be calculated if the contact domain Ω is given. For example, if Ω is the circle of the radius a, we may change to polar coordinates, and find

$$\iint\limits_{x'^2+y'^2\leq a^2}\frac{(x-x')\,dx'dy'}{(x-x')^2+(y-y')^2}=\int\limits_0^a\int\limits_0^{2\pi}\frac{(r\cos\theta-r'\cos\theta')r'dr'd\theta'}{r^2+r'^2-2rr'\cos(\theta-\theta')}.$$

Using the relationship

$$\int\limits_0^{2\pi}\frac{(r\cos\theta-r'\cos\theta')\,d\theta'}{r^2+r'^2-2rr'\cos(\theta-\theta')}=\begin{cases}\dfrac{2\pi}{r}\cos\theta, & \left|\dfrac{r'}{r}\right|<1,\\[2ex] 0, & \left|\dfrac{r'}{r}\right|>1,\end{cases}$$

and the result of integration

$$\int\limits_0^r\frac{2\pi}{r}\cos\theta r'\,dr'=\pi r\cos\theta=\pi x,$$

we reduce Eq. (3.44) to

$$\iint\limits_{\Omega}p(x',y')\left[\frac{1}{\sqrt{(x-x')^2+(y-y')^2}}+\frac{\mu\vartheta(x-x')}{(x-x')^2+(y-y')^2}\right]dx'dy'$$
$$=\frac{2}{K}\left[D-f(x,y)\right]-\pi\vartheta\tau_0x. \tag{3.45}$$

Eq. (3.45) differs from Eq. (3.39) only by the right side. The method of expansion with respect to the small parameter $\varepsilon=\mu\vartheta$ can again be used to solve Eq. (3.45).

Let us analyze the influence of the parameter $\dfrac{\tau_0}{E}$ on the solution of Eq. (3.45). At first, we consider the case of a smooth punch with surface described by the function $f(x,y) = \dfrac{x^2+y^2}{2R}$. Then the right side of Eq. (3.45) can be rewritten in the form

$$\frac{2}{K}[D - f(x,y)] - \pi\vartheta\tau_0 x = \frac{2}{K}\left[D_1 - \frac{(x+e)^2+y^2}{2R}\right], \tag{3.46}$$

where

$$e = \frac{\vartheta\tau_0 R\pi K}{2}, \quad D_1 = D + \frac{R\vartheta^2\tau_0^2\pi^2 K^2}{8}. \tag{3.47}$$

The relationships (3.47) indicate that the shift of the contact region e and the indentation of the punch D depend on the value of $\dfrac{\tau_0}{E}$.

Then let us consider the sliding contact of a punch with a flat base $\left(f(x,y) = 0,\ x^2+y^2 \le a^2\right)$. In this case the right-hand side of Eq. (3.45) has the form

$$\frac{2}{K}[D - f(x,y)] - \pi\vartheta\tau_0 x = \frac{2}{K}D - \pi\vartheta\tau_0 x.$$

In this case the contact pressure distribution corresponds to the solution of Eq. (3.39) for the punch with inclined flat base; the angle of inclination is proportional to $\pi\vartheta\tau_0$.

This conclusion about the influence of τ_0 on the contact characteristics is in a good agreement with that made in the two-dimensional problem (see § 3.2).

3.4 Sliding contact of viscoelastic bodies

We consider a rigid cylinder moving over a viscoelastic base with a constant velocity V (Fig. 3.7). We assume that the velocity V is much smaller than the speed of sound in the viscoelastic body, which permits the inertial terms to be neglected in the equilibrium equations. Note that the typical values of the speed of sound (V_s) are $V_s \approx 5 \cdot 10^3$ m/s (for steels), $V_s \approx 10^3$ m/s (for polymer materials), $V_s \approx 30 - 50$ m/s (for soft rubbers).

3.4.1 Constitutive equations for the viscoelastic body

The relationships between the strain and stress components in an isotropic viscoelastic body are taken in the following form:

$$\begin{aligned}
\varepsilon_{x^0} + T_\varepsilon \frac{\partial \varepsilon_{x^0}}{\partial t} &= \frac{1-\nu^2}{E}\left(\sigma_{x^0} + T_\sigma \frac{\partial \sigma_{x^0}}{\partial t}\right) \\
&\quad - \frac{\nu(1+\nu)}{E}\left(\sigma_{y^0} + T_\sigma \frac{\partial \sigma_{y^0}}{\partial t}\right), \\
\varepsilon_{y^0} + T_\varepsilon \frac{\partial \varepsilon_{y^0}}{\partial t} &= \frac{1-\nu^2}{E}\left(\sigma_{y^0} + T_\sigma \frac{\partial \sigma_{y^0}}{\partial t}\right) \\
&\quad - \frac{\nu(1+\nu)}{E}\left(\sigma_{x^0} + T_\sigma \frac{\partial \sigma_{x^0}}{\partial t}\right), \\
\gamma_{x^0y^0} + T_\varepsilon \frac{\partial \gamma_{x^0y^0}}{\partial t} &= \frac{1+\nu}{E}\left(\tau_{x^0y^0} + T_\sigma \frac{\partial \tau_{x^0y^0}}{\partial t}\right).
\end{aligned} \tag{3.48}$$

Here T_ε and T_σ are quantities characterizing the viscous properties of the medium, E and ν are the Young's modulus of elasticity and Poisson's ratio, respectively. Plane strain is considered here; plane stress can be considered in the similar way.

Eqs. (3.48) constitute the two-dimensional extension of the Maxwell–Thomson model, for which $H = \frac{T_\varepsilon E}{T_\sigma}$ is the instantaneous modulus of elasticity, $T_\varepsilon > T_\sigma$. The parameter $\frac{T_\varepsilon}{T_\sigma}$ is equal to $10^5 - 10^7$ for amorphic polymer materials, $10 - 10^2$ for high level crystalline polymer materials, $1.1 - 1.5$ for black metals; $\frac{1}{T_\varepsilon}$ is the coefficient of retardation.

Let us introduce a coordinate system (x, y) connected with the center of the cylinder (Fig. 3.7)

$$x = x^0 - Vt, \quad y = y^0.$$

The state of the viscoelastic medium is steady with respect to this coordinate system. The displacements and stresses depend on the coordinates (x, y) and are independent of time. i.e. $u^0(x + Vt, t) \equiv u(x)$, $\quad v^0(x + Vt, t) \equiv v(x)$ etc. After differentiating the first identity with respect to t and x, we obtain

$$\frac{\partial u^0(x^0,t)}{\partial x^0} V + \frac{\partial u^0}{\partial t} = 0, \quad \frac{\partial u^0(x^0,t)}{\partial x^0} = \frac{\partial u(x)}{\partial x},$$

or

$$\frac{\partial u^0(x^0,t)}{\partial t} = -V \frac{\partial u(x)}{\partial x}.$$

The time derivative of the function $v^0(x^0, t)$ and all components of stresses and strains in (3.48) can be found by the same procedure. Let us introduce the nota-

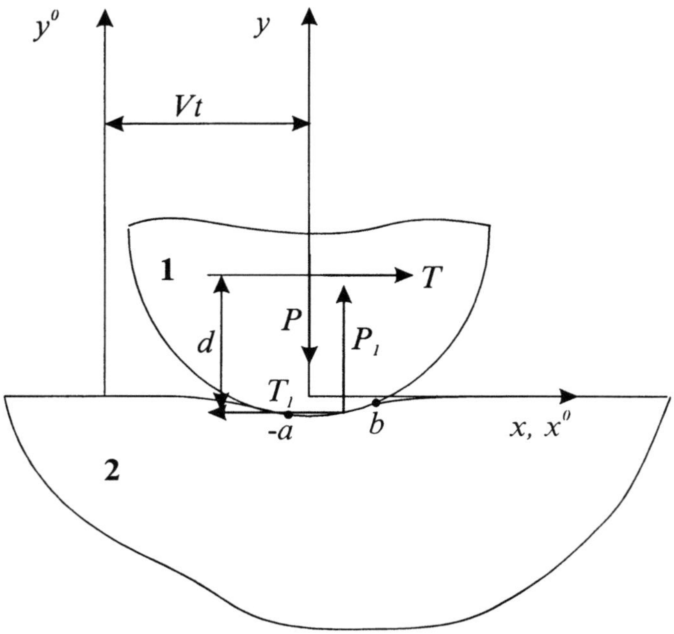

Figure 3.7: Scheme of the sliding contact of a cylinder and a viscoelastic half-space.

tions

$$\varepsilon^0_{ij} + T_\varepsilon \frac{\partial \varepsilon^0_{ij}}{\partial t} = \varepsilon_{ij} - T_\varepsilon V \frac{\partial \varepsilon_{ij}}{\partial x} = \varepsilon^*_{ij},$$

$$\sigma^0_{ij} + T_\sigma \frac{\partial \sigma^0_{ij}}{\partial t} = \sigma_{ij} - T_\sigma V \frac{\partial \sigma_{ij}}{\partial x} = \sigma^*_{ij}, \qquad (3.49)$$

$$u - T_\varepsilon V \frac{\partial u}{\partial x} = u^*, \quad v - T_\varepsilon V \frac{\partial v}{\partial x} = v^*.$$

The functions ε^*_x, ε^*_y, γ^*_{xy}, σ^*_x, σ^*_y, τ^*_{xy} introduced in this manner satisfy the equations equivalent to the equilibrium, strain compatibility and Hooke's law equations for an isotropic elastic body.

3.4.2 Problem formulation

Since the deformations are small, we describe the shape of the cylinder by the function $f(x) = \dfrac{x^2}{2R}$, and refer the boundary conditions to the undeformed surface ($y = 0$). The relationship $v = f(x) + \text{const}$ for the normal displacement v of the half-plane ($y = 0$) holds within the contact zone $(-a, b)$, hence

$$\frac{\partial v}{\partial x} = \frac{x}{R}, \quad (y = 0). \tag{3.50}$$

We suppose that there is limiting friction in the contact region. So the following relationship between the normal σ_y and tangential τ_{xy} stresses (Amontons' law of friction, see Amontons, 1699) holds within the contact zone

$$\tau_{xy} = -\mu\sigma_y \operatorname{sgn} V, \quad (y = 0), \tag{3.51}$$

where μ is the coefficient of sliding friction. The surface of the contacting bodies is stress free outside the contact area:

$$\sigma_y = \tau_{xy} = 0 \quad (-\infty < x < -a, \quad b < x < +\infty).$$

Using the notations (3.49), we find that Eqs. (3.50) and (3.51) give the following boundary conditions ($y = 0$)

$$\begin{gathered} \sigma_y^* = \tau_{xy}^* = 0, \quad (-\infty < x < -a, \quad b < x < +\infty), \\ \frac{\partial v^*}{\partial x} = \frac{x - T_\varepsilon V}{R}, \quad \tau_{xy}^* + \mu\sigma_y^* \operatorname{sgn} V = 0, \quad (-a \leq x \leq b). \end{gathered} \tag{3.52}$$

3.4.3 Analytical results

This boundary problem can be reduced to a Riemann-Hilbert problem by the method described by Galin (1953) and used in § 3.2. Then the real stresses and displacement can be found by solving of the differential equations (3.49). The solution of this problem in detail is published (Goryacheva, 1973) We give here only the final expressions.

The normal pressure $p(x)$ at any point of contact zone is defined by the formula

$$p(x) = -\sigma_y(x,0) = -\frac{\varrho \exp(x/(T_\sigma V))}{T_\sigma V \pi K R} \times$$

$$\times \int_{-a}^{x} \left[\frac{(a+b)^2}{2} \left(\frac{1}{4} - \eta^2 \right) + PKR + (x' - T_\varepsilon V)(a+b) \left(\frac{1}{2} - \eta \right) - \right. \tag{3.53}$$

$$\left. (x' - T_\varepsilon V)(a + x') \right] \frac{\exp(-x'/(T_\sigma V))dx'}{(a+x')^{1/2+\eta}(b-x')^{1/2-\eta}},$$

where K, ϱ and η are determined by Eqs. (3.7) and (3.9), P is the normal force applied to the cylinder

$$P = \int_{-a}^{b} p(x)\, dx. \tag{3.54}$$

A tangential stress τ_{xy} at the surface of the half-plane is determined by Eq. (3.51).

The width of the contact zone $l = a + b$ is found as the solution of the following equation

$$\left[1-\left(\frac{l}{l_E}\right)^2\right]\left[\Psi\left(\frac{3}{2}+\eta,3;2\zeta\right)\Phi\left(\frac{1}{2}+\eta,1;2\alpha\zeta\right)+\right.$$
$$\left.\frac{1}{2}\alpha\left(\frac{1}{2}-\eta\right)\Psi\left(\frac{1}{2}+\eta,1;2\zeta\right)\Phi\left(\frac{3}{2}+\eta,3;2\alpha\zeta\right)\right]+ \tag{3.55}$$
$$\left(\frac{l}{l_E}\right)^2(1-\alpha)\Psi\left(\frac{3}{2}+\eta,3;2\zeta\right)\Phi\left(\frac{3}{2}+\eta,3;2\alpha\zeta\right)=0,$$

where $\zeta = \dfrac{l}{2T_\varepsilon V}$ represents the ratio of the time taken an element to travel through the semi-contact width $\dfrac{l}{2}$ to the retardation time T_ε, $l_E = \sqrt{2PKR\Big/\left(\frac{1}{4}-\eta^2\right)}$ is the contact width in sliding of the cylinder over the elastic half-plane under the normal force P if the elastic properties of the half-plane are characterized by the parameters K and ϑ (see Eq. (3.7)),

$$\alpha = \frac{T_\varepsilon}{T_\sigma},$$

and $\Phi(\beta,\gamma;z)$ and $\Psi(\beta,\gamma;z)$ are the confluent hypergeometric functions (see Gradshteyn and Ryzhik (1963, 9.210) or Janke and Emde (1944))

$$\Phi(\beta,\gamma;z) = 1+\frac{\beta}{\gamma}\cdot\frac{z}{1!}+\frac{\beta(\beta+1)}{\gamma(\gamma+1)}\cdot\frac{z^2}{2!}+\frac{\beta(\beta+1)(\beta+2)}{\gamma(\gamma+1)(\gamma+2)}\cdot\frac{z^3}{3!}+\cdots,$$

$$\Psi(\beta,\gamma;z) = \frac{\Gamma(1-\gamma)}{\Gamma(\beta-\gamma+1)}\Phi(\beta,\gamma;z)+\frac{\Gamma(\gamma-1)}{\Gamma(\beta)}z^{1-\gamma}\Phi(\beta-\gamma+1,2-\gamma;z).$$

Eq. (3.55) shows that the contact width l depends on the viscoelastic properties of the half-plane, the normal force P applied to the cylinder, its radius R and also on the coefficient of friction μ. Since the last term in Eq. (3.55) is negative $\left(\alpha > 1, \quad |\eta| < \dfrac{1}{2}\right)$, the first one is positive, and $l^2 \leq l_E^2$.

The shift ϵ of the contact zone relative to the point $(0,0)$ can be found as

$$\epsilon = \frac{b-a}{b+a} = -2\eta + \frac{\left[1-\left(\frac{l}{l_E}\right)^2\right]\left(\frac{1}{2}-\eta\right)\Psi\left(\frac{1}{2}+\eta,1;2\zeta\right)}{2\zeta\left(\frac{l}{l_E}\right)^2\Psi\left(\frac{3}{2}+\eta,3;2\zeta\right)}. \tag{3.56}$$

The ends of the contact zone $-a$ and b can be found from Eqs. (3.55) and (3.56).

Fig. 3.7 illustrates the forces applied to the cylinder. The vertical component P_1 of the reaction of the viscoelastic half-plane does not pass through the cylinder center. Hence, the moment

$$M_1 = \int_{-a}^{b} xp(x)dx = -\int_{-a}^{b} x\sigma_y^*(x,0)dx - T_\sigma VP \tag{3.57}$$

resists the cylinder motion. To calculate the moment M_1 we use

$$\int_{-a}^{b} x\frac{\partial\sigma_y(x,0)}{\partial x}dx = P.$$

The last relation holds because of the continuity of the stresses at the boundary of the contact zone, Eq. (3.54) and the relation $\sigma_y(x,0) = -p(x)$. The following expression for the moment M_1 can be obtained by substituting Eq. (3.53) into Eq. (3.57)

$$\frac{M_1}{Pl} = \left(\frac{l}{l_E}\right)^2\left(\frac{1}{2\zeta} - \frac{\epsilon}{2} - \frac{\eta}{3}\right) + \left(\frac{\epsilon}{2} - \frac{1}{2\zeta\alpha} - \eta\right). \tag{3.58}$$

The tangential forces $T_1 = \mu P$ and T $(|T| = |T_1|)$ give rise to the moment $M_2 = \mu Pd$, $(0,d)$ is the point of application of the force T (see Fig. 3.7). The relations $M_1 = M_2$ $\left(\text{or } d = \dfrac{M_1}{\mu P}\right)$ must hold, to provide the steady motion of the cylinder.

3.4.4 Some special cases

If we assume $\eta = 0$ in the previous equations we obtain the solution of the frictionless problem for sliding of the rigid cylinder over the viscoelastic half-plane ($\mu = 0$).

If we put $\eta = 0$ in Eq. (3.53), we obtain the following expression for the contact pressure

$$p(x) = -\frac{\exp(x/(T_\sigma V))}{T_\sigma V\pi KR}\int_{-a}^{x}\frac{\frac{1}{8}(a+b)^2 + \frac{1}{2}(x' - T_\varepsilon V)(b-a-2x') + PKR}{\exp(x'/(T_\sigma V))\sqrt{(a+x')(b-x')}}dx'. \tag{3.59}$$

Since there is no friction, we have $\tau_{xy} = 0$.

If we put $\eta = 0$ in Eqs. (3.55) and (3.56) we obtain

$$\left[1 - \left(\frac{l}{l_0}\right)^2\right][I_0(\alpha\zeta)K_1(\zeta) + K_0(\zeta)I_1(\alpha\zeta)] + \frac{2}{\zeta}\left(\frac{1}{\alpha} - 1\right)\cdot\left(\frac{l}{l_0}\right)^2 I_1(\alpha\zeta)K_1(\zeta) = 0, \tag{3.60}$$

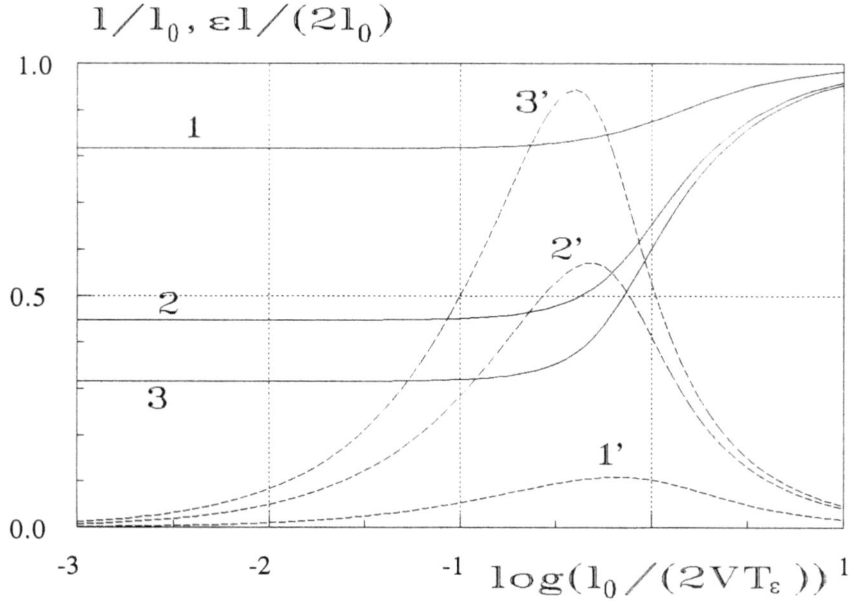

Figure 3.8: The contact width (solid lines) and the contact displacement (broken lines) in sliding/rolling contact ($\mu\vartheta = 0$) of a cylinder and a viscoelastic half-space for various values of $\alpha = T_\varepsilon/T_\sigma$: $\alpha = 1.5$ (curves 1, 1′), $\alpha = 5$ (curves 2, 2′), $\alpha = 10$ (curves 3, 3′).

$$\epsilon = \frac{1}{2}\left(\left(\frac{l_0}{l}\right)^2 - 1\right)\frac{K_0(\zeta)}{K_1(\zeta)}, \tag{3.61}$$

where $l_0 = \sqrt{8KRP}$ is the contact width in the corresponding problem for the elastic body, characterized by the parameter K (see Eq. (3.7)), $I_\nu(x)$ and $K_\nu(x)$ are modified Bessel functions. The following relationships (see Gradshteyn and Ryzhik (1963, § 8.4-8.5) or Janke and Emde (1944)) have been used to derive Eqs. (3.60) and (3.61)

$$\Phi\left(\frac{1}{2}+\nu, 1+2\nu; 2x\right) = 2^\nu\Gamma(\nu+1)x^{-\nu}\exp(x)I_\nu(x),$$

$$\Psi\left(\frac{1}{2}+\nu, 1+2\nu; 2x\right) = \frac{1}{\sqrt{\pi}}\exp(x)(2x)^{-\nu}K_\nu(x).$$

The dependence of dimensionless contact width $\dfrac{l}{l_0}$ and contact shift $\dfrac{\epsilon l}{2l_0}$ on the parameter $\zeta_0 = \dfrac{l_0}{2T_\varepsilon V}$ have been calculated based on Eqs. (3.60) and (3.61).

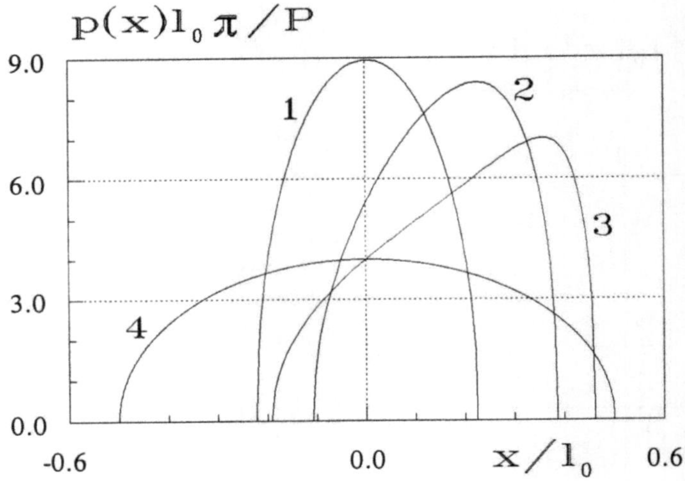

Figure 3.9: The pressure in sliding/rolling contact ($\mu\vartheta = 0$) of a cylinder and a viscoelastic half-space ($\alpha = 5$), for various values of ζ_0: $\zeta_0 = 10^{-3}$ (curve 1), $\zeta_0 = 0.4$ (curve 2), $\zeta_0 = 1$ (curve 3), $\zeta_0 = 10^4$ (curve 4).

The parameter ζ_0 is the ratio of the contact duration at any point of the half-plane to the double retardation time T_ε. Fig. 3.8 illustrates the results calculated for the cases $\alpha = 1.5$ (curve 1), $\alpha = 5$ (curve 2) and $\alpha = 10$ (curve 3). The results show that the contact width l changes within the limits $l_H \leq l \leq l_0$, where $l_H = \sqrt{\frac{8KRP}{\alpha}}$, l_H is the contact width in the corresponding problem for the elastic body, having the instantaneous modulus of elasticity $H = \alpha E$. The contact shift ϵ is a nonmonotonic function of the parameter ζ_0, with its maximum lying in the range (0.1, 1).

Fig. 3.9 illustrates the contact pressure distribution (Eq. (3.59)) for various parameters $\zeta_0 = \frac{l_0}{2T_\varepsilon V}$. For small values of this parameter ($\zeta_0 = 10^{-3}$, curve 1) the contact pressure is distributed symmetrically within the contact zone and it corresponds to the solution for elastic bodies having modulus H. For large values of the parameter ($\zeta_0 = 10^3$, curve 4), the contact pressure coincides with that for contact of elastic bodies having modulus E. If $\zeta_0 \in (10^{-3}, 10^3)$, the contact pressure becomes unsymmetrical (curves 2 and 3). The maximum contact pressure decreases as the parameter ζ_0 increases.

Equations (3.53), (3.55), (3.56) for $T_\varepsilon = T_\sigma$ give the solution of the contact problem with limiting friction, for a rigid cylinder and an elastic half-plane (with elastic modulus E). The following expressions can be obtained

$$p(x) = \frac{\varrho}{\pi KR}(a+x)^{1/2-\eta}(b-x)^{1/2+\eta}, \tag{3.62}$$

$$l = \sqrt{2KRP \Big/ \left(\frac{1}{4} - \eta^2\right)}, \quad \epsilon = -2\eta. \qquad (3.63)$$

The following relationship has been used to deduce Eq. (3.62)

$$I_0(x)K_1(x) + K_0(x)I_1(x) = \frac{1}{x}.$$

Eqs. (3.62) and (3.63) coincide with the results obtained in § 3.2 and in Galin (1980) and Johnson (1987).

3.5 Rolling contact of elastic and viscoelastic bodies

Contact problems for an elastic cylinder rolling along an elastic half-plane under the assumption of partial slip in the contact zone have been investigated by Carter (1926), Fromm (1927), Glagolev (1945), Poritsky (1950), Ishlinsky (1956), Johnson (1962), Mossakovsky and Mishchishin (1967), Kalker (1990), etc.

The effect of imperfect elasticity of the contacting bodies has been investigated by Hunter (1961), Morland (1962), Kalker (1991), etc. They considered a rolling contact of a rigid or a viscoelastic cylinder and a viscoelastic half-plane.

We consider the simultaneous effect of sliding in contact and imperfect elasticity affecting the resistance to rolling.

3.5.1 Problem formulation

We consider this problem as two-dimensional and quasistatic. Suppose that a viscoelastic cylinder (1) of radius R rolls with a constant velocity V and angular velocity ω over a base (2) of the same material (Fig. 3.7). As in the previous section, we consider a coordinate system (x, y) moving with the rolling cylinder. The relationship (3.50) holds within the contact zone $(-a, b)$. We assume that the contact zone $(-a, b)$ consists of two parts: a slip region $(-a, c)$ and a stick region (c, b). The validity of this assumption in rolling contact problems for bodies of the same mechanical properties has been proved by Goryacheva (1974) and by Goldstein and Spector (1986).

The velocities of the tangential displacements of points of the cylinder and of the half-plane are equal within the stick zone (c, b), i.e.

$$V - \omega R + \frac{\partial u_1^0}{\partial t} = \frac{\partial u_2^0}{\partial t}.$$

In the coordinate system (x, y) connected with the cylinder, this relation is written in the form:

$$\frac{du_2}{dx} - \frac{du_1}{dx} = \delta, \ y = 0, \ c < x < b, \quad \delta = \frac{\omega R - V}{V}. \qquad (3.64)$$

Within the slip zone $(-a, c)$ the Coulomb-Amontons' law of friction holds

$$\tau_{xy} = -\mu p(x) \operatorname{sgn} s_x, \quad p(x) > 0.$$

Here μ is the coefficient of sliding friction, and s_x is the difference in velocities of tangential displacement of boundary points of a half-plane and cylinder:

$$s_x = \frac{\partial u_2}{\partial t} - \frac{\partial u_1}{\partial t} - V + \omega R = V\left(\delta + \frac{du_1}{dx} - \frac{du_2}{dx}\right).$$

The surface of the viscoelastic body is stress free outside the contact zone $(-a, b)$. The relations between the strain and stress components are taken in the form (3.48).

3.5.2 Solution

In the coordinate system (x, y), the displacements and stresses do not depend explicitly on time and are functions only of the coordinates. As in § 3.4, we introduce the functions ϵ_x^*, ϵ_y^*, γ_{xy}^*, σ_x^*, σ_y^*, τ_{xy}^* (3.49) which satisfy the equations equivalent to the equilibrium, strain compatibility and Hooke's law. To find these functions we use the method developed by Galin (1980). We introduce two functions of a complex variable $w_1(z)$ and $w_2(z)$ in the lower half-plane, which are Cauchy type integrals $(z = x + iy)$

$$w_1(z) = \int_{-a}^{b} \sigma_y^*(t, 0)\frac{dt}{t - z} = U_1(x, y) - iV_1(x, y),$$

$$w_2(z) = \int_{-a}^{b} \tau_{xy}^*(t, 0)\frac{dt}{t - z} = U_2(x, y) - iV_2(x, y).$$

Expressing the functions

$$\sigma_y^*(x, 0), \quad \tau_{xy}^*(x, 0), \quad \frac{du_i^*(x, 0)}{dx}, \quad \frac{dv_i^*(x, 0)}{dx}, \quad (i = 1, 2)$$

in terms of the real and imaginary parts of the functions $w_1(z)$ and $w_2(z)$ (see Galin, 1980) and substituting them into the boundary conditions, modified somewhat, taking account of (3.49), we obtain a conjugate problem: to find two functions $w_1(z)$ and $w_2(z)$ which are analytic in the lower half-plane and satisfy

$$\begin{aligned}
&V_1 = V_2 = 0, && x \notin (-a, b), \\
&U_1 = -\frac{x - T_\varepsilon V}{KR}, \quad V_2 + \mu V_1 = 0, && x \in (-a, c), \\
&U_1 = -\frac{x - T_\varepsilon V}{KR}, \quad U_2 = \frac{\delta}{K}, && x \in (c, b), \\
&K = \frac{4(1 - \nu^2)}{\pi E}.
\end{aligned} \tag{3.65}$$

The functions satisfying the boundary conditions (3.65) are

$$w_1(z) = -\frac{1}{X_1(z)} \int_{-a}^{b} \frac{t - T_\varepsilon V}{\pi K R} X_1^+(t) \frac{dt}{t-z} - \frac{P}{X_1(z)},$$

$$w_2(z) = -\frac{X_2(z)}{\pi} \left[\mu \int_{-a}^{c} V_1(t,0) X_2^+(t) \frac{dt}{t-z} - \frac{\delta}{K} \int_{c}^{b} X_2^+(t) \frac{dt}{t-z} \right] - \tag{3.66}$$

$$\frac{X_2(z)}{z-c} C_2.$$

Here C_2 is some constant and

$$X_1(z) = \sqrt{(z+a)(z-b)},$$

$$X_1^+(t) = \sqrt{(a+t)(b-t)},$$

$$X_2(z) = \sqrt{\frac{z-c}{z-b}},$$

$$X_2^+(t) = \begin{cases} \sqrt{\dfrac{b-t}{c-t}}, & t \in (-a, c), \\ \sqrt{\dfrac{b-t}{t-c}}, & t \in (c, b), \end{cases}$$

$$P = \int_{-a}^{b} p(x)\, dx = -\int_{-a}^{b} \sigma_y(x,0)\, dx = -\int_{-a}^{b} \sigma_y^*(x,0)\, dx.$$

The last relation follows from Eq. (3.49) subject to the conditions $\sigma_y(-a,0) = \sigma_y(b,0) = 0$.

We can find $\sigma_y^*(x,0)$, $\tau_{xy}^*(x,0)$ by calculating the imaginary parts of the functions $w_1(z)$ and $w_2(z)$ on the real axis. Then true stresses $p(x) = -\sigma_y(x,0)$, $\tau_{xy}(x)$ within the contact zone are found by solving the equations (3.49).

The function $w_1(z)$ (3.66) shows that the tangential stress does not influence the pressure distribution for the contact of bodies having similar mechanical properties. The contact pressure in the problem under consideration is determined by Eq. (3.59) and can be represented by the curves in the Fig. 3.9.

Using the following relationships for the imaginary part $V_1(x,0)$ of the function $w_1(z)$ as $z \to x - i0$

$$V_1(x,0) = \frac{F(x)}{\sqrt{(a+x)(b-x)}}, \quad x \in (-a, b),$$

where

$$F(x) = \frac{(a+b)^2}{8KR} + \frac{(x - T_\varepsilon V)(b - a - 2x)}{2KR} + P \tag{3.67}$$

and the result of integration (see Gradshteyn and Ryzhik, 1963)

$$\int_c^b \sqrt{\frac{b-t}{t-c}} \cdot \frac{dt}{t-x} = \begin{cases} -\pi \left[1 - \sqrt{\left| \dfrac{b-x}{c-x} \right|} \right], & x \notin [c,b], \\ -\pi, & x \in (c,b), \end{cases} \tag{3.68}$$

we obtain the relationships for the imaginary part $V_2(x,0)$ of the function $w_2(z)$ (3.66) as $z \to x - i0$

$$V_2(x,0) = -\mu V_1(x,0), \quad x \in (-a,c),$$

$$V_2(x,0) = -\mu V_1(x,0) + \left[\frac{\mu}{2KR}(b - c - 2x + 2T_\varepsilon V) + \frac{\delta}{K} \right] \times \tag{3.69}$$

$$\times \sqrt{\frac{x-c}{b-x}} + \frac{C_2}{\sqrt{(x-c)(b-x)}} = \pi \tau_{xy}^*(x,0), \quad x \in (c,b).$$

Then the tangential stresses $\tau_{xy}(x)$ can be found by solving Eq. (3.49) (see Goryacheva, 1973):

- in the slip zone $(-a,c)$

$$\tau_{xy}(x) = \mu p(x),$$

- in the stick zone (c,b)

$$\tau_{xy}(x) = \mu p(x) - \frac{\mu \exp(x/(T_\sigma V))}{2\pi K R T_\sigma V} \times$$

$$\int_c^x \left(b - c - 2t + 2T_\varepsilon V + \frac{2R\delta}{\mu} \right) \sqrt{\frac{t-c}{b-t}} \exp\left(-\frac{t}{T_\sigma V} \right) dt - \tag{3.70}$$

$$\frac{C_2}{\pi T_\sigma V} \exp\left(\frac{x}{T_\sigma V} \right) \int_c^x \frac{\exp(-t/(T_\sigma V))}{\sqrt{(t-c)(b-t)}} dt.$$

For determining the constant C_2 and the point c of transmission of slip to stick zone we use two conditions. The first one is the relation (3.64), which can be written at $x = b$ in the form

$$\int_b^{+\infty} U_2(x,0) \exp\left(-\frac{x}{T_\varepsilon V} \right) dx = \frac{\delta T_\varepsilon V}{K} \exp\left(-\frac{b}{T_\varepsilon V} \right), \tag{3.71}$$

where $U_2(x,0)$ is a real part of the function $w_2(z)$ as $z \to x - i0$.

$$U_2(x,0) = -\left[\frac{\mu}{2KR}(b - c - 2x + 2T_\varepsilon V) + \frac{\delta}{K} \right] \sqrt{\frac{x-c}{x-b}} +$$

$$\frac{\delta}{K} + \frac{\mu F(x)}{\sqrt{(x+a)(x-b)}} - \frac{C_2}{\sqrt{(x-c)(x-b)}}, \quad x \in (b, +\infty),$$

where $F(x)$ is defined by Eq. (3.67). The second one is the relation $\tau_{xy}(b,0) = 0$, which holds because of the continuity of the stresses at the ends of the contact zone, and gives the equation

$$\int_{-a}^{b} V_2(x,0)\exp\left(-\frac{x}{T_\sigma V}\right) dx = 0. \tag{3.72}$$

Using Eqs. (3.69), (3.71) and (3.72) we obtain

$$C_2 = \frac{\mu(b-c)}{4KR}\left[b + c - 2T_\varepsilon V - \frac{2R\delta}{\mu} + \left(b + c + \frac{2R\delta}{\mu}\right)\frac{K_1\left(\dfrac{b-c}{2T_\varepsilon V}\right)}{K_0\left(\dfrac{b-c}{2T_\varepsilon V}\right)}\right].$$

3.5.3 The contact width and the relation between the slip and stick zones

The unknown ends of the contact $-a$ and b, and the transition point c can be determined by satisfying the conditions for the real stresses and displacements at the boundary of the elastic bodies. Goryacheva (1973) showed that the relationships for the contact width $l = a + b$ and the contact shift $\epsilon = \dfrac{b-a}{b+a}$ are the same as (3.60) and (3.61) which hold in the sliding problem for the rigid cylinder and the viscoelastic half-plane. The plots of these functions are presented in Fig. 3.8.

Using Eqs. (3.71) and (3.72) we can derive the following equation for determining the width $\beta = \dfrac{b-c}{b+a}$ of the stick zone

$$\begin{aligned}&\left(1 + \epsilon - \beta - \frac{2R\delta}{\mu l}\right)[I_0(\alpha\beta\zeta)K_1(\beta\zeta) + I_1(\alpha\beta\zeta)K_0(\beta\zeta)] + \\ &\qquad \frac{1}{\zeta}\left(\frac{1}{\alpha} - 1\right) I_1(\alpha\beta\zeta)K_0(\beta\zeta) = 0.\end{aligned} \tag{3.73}$$

3.5.4 Rolling friction analysis

The cylinder is subjected to the normal load P, tangential load T and moment M. The reaction forces P_1 and T_1 are due to the normal and tangential stress distributions caused by the contact of the cylinder with the viscoelastic body (see Fig. 3.7). The condition for moment equilibrium about the center of the cylinder is

$$M + M_1 + T_1 R = 0,$$

where

$$M_1 = \int_{-a}^{b} xp(x)\, dx$$

and

$$T_1 = \int_{-a}^{b} \tau_{xy} dx = \int_{-a}^{b} \tau^*_{xy} dx = \frac{1}{\pi} \int_{-a}^{b} V_2(x,0)\, dx,$$

$$(\tau_{xy}(-a) = \tau_{xy}(b) = 0).$$

Eqs. (3.57), (3.67) and (3.69) show that the equations for M_1 and T_1 can be transformed to the following expressions

$$\frac{M_1}{Pl} = \frac{\epsilon}{2}\left(1 - \left(\frac{l}{l_0}\right)^2\right) + \frac{1}{2\zeta}\left(\left(\frac{l}{l_0}\right)^2 - \frac{1}{\alpha}\right), \tag{3.74}$$

$$\frac{T_1}{\mu P} = 1 - \left(\frac{\beta l}{l_0}\right)^2 - 2\beta\left(\frac{l}{l_0}\right)^2\left(1 + \epsilon - \beta - \frac{2\delta R}{\mu l}\right)\frac{K_1(\beta\zeta)}{K_0(\beta\zeta)}. \tag{3.75}$$

Provided that the contact width l is in the limits $l_H \leq l \leq l_0$ $\left(l_H^2 = \frac{l_0^2}{\alpha}\right)$, both terms in the side of (3.74) are nonnegative and so $M_1 \geq 0$. The sum of the moments of the normal and tangential contact stress with respect to the center of the cylinder gives the rolling friction moment $M^* = M_1 + T_1 R$.

The rolling friction is characterized by the rolling friction coefficient, which gives the relation between the moment of friction M^* and the normal load P. Using Eqs. (3.62), (3.73), (3.74) and (3.75) we obtain

$$\mu_r = \frac{M^*}{Pl} = \left(\left(\frac{l_0}{l}\right)^2 - 1\right)\frac{K_0(\zeta)}{4K_1(\zeta)} +$$

$$\frac{1}{2\zeta}\left(\left(\frac{l}{l_0}\right)^2 - \frac{1}{\alpha}\right) + \frac{\mu R}{l}\left(1 - \frac{\beta^2 l^2}{l_0^2}\right) -$$

$$\frac{2\mu\beta R l \left(1 - \frac{1}{\alpha}\right) I_1(\alpha\beta\zeta)\, K_0(\beta\zeta)}{l_0^2 \zeta \left[I_0(\alpha\beta\zeta)\, K_1(\beta\zeta) + I_1(\alpha\beta\zeta)\, K_0(\beta\zeta)\right]}.$$

Free rolling occurs if $T = 0$ and $M = M_1$. Fig. 3.10 illustrates the dependence of the coefficient μ_r of a rolling friction on the parameter $\zeta_0 = \frac{l_0}{2T_\epsilon V}$ for free rolling. The results indicate that the maximum value of the friction coefficient takes place for $\zeta_0 \approx 1$. The maximum value of μ_r depends essentially on the parameter α characterizing viscous properties of contacting bodies.

The analysis of Eqs. (3.74), (3.75) and the equilibrium conditions show that tangential contact stresses acting on the half-plane are parallel to the velocity V ($\mu > 0$) if $M > M_1$. If $M < M_1$ the tangential stresses have the opposite direction ($\mu < 0$), in this case the active tangential force T in the direction of motion is applied to the cylinder. Eqs. (3.73) and (3.75) show that the width of stick zone

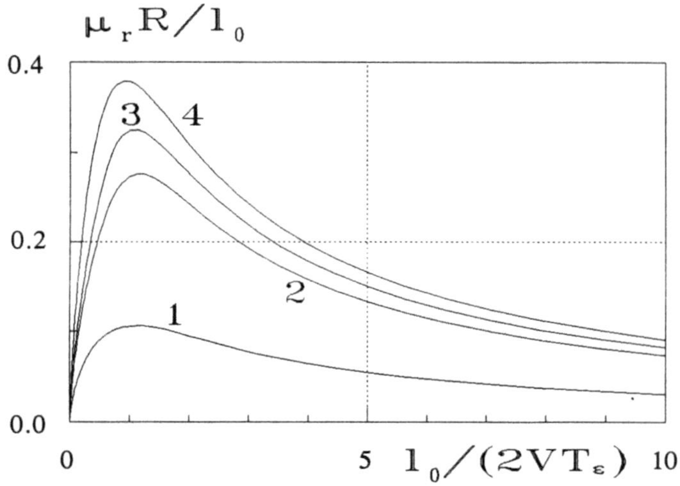

Figure 3.10: The rolling resistance of a viscoelastic cylinder on a viscoelastic half-space (similar materials, $\mu = 0$) for various value of the parameter $\alpha = T_\varepsilon/T_\sigma$: $\alpha = 1.5$ (curve 1), $\alpha = 5$ (curve 2), $\alpha = 10$ (curve 3), $\alpha = 100$ (curve 4).

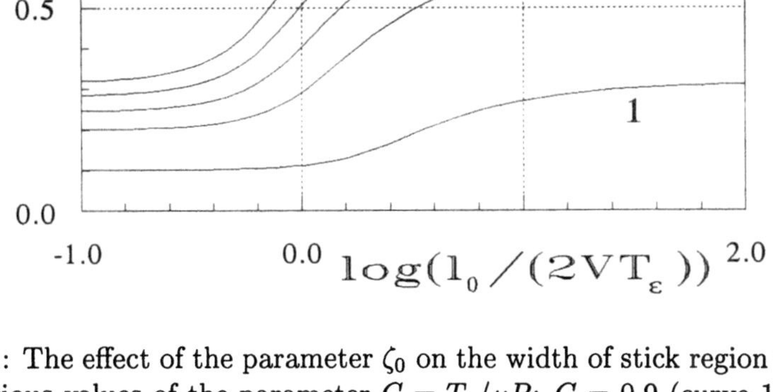

Figure 3.11: The effect of the parameter ζ_0 on the width of stick region for $\alpha = 10$ and for various values of the parameter $C = T_1/\mu P$: $C = 0.9$ (curve 1), $C = 0.6$ (curve 2), $C = 0.4$ (curve 3), $C = 0.2$ (curve 4), $C = 0$ (curve 5).

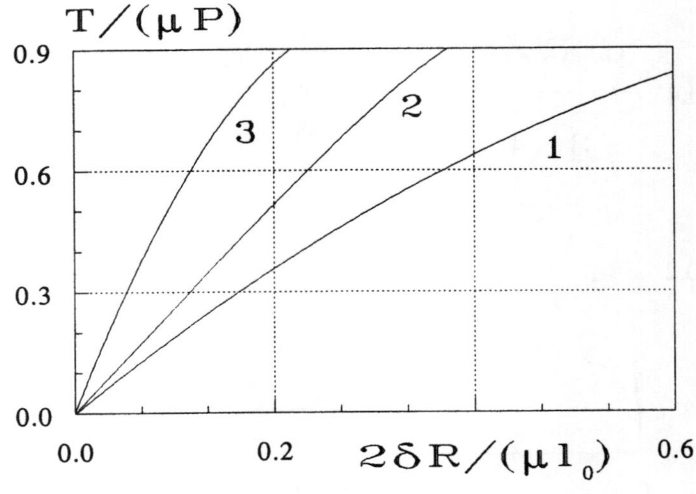

Figure 3.12: Creep curves for a tractive rolling contact of a viscoelastic cylinder on a viscoelastic half-space (similar materials, $\alpha = 10$) for various values of the parameter $\zeta_0 = l_0/2T_\varepsilon V$: $\zeta_0 = 10^2$ (curve 1), $\zeta_0 = 10^{-1}$ (curve 2), $\zeta_0 = 10^{-4}$ (curve 3).

depends on the ratio $C = \dfrac{T_1}{\mu P}$. Eq. (3.73) has been solved for various parameters C. The plots are shown in Fig. 3.11. The width of the stick zone increases as the parameter C decreases. For $C = 0$, the stick region is spread within the whole of the contact zone.

The creep ratio δ for the rolling cylinder can be found from Eq. (3.75). Fig. 3.12 illustrates the dependence of the parameter C on the creep ratio for various parameters ζ_0. The results show that for a fixed value of the parameter C, the creep ratio decreases as the parameter ζ_0 decreases (the velocity V increases).

3.5.5 Some special cases

If $\alpha = 1$, then the equations obtained above yield the solution of the problem of rolling of an elastic cylinder over a base of the same material, with elastic modulus E.

We obtain the following expressions for the normal and tangential stresses within the contact zone $(-a, a)$ which is symmetrical in this case ($\epsilon = 0$)

$$p(x) = \frac{\sqrt{a^2 - x^2}}{\pi K R} \quad (-a < x < a),$$

$$\tau_{xy}(x) = \begin{cases} \dfrac{\mu}{\pi K R}\sqrt{a^2 - x^2}, & (-a < x < c), \\ \dfrac{\mu}{\pi K R}\left[\sqrt{a^2 - x^2} - \sqrt{(a-x)(x-c)}\right], & (c < x < a). \end{cases}$$

The contact width is $l = 2a = \sqrt{8KRP}$, the width of the stick zone is $\beta = 1 - \dfrac{\delta R}{\mu a}$. The contact pressure distribution is symmetrical ($M_1 = 0$). The tangential force T is calculated by the formula

$$\frac{T}{\mu P} = \frac{\delta R}{\mu a}\left(2 - \frac{\delta R}{\mu a}\right).$$

Note that the relative width of the stick zone does not depend on the elastic properties of contacting bodies, and it is calculated by

$$\beta = \sqrt{1 - \frac{T}{\mu P}}.$$

If $\alpha \neq 1$, the contact characteristics for viscoelastic bodies approach those for elastic bodies with the elasticity moduli E and $H = \alpha E$, as $T_\varepsilon V \to 0$ and $T_\varepsilon V \to +\infty$, respectively.

3.6 Mechanical component of friction force

We investigated the sliding contact of a rigid cylinder and an viscoelastic half-space in § 3.4. The results show that there is a resistance to the motion of the cylinder, even though we assume that the tangential stresses are zero at the interface. Under the same boundary conditions, there is no resistance to the motion in sliding contact of elastic bodies (see § 3.2 and § 3.3). The reason is that the deformation is reversible for elastic bodies so that both the contact region and stress distribution are symmetrical relative the axis of symmetry of the cylinder. This is not so for viscoelastic bodies. The center of the contact region, and the point where the maximum pressure takes place, are shifted towards the leading edge of the contact (see § 3.2). It is precisely these phenomena that are responsible for the resistance in sliding.

Let us calculate the tangential force T that has to be applied to the cylinder to provide its steady motion (Fig. 3.13). We assume that the tangential stress is negligible within the contact zone ($\tau_{xy} = 0$). This enables us to study the mechanical friction component alone. Since the normal stress is directed to the center of the cylinder, the reaction force F is also directed to the center (see Fig. 3.13(a)). Let us calculate the x- and y- components T_d and P_1 of the reaction force F. Taking into account that the contact width $l = a + b$ is much less than the radius R, we can write

$$P_1 = \int_{-a}^{b} p(x)\cos\varphi(x)dx \approx \int_{-a}^{b} p(x)\,dx, \qquad (3.76)$$

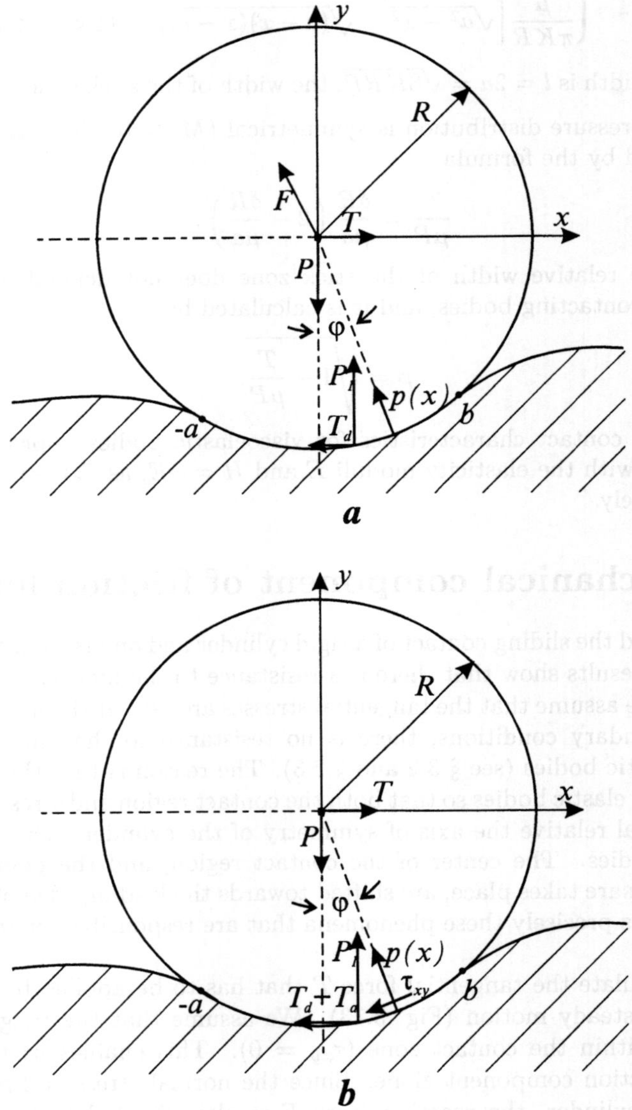

Figure 3.13: Scheme of the forces applied to the cylinder in sliding contact: frictionless contact (a), contact with friction (b).

$$T_d = \int_{-a}^{b} p(x) \sin \varphi(x)\, dx \approx \frac{M}{R}, \tag{3.77}$$

where

$$M = \int_{-a}^{b} x p(x)\, dx.$$

The equations of equilibrium show that $T_d = T$ and $P_1 = P$. The force T_d is called the mechanical friction component. The mechanical friction coefficient μ_d can be obtained by dividing the equation (3.77) by the equation (3.76), with the result

$$\mu_d = \frac{T_d}{P} = \frac{M}{PR}, \tag{3.78}$$

where M is estimated from (3.58) provided that $\eta = 0$ ($l_E = l_0$ if $\eta = 0$). Hence the expression for μ_d can be written in the form

$$\mu_d = \alpha_h \frac{l_0}{R}, \tag{3.79}$$

where

$$\alpha_h = \frac{\varepsilon l}{2l_0}\left(1 - \left(\frac{l}{l_0}\right)^2\right) + \frac{\left(\frac{l}{l_0}\right)^2 - \frac{1}{\alpha}}{2\zeta_0}. \tag{3.80}$$

It is worth noting that the mechanical friction coefficient μ_d (Eq (3.79)) coincides with the coefficient of rolling friction for free rolling of a viscoelastic cylinder over a viscoelastic half-space. This conclusion follows from the fact that Eq.(3.80) is similar to Eq. (3.74) divided by $\frac{l_0}{l}$. So the curves in Fig. 3.10 illustrate the dependence of the mechanical friction coefficient μ_d on the parameter ζ_0. The dependence is not monotonic, and has a maximum when $\zeta_0 \approx 1$, i.e. the semi-contact time is roughly equal to the retardation time. The mechanical component of friction force tends to zero for small or large values of the parameter ζ_0.

Tabor (1952) was the first who proposed to determine the mechanical friction coefficient from a rolling contact test. Later experiments supported his idea. Fig. 3.14 illustrates the experimental results obtained by Greenwood and Tabor (1958). The rolling and sliding contact of steel balls over high-hysteresis rubber specimens was investigated. A soap was used as lubricant in sliding contact to decrease the adhesive component of the friction force. The results in sliding (solid symbols) and in rolling (open symbols) agree very closely. For a nominal pressure less than $3 \cdot 10^4 Pa$, they are in a good agreement with the theoretical curve based on the hysteresis theory of friction. According to this theory elaborated for the rolling friction, the coefficient of rolling friction is determined from the expression (3.79). It is supposed that the coefficient α_h is dependent on the viscoelastic properties of material and the rolling velocity. The value of the coefficient α_h is determined from experiments of cyclic loading of the material.

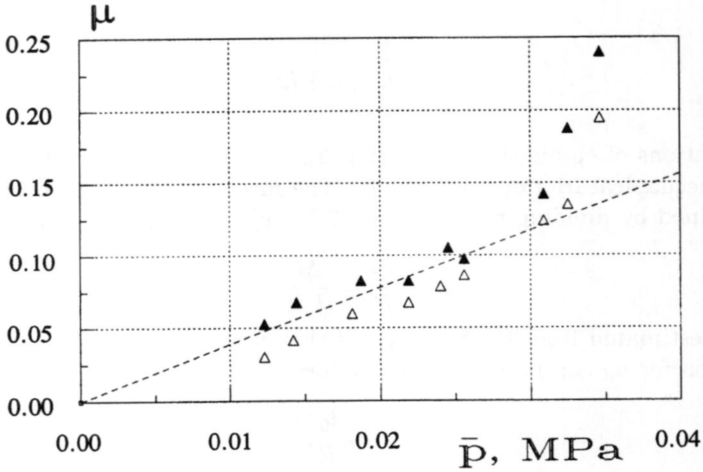

Figure 3.14: The friction coefficient of a steel sphere on well-lubricated rubber, as a function of the average contact pressure in rolling contact (open symbols) and in sliding contact (solid symbols) (the experimental results, Greenwood and Tabor, 1958). The broken line is a theoretical curve obtained from the hysteresis theory of friction (Tabor, 1955).

The investigation of contact problems for a cylinder and a viscoelastic half-space (see § 3.4 and § 3.5) makes it possible to analyze the dependence of the coefficient α_h (3.80) in a sliding/rolling contact on the viscoelastic characteristics of the material (E, ν, T_σ, T_ε) and the sliding/rolling velocity. An analysis of the equation (3.80) shows that the magnitude of α_h also depends on the normal load P because of $l_0 \sim \sqrt{P}$. The discrepancy between the theoretical and experimental results (see Fig. 3.14) may be explained by the neglect in the calculations of the dependence of α_h on pressure (the theoretical curve corresponds to $\alpha_h = 0.35$).

It was suggested in the previous analysis that the energy dissipation due to irreversible deformation is the only reason for the friction force. Considering that both of the causes of energy dissipation (adhesion and deformation) are simultaneously realized in sliding contact, it is important to investigate their joint influence on the friction force. Are there mutual influences between the adhesive and mechanical components of the friction force? Some results obtained in this chapter (see § 3.4) make it possible to answer this question.

We consider the cylinder of radius R sliding with friction ($\tau_{xy}(x) = \mu_a p(x)$) over viscoelastic body (Fig. 3.13(b)). In this case the adhesive component of the

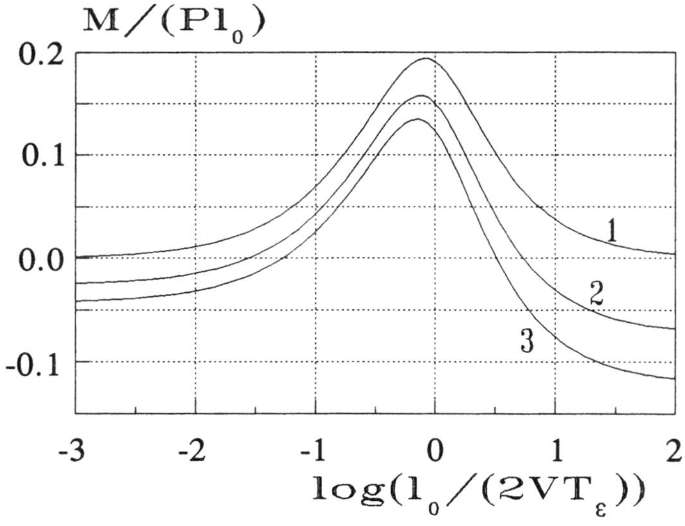

Figure 3.15: The mechanical component of friction force for different friction coefficients μ_a: $\mu_a = 0$ (curve 1), $\mu_a = 0.3$ (curve 2), $\mu_a = 0.6$ (curve 3), $\alpha = 5$.

friction force T_a can be written as

$$T_a = \int_{-a}^{b} \tau_{xy} \cos \varphi(x)\, dx \approx \mu_a \int_{-a}^{b} p(x) dx = \mu_a P. \tag{3.81}$$

The equation of equilibrium shows that

$$T = T_a + T_d = \mu_a P + \frac{M}{R}.$$

Hence the total friction coefficient is given by the expression

$$\mu = \frac{T}{P} = \mu_a + \frac{M}{PR}. \tag{3.82}$$

The second term in (3.82) is generally classified as the coefficient of the mechanical component of the friction force. The moment M is given by Eq. (3.58). Since the moment M depends on the parameter η (see Eqs. (3.55), (3.56) and (3.58)), and η in turn is a function of the adhesive friction coefficient μ_a (see Eq. (3.9)), the mechanical component is governed by the adhesive one. Fig. 3.15 illustrates the dependence of the dimensionless moment $\frac{M}{Pl_0}$, which is proportional to the mechanical component of friction force, on the parameter ζ_0 for different friction coefficients μ_a. The results show that the coefficient μ_a decreases the mechanical component. For small values of the parameter ζ_0, the mechanical component becomes negative as the coefficient μ_a increases.

Chapter 4

Contact of Inhomogeneous Bodies

The use of surface treatment of different types leads to the changes in the surface properties relative to the bulk ones. This chapter is devoted to contact problems for bodies with specific surface properties, and to the analysis of the influence of mechanical properties of the surface layer (i.e. coating, boundary lubricant, etc.) on contact characteristics and internal stresses that govern the surface fracture of contacting bodies.

4.1 Bodies with internal defects

In solving certain applied problems, the influence of systems of defects (such as microcracks and microvoids) on the stress-strain state of elastic bodies has to be taken into account. From the standpoint of evaluating the surface strength of bodies in their contact interaction, of special interest is the study of the stress fields in sub-surface layers, where the manifestation of microcracks and other kinds of defects is, as a rule, associated with various kinds of mechanical and thermal treatment of surfaces (e.g., coating, hardening, etc.).

Such a study necessarily involves solving a boundary value problem of elasticity in a very complicated domain, which admits exact solution only in a few idealized cases. One of the widespread idealizations is the assumption that the domain where defects are arranged is unbounded. The approaches to stress analysis in the vicinity of the internal stress concentrators such as cracks, cuts, and thin inclusions, with this assumption are described in monographs by Muskhelishvili (1949), Savin (1968), Popov (1982).

The case of defects localized near the boundary of the elastic body can not be analyzed in the framework of this idealization. In Mozharovsky and Starzhinsky (1988), a method is proposed for solving a plane elasticity problem for a strip discretely soldered to the foundation (i.e., having finitely many cuts at the interface). Savin (1968) considers system of circular holes localized near the half-plane boundary. However, the algorithms developed there and in some other works are too complicated to be used in specific problems (especially, in contact problems) and moreover assume that the domain contains a finite number of defects.

There are many cases, in which the stress field some distance away from defects is the matter of consideration. And the question to be answered is: "How can we take into account the total influence of a system of defects without solving the original problem exactly?"

In what follows we propose an approach to solving the problem in question, which is based on the theory of differential operators in domains with fine-structured boundary. It is assumed that the system of defects is localized near some internal surface Γ in the domain Ω occupied by an elastic body. The key idea of the method proposed is to introduce characteristics of the defect layer (the layer with the system of defects) which represent, on the average, the behavior of the layer under deformation. This permits us to reduce the original exact statement of the boundary conditions on the surface of defects to the matching condition on Γ. The method allows us to calculate the averaged stresses at some distance away from the system of defects.

4.1.1 Boundary problem for elastic bodies with an internal system of defects

We consider the elastic body in the region $\Omega \subset R^n$. There is an internal system of defects, i.e. the set $F^{(s)} = \bigcup_i F_i^{(s)} \subset \Omega$ (see Fig. 4.1). All elements $F_i^{(s)}$ are localized near some internal surface Γ (or near an internal curve in the two-dimensional case).

The problem is reduced to investigation of the Lamé equations in $\Omega \setminus F^{(s)}$, i.e. Ω excluding $F^{(s)}$. When body forces are neglected, Lamé's equations are

$$(\lambda + \mu)u_{j,jk} + \mu \Delta u_k = 0, \tag{4.1}$$

where λ and μ are the Lamé parameters, u_j is the displacement along the j-th axis of coordinates ($j = 1,2,3$ in 3-D case, $n = 3$, or $j = 1,2$ in 2-D case, $n = 2$). We assume that the boundary $\partial F^{(s)}$ of defects is free of loading, i.e.

$$\mathbf{T}_\nu \Big|_{\partial F^{(s)}} = 0, \tag{4.2}$$

where $\mathbf{T}_\nu$ is the vector of load applied to a unit area element of $\partial F^{(s)}$ with normal ν. We use Einstein's summation convention: a repeated suffix j as in Eq. (4.1) means summation over $j = 1, 2, \ldots, n$.

The condition at the boundary of the body $\partial\Omega$ can be arbitrary, in terms of stresses and/or displacements, etc.

The solution of this boundary problem is denoted as the vector $\mathbf{u}^{(s)}$ with components $u_i^{(s)}$.

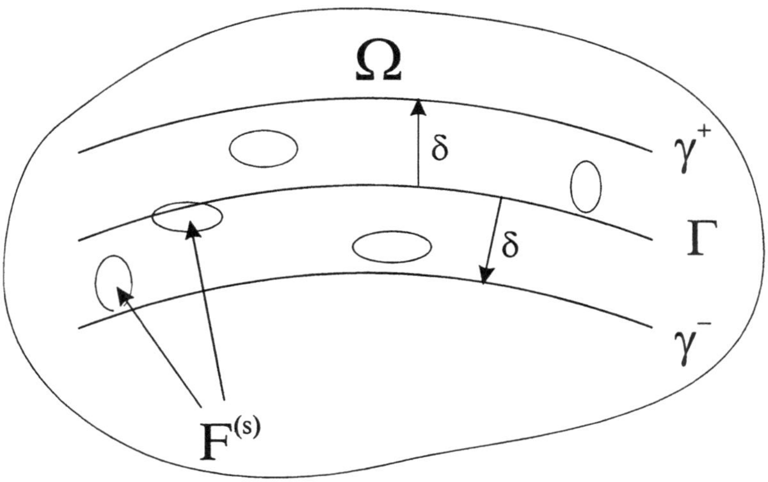

Figure 4.1: Location of the system of defects $F^{(s)}$ within an elastic body Ω.

4.1.2 The tensor of influence

We introduce here some characteristics of influence of the system of defects $F^{(s)}$ on the solution of the Lamé equations in the region Ω.

We suppose that γ is an open connected set on Γ; γ^+ and γ^- are surfaces formed by the ends of the vectors of length δ normal to γ (see Fig. 4.1); $T(\gamma, \delta)$ is a layer of thickness 2δ whose mid-surface is γ, $F^{(s)} \subset T(\gamma, \delta)$. We define the domain $T(\gamma, \delta, s)$ as $T(\gamma, \delta, s) = T(\gamma, \delta) \setminus F^{(s)}$.

We determine u_k^α as a solution of the Lamé equations (4.1) in the domain $T(\gamma, \delta, s)$ with the following boundary conditions:

$$\begin{aligned} \gamma^-: &\quad u_i^\alpha = 0, \\ \gamma^+: &\quad u_i^\alpha = \delta_{i\alpha}, \end{aligned} \tag{4.3}$$

$$T_\nu \Big|_{\partial F^{(s)}} = 0,$$

where index α denotes that only displacement in the α-axis direction on γ^+ is not equal to zero ($\alpha = 1, 2, (3)$), $\delta_{i\alpha}$ is the Kronecker delta.

The influence of the set $F^{(s)}$ is characterized by the tensor $P(\gamma, \delta, s)$ with components

$$P^{\alpha\beta}(\gamma, \delta, s) = \int\limits_{T(\gamma,\delta,s)} \sigma_{ki}^\alpha u_{k,i}^\beta dv, \tag{4.4}$$

$$\alpha, \beta = 1, 2, (3), \quad dv = \prod_{i=1}^{n} dx_i, \quad n = 2, (3),$$

where σ_{ki}^{α} is the component of the stress tensor, corresponding to the solution of the boundary problem (4.3)

$$\sigma_{ki}^{\alpha} = \lambda \Theta^{\alpha} \delta_{ik} + \mu \left(u_{k,i}^{\alpha} + u_{i,k}^{\alpha} \right), \tag{4.5}$$

$$\Theta^{\alpha} = u_{j,j}^{\alpha}, \quad \left(u_{k,i}^{\alpha} \equiv \frac{\partial u_k^{\alpha}}{\partial x_i} \right),$$

u_k^{β} is the solution of the boundary problem (4.3) where index α is replaced by β ($\beta = 1, 2, (3)$).

The integral term in Eq. (4.4) may be written as

$$\int_{T(\gamma,\delta,s)} \sigma_{ki}^{\alpha} u_{k,i}^{\beta} dv = \int_{T(\gamma,\delta,s)} \left(\sigma_{ki}^{\alpha} u_k^{\beta} \right)_{,i} dv - \int_{T(\gamma,\delta,s)} \sigma_{ki,i}^{\alpha} u_k^{\beta} dv. \tag{4.6}$$

The second integral in the right-hand side of Eq. (4.6) is equal to zero due to equilibrium equation in the absence of the body and inertial forces ($\sigma_{ki,i}^{\alpha} = 0$), and the first integral may be transformed into a surface integral using Gauss' theorem

$$\int_{T(\gamma,\delta,s)} \left(\sigma_{ki}^{\alpha} u_k^{\beta} \right)_{,i} dv = \int_{\gamma^-} \sigma_{ki}^{\alpha} u_k^{\beta} \nu_i \, d\Sigma + \int_{\gamma^+} \sigma_{ki}^{\alpha} u_k^{\beta} \nu_i \, d\Sigma + \int_{\partial F^{(s)}} \sigma_{ki}^{\alpha} u_k^{\beta} \nu_i \, d\Sigma.$$

Thus, using the boundary conditions (4.3), we reduce Eq. (4.4) to the following form

$$P^{\alpha\beta}(\gamma, \delta, s) = \int_{\gamma^+} T_k^{\alpha} \delta_{k\beta} \, d\Sigma = \int_{\gamma^+} T_{\beta}^{\alpha} \, d\Sigma, \tag{4.7}$$

$$\alpha, \beta = 1, 2, (3), \quad d\Sigma = \prod_{i=1}^{n-1} dx_i, \quad n = 2, (3)$$

where $T_k^{\alpha} = \sigma_{ki}^{\alpha} \nu_i$ is the k-th component of the vector of the load $\mathbf{T}_{\nu}^{(\alpha)}$ acting at the boundary γ^+ of the domain $T(\gamma, \delta, s)$ on a unit area element with the normal ν, with components ν_i, in the problem with the boundary conditions (4.3). Thus, the component $P^{\alpha\beta}(\gamma, \delta, s)$ of the tensor of influence $\mathbf{P}(\gamma, \delta, s)$ is equal to the work done by the force $T_{\nu}^{(\alpha)}$ on the β-displacement of the boundary γ^+ satisfying the boundary conditions (4.3). By Betti's reciprocal theorem (Gladwell, 1980), the tensor $\mathbf{P}(\gamma, \delta, s)$ is symmetric.

4.1.3 The auxiliary problem

Together with the main boundary problem formulated in § 4.1.1, we consider the boundary problem for the Lamé equations (4.1) in the region $\Omega \setminus \Gamma$. We denote its solution by the vector $\mathbf{u}$ with components u_i, $i = 1, 2, (3)$. The functions u_i satisfy the same conditions at the boundary $\partial\Omega$, as the functions $u_i^{(s)}$, and the following condition at the surface Γ

$$\mathbf{T}_\nu^+\Big|_\Gamma = \mathbf{T}_\nu^-\Big|_\Gamma = \mathbf{k}(\mathbf{x})\left(\mathbf{u}^+ - \mathbf{u}^-\right)\Big|_\Gamma, \tag{4.8}$$

where $\mathbf{T}_\nu^+$ and $\mathbf{T}_\nu^-$ are load vectors on unit area elements with normal ν on different sides of Γ, $(\mathbf{u}^+ - \mathbf{u}^-)$ is the jump of the vector of displacements on Γ, $\mathbf{k}(\mathbf{x})$ is a tensor with nonnegative components.

The relations between the tensors $\mathbf{k}(\mathbf{x})$, $\mathbf{P}(\gamma, \delta, s)$ and the solutions of the main and auxiliary problems, i.e. functions $u_i^{(s)}$ and u_i, can be established based on the following *theorem*:

Let the following conditions be satisfied as $s \to +\infty$:

1. All the elements $F_i^{(s)}$ are in an arbitrarily small vicinity of Γ.

2. For any $\gamma \subset \Gamma$, the following limits and the function $\mathbf{k}(\mathbf{x})$ exist such that

$$\lim_{\delta\to 0} \varliminf_{s\to+\infty} \mathbf{P}(\gamma, \delta, s) = \lim_{\delta\to 0} \varlimsup_{s\to+\infty} \mathbf{P}(\gamma, \delta, s) = \int_\gamma \mathbf{k}(\mathbf{x})\, d\Sigma, \tag{4.9}$$

$$\mathbf{x} = \{x_1, x_2, (x_3)\}.$$

Then the sequence $\mathbf{u}^{(s)}$ of solutions of the main boundary problem formulated in § 4.1.1 converges to the solution $\mathbf{u}$ of the auxiliary problem, the convergence occurring not only with respect to the functions $\mathbf{u}^{(s)}$ but also with respect to their first and second derivatives.

This theorem is a particular case of a more general theorem, stated and proved in Marchenko and Khruslov (1974) (see also Marchenko, 1971). A discussion of its application for the problem under consideration may be found in Goryacheva and Feldstein (1995, 1996).

For applications, the theorem yields an asymptotic analysis of the stress-strain state at some distance away from Γ. Averaging methods in continuous media mechanics are also discussed in Sanchez-Palensia (1980).

By comparing Eqs. (4.7) and (4.9), we find that the components of the tensor $\mathbf{k}(\mathbf{x})$ are numerically equal to the limit values as $\delta \to 0$, $s \to +\infty$ of components of the force vectors $T_\nu^{(\alpha)}$ $(\alpha = 1, 2, (3))$ acting on a unit area element of the boundary γ^+ (in the limit, γ^+ coincides with γ) in the problem with the boundary conditions (4.3)).

Thus, the tensor $\mathbf{k}(\mathbf{x})$, used in the formulation of the auxiliary problem, characterizes on the average the deformation properties of the thin layer with defects.

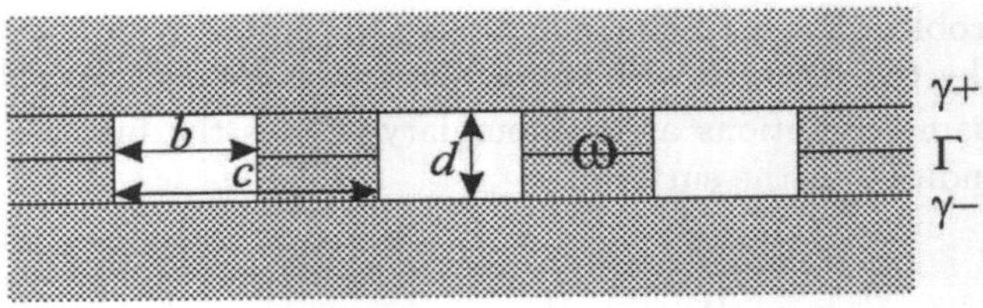

Figure 4.2: Schematic of the arrangement of the system of defects in some special case.

4.1.4 A special case of a system of defects

As an example, we determine the tensor $\mathbf{k}(\mathbf{x})$ in a special case. Let $\Omega \subset R^2$, Γ be a line parallel to the x-axis, and let $F^{(s)}$ be a set of similar rectangles, with sides b and d, uniformly distributed along Γ (see Fig. 4.2). We choose γ^+ and γ^- so that the layer $T(\gamma, \delta)$ has the thickness 2δ which is equal to the side of the rectangle d. The reason for this choice is as follows. The theorem provides a method for analyzing the asymptotic behavior of the solution of the problem (4.1)–(4.2) at some distance away from Γ that increases with the thickness of $T(\gamma, \delta)$. Any other choice of γ^+ and γ^- gives rise to a worse asymptotic approximation. This can be illustrated by the following limit case: if δ is much greater than the characteristic size of $F_i^{(s)}$, then the solution of the problem (4.3) and the components of $\mathbf{P}(\gamma, \delta, s)$ do not feel the set $F_i^{(s)}$; this situation corresponds to the solution of the main boundary problem at infinity.

We consider the element ω bounded by γ^+, γ^- and sides of two adjacent defects so that ω is a rectangle with sides $c - b$ and d (see Fig. 4.2). We assume that the deformation of each element ω is independent of the deformation of the neighboring elements. To satisfy the boundary conditions (4.3) on γ^+ and γ^-, we consider the solution of the Lamé equations (4.1) in the form ($x = x_1$ and $y = x_2$)

$$\mathbf{u}^1 = \begin{pmatrix} \dfrac{y}{d} \\ 0 \end{pmatrix}, \quad \mathbf{u}^2 = \begin{pmatrix} 0 \\ \dfrac{y}{d} \end{pmatrix}. \tag{4.10}$$

The displacements $\mathbf{u}^1$ ($\alpha = 1$) and $\mathbf{u}^2$ ($\alpha = 2$) provide the uniform stress field inside the element ω

$$\sigma_{11}^1 = \sigma_{22}^1 = 0, \quad \sigma_{12}^1 = \frac{\mu}{d}, \quad (\alpha = 1), \tag{4.11}$$

$$\sigma_{11}^2 = \frac{\lambda}{d}, \quad \sigma_{22}^2 = \frac{(\lambda + 2\mu)}{d}, \quad \sigma_{12}^2 = 0, \quad (\alpha = 2). \tag{4.12}$$

So the boundary conditions on the sides $x = b$ and $x = c$ of the rectangle (unloaded sides) are more nearly satisfied, the more elongated the element ω is along the y-axis.

Since the elements of $F^{(s)}$ (defects) are assumed to be distributed uniformly , the value of $\mathbf{P}(\gamma, \delta, s)$ depends only on the length of γ. We consider N elements ω of length l on γ, so that $N = l/c$, $N \gg 1$. Then we obtain from Eq. (4.7)

$$P^{\alpha\beta}(\gamma, \delta, s) = \frac{l}{c} \int\limits_b^c T^{\alpha}_{\beta}\, dx. \tag{4.13}$$

The components T^{α}_{β} on the boundary γ^+ having the normal vector $\nu = \{0; 1\}$ are

$$T^1_1 = \sigma^1_{12}, \quad T^2_2 = \sigma^2_{22}, \quad T^1_2 = T^2_1 = 0.$$

Then using Eqs. (4.11), (4.12) and (4.13), we obtain

$$\mathbf{P}(\gamma, \delta, s) = \begin{pmatrix} \mu & 0 \\ 0 & \lambda + 2\mu \end{pmatrix} \frac{(c-b)l}{dc}.$$

Since the distribution of ω is uniform, Eq. (4.9) gives

$$\mathbf{k} = \begin{pmatrix} \mu & 0 \\ 0 & \lambda + 2\mu \end{pmatrix} \frac{c-b}{dc} = \begin{pmatrix} k_1 & 0 \\ 0 & k_2 \end{pmatrix}. \tag{4.14}$$

Eq. (4.8) shows that, for the defects localized near the line Γ, parallel to the x-axis, the following relations must hold:

$$\begin{aligned} \tau^+_{xy} = \tau^-_{xy} &= k_1 \left(u^+_x - u^-_x\right), \\ \sigma^+_y = \sigma^-_y &= k_2 \left(u^+_y - u^-_y\right), \end{aligned} \tag{4.15}$$

where $\tau^{\pm}_{xy}$, $\sigma^{\pm}_y$, $u^{\pm}_x$, $u^{\pm}_y$ are the tangential and normal stresses and displacements on Γ.

It is worth noting that if the shape of defects tends to a cutset ($d \to 0$) then $k_i \to \infty$. For $d \neq 0$ the parameters $k_i \to 0$, if $b \to c$. In this case the upper layer lies on the substrate without friction.

4.1.5 Half-plane weakened by a system of defects

We investigate now the effect of the system of defects disposed near the boundary of elastic body when there is a non-uniform stress field due to the contact with counterbody.

We consider a problem in the 2-D formulation for the punch with the shape function $y = f(x)$ which penetrates without friction into the elastic half-plane $\Omega = \{(x, y) : y > 0\}$ under the load P (Fig. 4.3). The boundary conditions at the

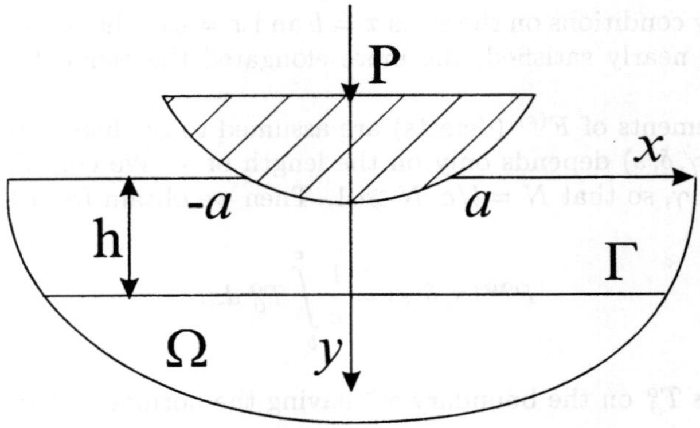

Figure 4.3: Scheme of contact of an indenter and the elastic semi-infinite plane weakened by the system of defects localized along the line Γ.

surface $y = 0$ have the form

$$\begin{aligned} u_y &= D - f(x), & |x| &\leq a, \\ \sigma_y &= 0, & |x| &> a, \\ \tau_{xy} &= 0, & -\infty &< x < +\infty, \end{aligned} \tag{4.16}$$

where a is a half-width of the contact zone, and D is a punch penetration.

The system of defects $F^{(s)}$ is uniformly distributed in the half-plane along the line $\Gamma : \{y = h\}$ (see Fig. 4.3). We assume that the linear relationships between stresses and the jump of displacements have the form of Eqs. (4.15) on Γ. The values of k_1 and k_2 are determined from Eq. (4.14) and, according to Eq. (4.14),

$$k_2 = \frac{\lambda + 2\mu}{\mu} k_1 = \frac{2(1-\nu)}{1-2\nu} k_1.$$

We solve the problem by the Fourier transform method described in detail in Ufland (1965), Sneddon (1972) and Gladwell (1980). Petrishin, Privarnikov and Shavalyakov (1965), Braat and Kalker (1993) developed this method for studying boundary problems for a multilayer elastic half-plane. The Fourier transforms H_1 and H_2 of the Airy stress functions, which are biharmonic in the domains $\{(x,y) : 0 < y < h\}$ and $\{(x,y) : y > h\}$, have the form

$$\begin{aligned} H_1(\omega, y) &= [A_1(\omega) + B_1(\omega)y]\exp(y\omega) + [C_1(\omega) + D_1(\omega)y]\exp(-y\omega), \\ H_2(\omega, y) &= [A_2(\omega) + B_2(\omega)y]\exp(-y|\omega|). \end{aligned} \tag{4.17}$$

The Fourier transforms $\hat{u}_x(\omega, y)$, $\hat{u}_y(\omega, y)$, $\hat{\sigma}_y(\omega, y)$, $\hat{\tau}_{xy}(\omega, y)$ of displacements and stresses are expressed in terms of the function $H(\omega, y)$ which coincides with $H_1(\omega, y)$ for the domain $\{(x, y) : 0 < y < h\}$ and with $H_2(\omega, y)$ for the domain

$\{(x, y) : y > h\}$ by:

$$\hat{u}_x(\omega, y) = \frac{1-\nu^2}{E} \cdot \frac{\partial^3 H}{\partial y^3} + \omega^2 \frac{\nu(1+\nu)}{E} \cdot \frac{\partial H}{\partial y}, \tag{4.18}$$

$$\hat{u}_y(\omega, y) = i\omega^3 \frac{1-\nu^2}{E} H + i\omega \frac{\nu(1+\nu)}{E} \cdot \frac{\partial^2 H}{\partial y^2}, \tag{4.19}$$

$$\hat{\sigma}_y(\omega, y) = i\omega^3 \frac{\partial H}{\partial y}, \tag{4.20}$$

$$\hat{\tau}_{xy}(\omega, y) = \omega^2 \frac{\partial^2 H}{\partial y^2}. \tag{4.21}$$

The coefficients $A_1(\omega)$, $B_1(\omega)$, $C_1(\omega)$, $D_1(\omega)$, $A_2(\omega)$, $B_2(\omega)$ in Eq. (4.17) are determined by solving the system of equations obtained by substituting Eqs. (4.18), (4.19), (4.20), (4.21) into the transformed boundary conditions (4.15) at $y = h$ and (4.16) at $y = 0$. This procedure is described in detail in Braat and Kalker (1993). Then the numerical Fourier inverse transformation is applied to find the stresses and displacements within the elastic body.

4.1.6 Influence of defects on contact characteristics and internal stresses

We present here the results of calculation of contact characteristics and internal stresses within the half-plane weakened by the system of defects localized near the line $y = h$ and having the characteristics described in the example of § 4.1.4. In calculations, the following dimensionless parameters and functions were used:

$$\bar{y} = \frac{y}{a}, \quad \bar{h} = \frac{h}{a}, \quad L = \frac{a}{R}, \quad \bar{p}(x) = -\frac{\sigma_y(x, 0)}{E}, \quad \bar{k} = \frac{k_1 a}{E}.$$

Fig. 4.4 demonstrates the contact pressure distribution under the punch of parabolic shape, i.e. $f(x) = x^2/2R$ (R is the radius of curvature).

The calculations show that as the number of defects in the layer increases ($\bar{k}$ decreases), the size of the contact zone and the layer thickness remaining the same, the maximum values of the contact pressure $\bar{p}(x)$ decrease and the load required for the contact zone to attain a given size diminishes. For comparison, the dashed line in Fig. 4.4 presents the Hertz pressure distribution corresponding to the same value of L. The results also show that the influence of the defects on the contact pressure distribution is weaker for larger distances between the surface and the layer with defects.

The results of computations of the internal stresses along the y-axis for the pressure p_0 uniformly distributed within the region ($-a \leq x \leq a$, $y = 0$) are shown in Fig. 4.5.

Fig. 4.5(a) illustrates the variation of the principal shear stress $\tau_1(y)$ for different values of parameter $\bar{k}$ and $\bar{h} = 2$. For large values of $\bar{k}$, the shear stress distribution is similar to that corresponding to a homogeneous half-space (curve 1);

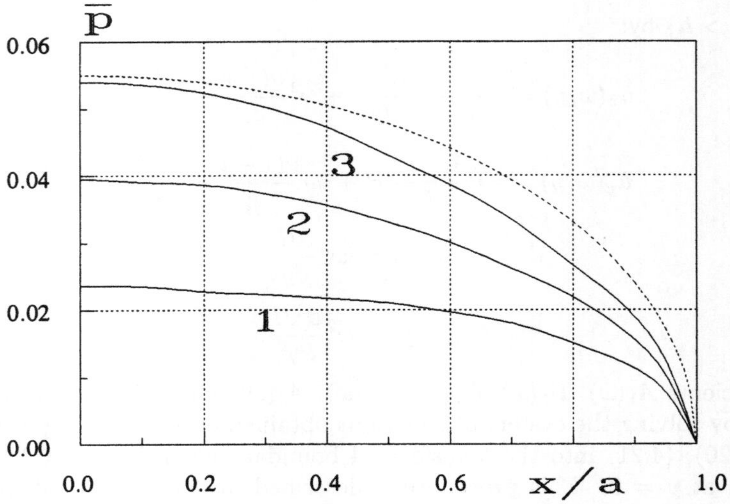

Figure 4.4: Contact pressure distribution under the punch of a parabolic shape for $\bar{h} = 1$, $L = 0.1$ and $\bar{k} = 0.1$ (curve 1), $\bar{k} = 1.0$ (curve 2), $\bar{k} = 3.0$ (curve 3), dashed line corresponds to the Hertz solution.

for smaller values of $\bar{k}$, τ_1 has a jump at Γ (i.e. $y = h$). The smaller the value of $\bar{k}$, the greater is the jump.

The stress σ_x also has a jump; that is illustrated in Fig. 4.5(b) (curves with the same numbers in Fig. 4.5 are constructed for the same values of the parameter $\bar{k}$). The behavior of σ_x has an interesting feature: it changes the sign at $y = h$ for small values of $\bar{k}$.

The results presented in Figs. 4.4 and 4.5 correspond to $\nu = 0.3$.

The study shows that there is a range of the parameters characterizing the amount of defects per unit area in the defect layer and the layer-boundary distance, within which the defect layer influences the contact characteristics substantially. The proposed approach enables one to allow for this influence in solving contact problems and in analysing the stress state of elastic bodies having internal systems of defects.

4.2 Coated elastic bodies

In normal contact of bodies with coatings, the model of a two-layered elastic body is usually used to analyze the stress field within the coating and substrate, and to calculate the contact characteristics. The method of integral transformations such as Fourier transform for 2-D case and Hankel transform for axisymmetric case is applied to solving the contact problems for two-layered elastic foundation (Nikishin and Shapiro, 1970, 1973, Makushkin, 1990a, 1990b, Kuo and Keer, 1992).

For coated bodies, there is a question which is very important from a tribologi-

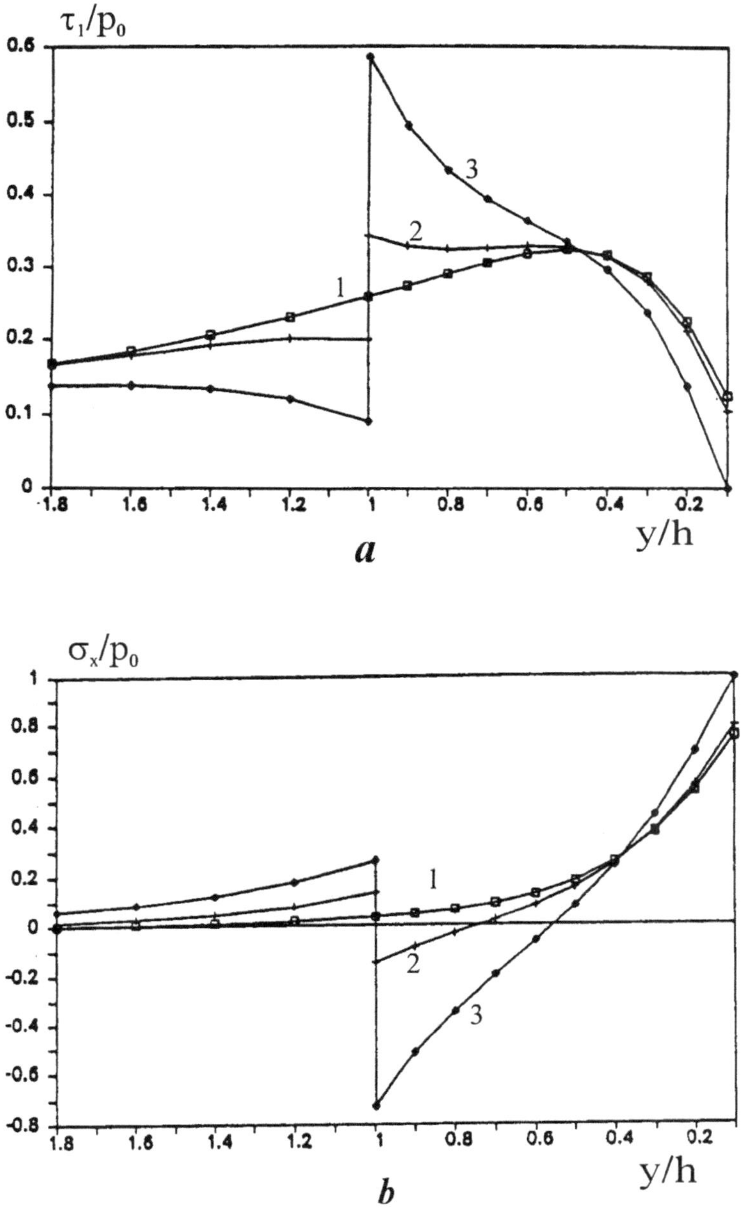

Figure 4.5: The principal shear stress τ_1 (a) and the component σ_x (b) along the y-axis for $\bar{h} = 2$ and $\bar{k} = 5000$ (curve 1), $\bar{k} = 0.5$ (curve 2), $\bar{k} = 0.05$ (curve 3).

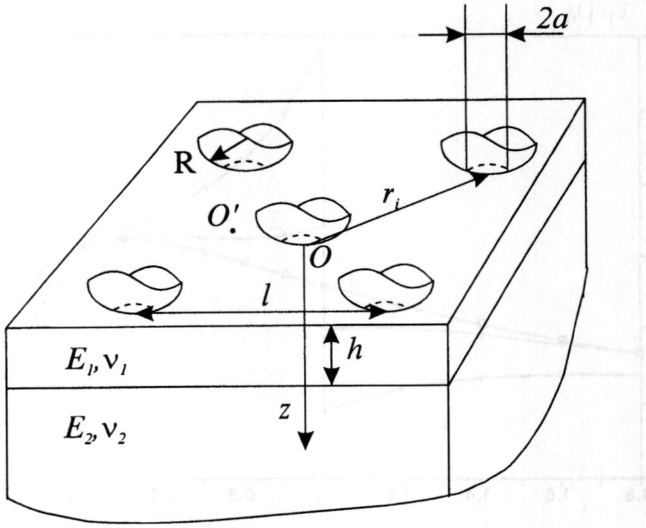

Figure 4.6: Scheme of contact of a periodic system of indenters and a coated elastic half-space.

cal point of view: What is the influence of discrete contact on the internal stresses of coated bodies? The answer to this question is essential due to the extensive use of thin coatings, and to the existence of thin films at the surfaces of contacting bodies with thickness comparable to the distance between asperities or to the size of each contact spot. Some results of numerical simulation of the contact of layered elastic bodies with real rough surfaces are discussed in Sainsot, Leroy and Villechase (1990) and in Cole and Sayles (1991). They showed that, for both soft and hard surface layers, the stresses in the layer and at the interface between the layer and the half-space are significantly affected by contact discreteness. The results are of interest for predicting the layer failure pattern, but it is difficult to analyze them because of the erratic character of the roughness.

In what follows we investigate the combined effect of surface roughness and coatings in normal contact using a simple model of discrete loading with spots arranged periodically on the surface of a two-layered elastic half-space.

4.2.1 Periodic contact problem

We consider a system of indenters, located on a hexagonal lattice with a constant pitch l. The system penetrates into the elastic layer of thickness h bonded with an elastic half-space (Fig. 4.6). The following conditions are satisfied at the interface between the layer and substrate ($z = h$):

$$\begin{aligned} &\sigma_z^{(1)} = \sigma_z^{(2)}, \quad \tau_{rz}^{(1)} = \tau_{rz}^{(2)}, \quad \tau_{\theta z}^{(1)} = \tau_{\theta z}^{(2)}, \\ &u_r^{(1)} = u_r^{(2)}, \quad u_\theta^{(1)} = u_\theta^{(2)}, \quad w^{(1)} = w^{(2)}, \end{aligned} \tag{4.22}$$

where $\sigma_z^{(i)}$, $\tau_{rz}^{(i)}$, $\tau_{\theta z}^{(i)}$ and $u_r^{(i)}$, $u_\theta^{(i)}$, $w^{(i)}$ are the components of stresses and displacements in the layer ($i = 1$) and in the half-space ($i = 2$).

The conditions on the upper surface ($z = 0$) of the layer are

$$\begin{aligned} w(\vec{r}) = f(\vec{r} - \vec{r}_i), \quad \vec{r} \in \omega_i = \{|\vec{r} - \vec{r}_i| < a\}, \\ \sigma_z^{(1)} = 0, \quad \vec{r} \notin \omega_i, \quad i = 1, 2, 3, \ldots, \\ \tau_{rz}^{(1)} = 0, \quad \tau_{\theta z}^{(1)} = 0, \end{aligned} \tag{4.23}$$

where $f(\vec{r})$ is the shape of the indenter, $\vec{r}_i$ is the radius-vector of the center of a contact spot ω_i from the origin of the system of coordinates. It is assumed that indenters are under identical conditions, so that the contact spots ω_i all have the same radius a. The load P acting on each indenter is related to the nominal contact pressure p_n by (see Chapter 2)

$$P = \frac{\sqrt{3}}{2} p_n l^2, \tag{4.24}$$

and the following equilibrium condition is satisfied

$$P = \int_0^a \int_0^{2\pi} p(r) r \, dr d\varphi. \tag{4.25}$$

4.2.2 Method of solution

We place the origin of the polar system of coordinates at the point where the axis of symmetry of any indenter intersects the plane $z = 0$. Using the principle of localization formulated in Chapter 2, we reduce the periodic problem under consideration to the following axisymmetric problem, in which the boundary conditions at the upper layer surface ($z = 0$) are in the form

$$\begin{aligned} w(r) &= f(r), & 0 < r < a, \\ \sigma_z^{(1)} &= 0, & a < r < R_1, \\ \sigma_z^{(1)} &= p_n, & R_1 < r < +\infty, \\ \tau_{rz}^{(1)} = \tau_{\theta z}^{(1)} &= 0, & 0 < r < +\infty, \end{aligned} \tag{4.26}$$

i.e. we consider the real contact condition for any fixed indenter with center at the point O, and replace the action of the remaining indenters by the nominal pressure p_n distributed uniformly in the region $r > R_1$. R_1 is chosen to satisfy the equilibrium condition within the circle $r \leq R_1$, i.e. $\pi R_1^2 p_n = P$. Taking Eq. (4.24) into account, we obtain

$$R_1 = l \left(\frac{\sqrt{3}}{2} \right)^{1/2} . \tag{4.27}$$

We use the Hankel transform for solving the axisymmetric problem for the elastic layer on the elastic substrate with boundary conditions (4.22) and (4.26). Following Nikishin and Shapiro (1970), we represent stresses and displacements within the layer ($i = 1$) and the half-space ($i = 2$) in the Love form

$$\begin{aligned}
\sigma_r^{(i)} &= \frac{\partial}{\partial z} \left[\nu_i \Delta \varphi^{(i)} - \frac{\partial^2 \varphi^{(i)}}{\partial r^2} \right], \\
\sigma_\theta^{(i)} &= \frac{\partial}{\partial z} \left[\nu_i \Delta \varphi^{(i)} - \frac{1}{r} \cdot \frac{\partial \varphi^{(i)}}{\partial r} \right], \\
\sigma_z^{(i)} &= \frac{\partial}{\partial z} \left[(2 - \nu_i) \Delta \varphi^{(i)} - \frac{\partial^2 \varphi^{(i)}}{\partial z^2} \right], \\
\tau_{rz}^{(i)} &= \frac{\partial}{\partial r} \left[(1 - \nu_i) \Delta \varphi^{(i)} - \frac{\partial^2 \varphi^{(i)}}{\partial z^2} \right], \\
u_r^{(i)} &= -\frac{1 + \nu_i}{E_i} \cdot \frac{\partial^2 \varphi^{(i)}}{\partial r \partial z}, \\
w^{(i)} &= \frac{1 + \nu_i}{E_i} \left[2 (1 - \nu_i) \Delta \varphi^{(i)} - \frac{\partial^2 \varphi^{(i)}}{\partial z^2} \right].
\end{aligned} \tag{4.28}$$

Here $\varphi^{(i)}(r, z)$ ($i = 1, 2$) are biharmonic functions in the layer and substrate. These functions are represented in the form

$$\begin{aligned}
\varphi^{(1)}(r, z) &= \int_0^{+\infty} [(A_1(\alpha) + z B_1(\alpha)) \exp(\alpha z) + \\
&\quad (C_1(\alpha) + z D_1(\alpha)) \exp(-\alpha z)] J_0(\alpha r) \, d\alpha, \\
\varphi^{(2)}(r, z) &= \int_0^{+\infty} (A_2(\alpha) + z B_2(\alpha)) \exp(-\alpha z) J_0(\alpha r) \, d\alpha,
\end{aligned} \tag{4.29}$$

where $J_0(x)$ is the Bessel function of the first kind of order 0.

The relations (4.22) and (4.26) expressed through Eq. (4.28) and the biharmonic functions (4.29) permit the problem to be reduced to a system of six linear equations for the coefficients $A_1(\alpha)$, $B_1(\alpha)$, $C_1(\alpha)$, $D_1(\alpha)$, $A_2(\alpha)$, $B_2(\alpha)$. Then the inverse integral transform is used to calculate stresses and displacements within the layer and substrate.

To simplify this procedure, we solve the contact problem in two stages. At first we determine the shape $g(r)$ of the upper surface of the layer ($z = 0$) within the circle $0 < r < R_1$ if the layered half-space is loaded by the nominal pressure p_n within the region $R_1 < r < +\infty$. To exclude the infinity from calculations, we solve the problem with the following boundary conditions at $z = 0$

$$\sigma_z^{(1)} = \begin{cases} -p_n, & 0 < r < R_1, \\ 0, & R_1 < r < +\infty. \end{cases} \tag{4.30}$$

The solution $\bar{g}(r)$ of the problem with boundary conditions (4.30) relates to the function $g(r)$ as follows

$$\bar{g} = g(r) + C. \tag{4.31}$$

In the second stage we use the function $\bar{g}(r)$ in formulating the contact conditions within the region $r < a$. We divide the circular region of radius a into N rings of thickness Δr. The contact pressure is presented as a piecewise function $p(r) = p_j$, ($r_{j-1} < r < r_j$, $r_j = j \cdot \Delta r$, $j = 1, 2, \ldots, N$) which is found from the following system of equations

$$p_1 \kappa_1^{(i)} + p_2 \kappa_2^{(i)} + \cdots + p_N \kappa_N^{(i)} = f_1(r_i) - g_1(r_i), \quad (i = 1, 2, \ldots, N-1), \tag{4.32}$$

where $f_1(r) = f(r) - f(a)$, $g_1(r) = \bar{g}(r) - \bar{g}(a)$ (this representation excludes the constant C in Eq. (4.31) from consideration). A coefficient $k_j^{(i)}$ determines the difference between the normal displacements of the rings with the external radii r_i and r_N when unit pressure acts within the ring with the radius r_j.

For a punch with flat base penetrating into the layered foundation, the contact radius a is given. To complete the system of equations (4.32), we must add the equilibrium condition in the form

$$\pi \sum_{i=1}^{N} p_i \left(r_i^2 - r_{i-1}^2 \right) = P,$$

or using the relationship $r_i = i\Delta r$, we have

$$\pi (\Delta r)^2 \sum_{i=1}^{N} p_i (2i - 1) = P. \tag{4.33}$$

If the shape of the indenter is described by the smooth function $f(r)$, there is an additional equation

$$p_N = 0. \tag{4.34}$$

For a smooth indenter, the radius a of the contact spot is unknown. To find the radius a, we can add Eq. (4.34) to the system of Eqs. (4.32), (4.33) and use iteration.

It was shown in Chapter 2 that the accuracy of the solution obtained from the principle of localization is higher if we consider the exact contact conditions also under the neighboring indenters. To evaluate the accuracy of the solution obtained above, we considered also the problem when two or more layers of indenters are taken into account (in the axisymmetric formulation we replace the action of the layer of indenters located at the radius l from the fixed indenter by an equivalent pressure applied within the ring of thickness $2a$). The results of calculations for a system of spherical indenters showed that the difference in the radii of contact region calculated both ways does not exceed 8%.

The stress field in the layer and substrate can also be calculated from the axisymmetric approach. We use the following conditions on the upper surface of the layer ($z = 0$)

$$\sigma_z^{(1)} = \begin{cases} p(r), & 0 < r < a, \\ 0, & a < r < R_1, \\ p_n, & R_1 < r < +\infty, \end{cases} \tag{4.35}$$

where $p(r)$ is the contact pressure obtained above. To exclude the infinity from calculations, we present the stress field inside the layered body as a superposition of the uniform stress field ($\tilde{\sigma}_z^{(1)}(z, r) = p_n$, $\sigma_r = \sigma_\theta = \tau_{rz} = \tau_{\theta z} = 0$) produced by the pressure distributed uniformly on the upper surface of the layer, and the stresses corresponding to the solution of the problem with the following boundary conditions ($z = 0$):

$$\bar{\sigma}_z^{(1)} = \begin{cases} p(r) - p_n, & 0 < r < a, \\ -p_n, & a < r < R_1, \\ 0, & R_1 < r < +\infty, \end{cases} \tag{4.36}$$

where R_1 is determined by Eq. (4.27). The solution of the axisymmetric problem with the boundary conditions (4.22) and (4.36) is found using Hankel transforms (Goryacheva and Torskaya, 1994).

To calculate the stresses under the unloaded zone with the center at the point O' (Fig. 4.6), we solve the axisymmetric problem with the following boundary conditions at $z = 0$

$$\sigma_z^{(1)} = \begin{cases} 0, & 0 < r < R_2, \\ p_c, & R_2 < r < R_3, \\ 0, & R_3 < r < R_4, \\ p_n, & R_4 < r < +\infty. \end{cases} \tag{4.37}$$

To obtain these conditions, we substitute the real contact pressure within three contact spots which are the nearest to the point O', by the pressure p_c uniformly distributed within the ring (R_2, R_3) where $R_2 = \dfrac{l}{\sqrt{3}} - a$, $R_3 = \dfrac{l}{\sqrt{3}} + a$. The pressure p_c is obtained from the equilibrium condition:

$$p_c \pi (R_3^2 - R_2^2) = 3P.$$

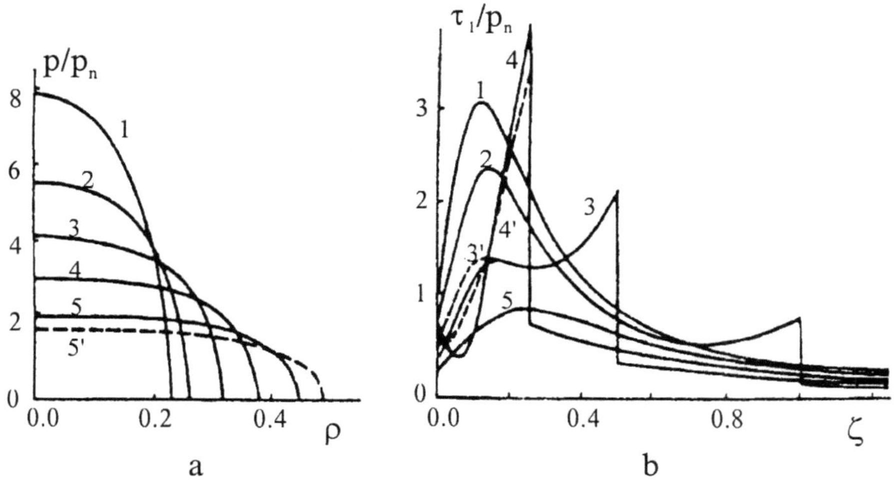

Figure 4.7: Contact pressure (a) and the principal shear stress along the z-axis (b) for hard coating ($\chi = 10$), for $p_n/E_2 = 0.1$, $\varrho = 2$ and $\tilde{h} = +\infty$ (curve 1), $\tilde{h} = 1$ (curve 2), $\tilde{h} = 0.5$ (curve 3), $\tilde{h} = 0.25$ (curve 4), $\tilde{h} = 0$ (curve 5); curves 3′ and 4′ correspond to the Hertz pressure distribution for the same values of parameters as curves 3 and 4; curve 5′ corresponds to $\tilde{h} = 0$ and $\varrho = 0$.

The radius R_4 is found from the condition that the average pressure within the circle of radius R_4 is equal to the nominal pressure p_n, so

$$R_4 = (3\sqrt{3}/2\pi)^{1/2} l.$$

The internal stresses in the axisymmetric problem with the boundary conditions in the form Eqs. (4.22) and (4.37) is found by the method described above (see also Goryacheva and Torskaya, 1995).

We have compared the solution of the axisymmetric problems with the boundary conditions at $z = 0$ in the form of Eqs. (4.35) and (4.37), with the exact solution obtained by the superposition of the stress fields produced by an each indenter. The results show that the maximum error (for the case $a/l = 0.5$) does not exceed of 5%.

4.2.3 The analysis of contact characteristics and internal stresses

We consider a system of spherical indenters ($f(r) = r^2/2R$).

It has been established that the solution of the problem depends on the following dimensionless parameters: the relative elasticity modulus of the surface layer

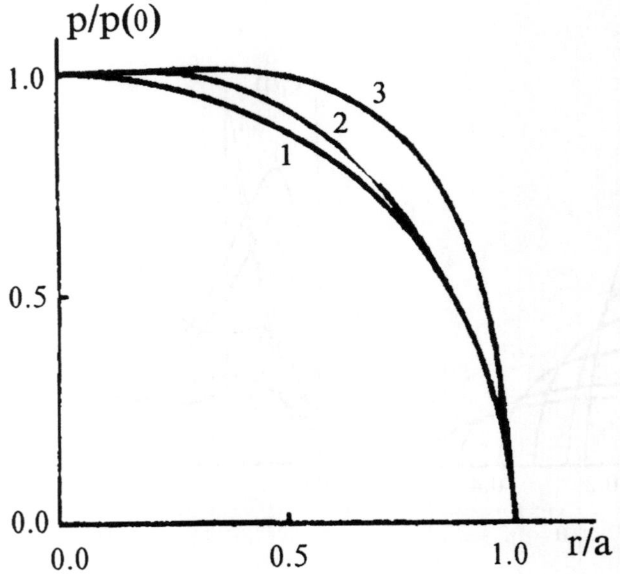

Figure 4.8: The contact pressure for thin hard coating, for $\tilde{h} = 0.25$ (curve 2), $\tilde{h} = 0.5$ (curve 3); $p_n/E_2 = 0.1$, $\chi = 10$, $\varrho = 2$; curve 1 corresponds to the Hertz solution.

$\chi = E_1/E_2$; the relative layer thickness $\tilde{h} = h/l$; the relative radius of curvature of the indenters $\varrho = R/l$, which characterizes also the density of arrangement of the indenters; the dimensionless nominal pressure p_n/E_2, and the Poisson ratio ν (in the calculation, we assumed that $\nu_1 = \nu_2 = \nu$).

We will analyze the influence of the relative mechanical and geometrical properties of the surface layer, and the density of indenter arrangement, on the contact pressure $\bar{p}(\rho) = p(\rho)/p_n$ ($\rho = r/l$), the relative radius of each contact spot a/l, and the internal stresses $\sigma_{ij}(\zeta)/p_n$ ($\zeta = z/l$) along the axis Oz and $O'z$ (Fig. 4.6).

It is convenient to consider separately two types of surface layer: hard ($\chi > 1$), and soft ($\chi < 1$) coatings.

Hard coatings

The results presented here have been calculated for $p_n/E_2 = 0.1$ and $\nu = 0.3$.

Figs. 4.7(a) and 4.8 illustrate the pressure distribution within a contact spot for different values of parameters $\tilde{h}$ and ϱ. The curves 1–5 in Fig. 4.7(a) correspond to the layer thickness changing from infinity to zero (uncoated substrate), respectively, and to a constant density of indenter arrangement, namely $\varrho = 2$. The results indicate that the maximum contact pressure decreases, and the contact radius increases, as the thickness of the coating decreases. However, for fixed thickness of the coating, the contact radius for the periodic problem is less than that calculated for one indenter penetrating the layered foundation. This conclu-

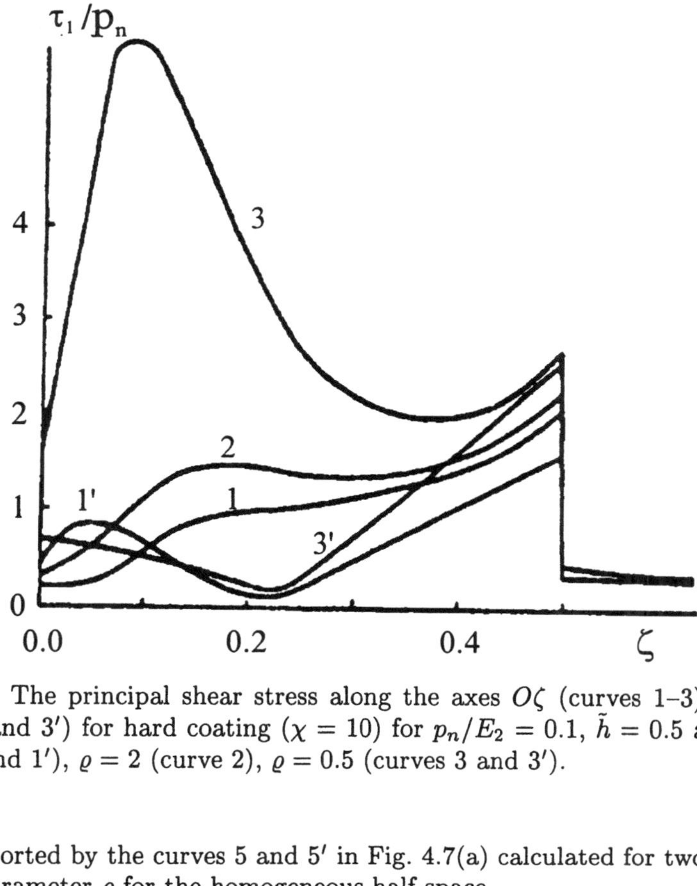

Figure 4.9: The principal shear stress along the axes $O\zeta$ (curves 1–3) and $O'\zeta$ (curves $1'$ and $3'$) for hard coating ($\chi = 10$) for $p_n/E_2 = 0.1$, $\tilde{h} = 0.5$ and $\varrho = 4$ (curves 1 and $1'$), $\varrho = 2$ (curve 2), $\varrho = 0.5$ (curves 3 and $3'$).

sion is supported by the curves 5 and $5'$ in Fig. 4.7(a) calculated for two different values of parameter ϱ for the homogeneous half-space.

Fig. 4.8 illustrates the distribution of the dimensionless pressure $p(\rho_1)/p(0)$ ($\rho_1 = r/a$) within the contact spot ($\rho_1 < 1$) for different values of the parameter $\tilde{h}$. The results show that the pressure distribution differs from the Hertz solution (curve 1) with the difference increasing as the parameter a/h increases beyond 1 (see $a/h = 1.2$ for curve 2 and $a/h = 3.2$ for curve 3).

The analysis of the influence of the parameter χ on contact characteristics shows that the radius a/l of the contact spot decreases, and the maximum contact pressure $p(0)$ increases, as the parameter χ increases.

We also investigated the influence of the parameters $\tilde{h}$ (Fig. 4.7(b)) and ϱ (Fig. 4.9) on the principal shear stress distribution $\tau_1(\zeta)$ along the axis Oz and $O'z$. The results show that it is specific for the hard coating to have a jump of values of τ_1 at the layer-substrate interface $\zeta = \tilde{h}$, so that $\tau_1^{(1)} - \tau_1^{(2)} > 0$, where $\tau_1^{(i)}$ is the value of τ_1 at the interface from the side of the layer ($i = 1$) and substrate ($i = 2$), respectively. As a rule, the function $\tau_1(\zeta)$ has two maxima: the first is inside the layer, or at the layer surface $\zeta = 0$ for very thin layers (curve 4 in Fig. 4.7(b)) and the other is at the layer-substrate interface $\zeta = \tilde{h}$. The relation between maxima changes depending on layer thickness (Fig. 4.7(b)): for relatively thick layers (curves 1 and 2) the maximum value of τ_1 occurs inside the layer, for thinner layers (curves 3 and 4) it is at the layer-substrate interface.

We compared the internal stresses produced by the contact pressure calculated from the periodic contact problem considered above, with internal stresses produced by the Hertz pressure applied within the contact spots of radius a. The results are presented in Fig. 4.7(b)) where curves $3'$ and $4'$ are constructed from the Hertz pressure distribution and the same values of parameters as curves 3 and 4. The difference between the curves is visible only for $\zeta \leq \zeta^*$, and the value of ζ^* decreases as the parameter a/h decreases ($a/h = 1.54$ for the curve 4 and $a/h = 0.64$ for the curve 3). So it is possible to simplify the calculations, changing the real pressure distribution to the Hertz pressure when we investigate the internal stresses at some distance away from the surface.

The dependence of the principal shear stress distribution along the $O\zeta$-axis on the parameter ϱ is illustrated by the curves 1–3 in Fig. 4.9. There are also the plots of the function $\tau_1(\zeta)$ along the $O'\zeta$ axis which crosses the plane $y = 0$ at point O', which is the center of unloaded zone (they are the curves $1'$ and $3'$ calculated for the same values of parameters as the curve 1 and 3 in this figure). Comparing the results, we can conclude that, for a fixed $\tilde{h}$, the maximum difference $\Delta\tau_1(\zeta)$ of the values of $\tau_1(\zeta)$ at the fixed depth ζ decreases as the parameter ϱ and, consequently, the parameter a/h, increases. The same conclusion was established in Chapter 2, where we analyzed the effect of the contact density parameter for a homogeneous half-space. For small values of the parameter ϱ, the function $\Delta\tau_1(\zeta)$ approaches the function $\tau_1(\zeta)$.

Soft coatings

This case ($\chi < 1$) has been calculated for $p_n/E_2 = 0.005$ and $\nu = 0.3$.

The results of calculations of contact pressure $p(\rho)/p_n$ and the principal shear stress $\tau_1(\zeta)/p_n$ for $\chi = 0.1$ are presented in Figs. 4.10 and 4.11. The analysis of the contact pressure distribution for the various layer thicknesses (Fig. 4.10(a)) shows that the radius a of the contact spot increases, and the maximum contact pressure decreases, as the layer thickness increases. It should be noted also that the influence of the substrate properties on the contact characteristics becomes negligible if the layer thickness $\tilde{h}$ is more than some critical value $\tilde{h}^*$ which depends on the parameters χ and ϱ. This conclusion follows from the comparison of the curves 2, 3 and 4 (the last one corresponds to the case $\tilde{h} \to +\infty$) in Fig. 4.10(a). The results of calculations of the contact pressure for various values of parameter $\chi < 1$ indicate that the critical value $\tilde{h}^*$ increases as the parameter χ decreases.

We also calculated the principal shear stress τ_1 along the axis $O\zeta$ for the same values of the parameters as we used in the contact pressure analysis (Fig. 4.10(b)). The results show that the maximum value of the principal shear stress can be achieved inside the layer, or inside the substrate, depending on the layer thickness

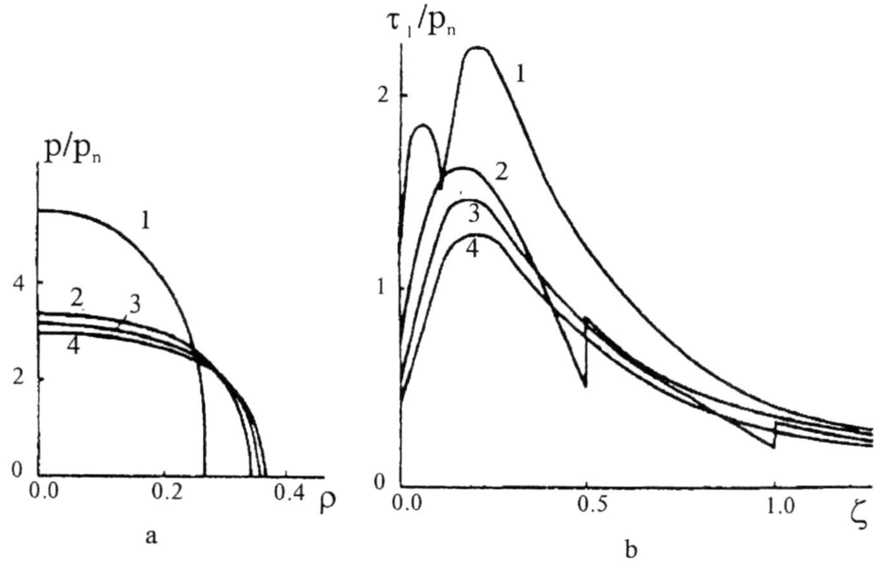

Figure 4.10: Contact pressure (a) and the principal shear stress along the axis $O\zeta$ (b) for soft coating, ($\chi = 0.1$) for $p_n/E_2 = 0.005$, $\varrho = 2$ and $\tilde{h} = 0.1$ (curve 1), $\tilde{h} = 0.5$ (curve 2), $\tilde{h} = 1$ (curve 3), $\tilde{h} = +\infty$ (curve 4).

$\tilde{h}$. For thick layers the maximum value of the principal shear stress occurs inside the layer, and for thin layers (curve 1) it is inside the substrate.

The results presented in Fig. 4.11 illustrate the dependence of the function $\tau_1(\zeta)$ along the axis $O\zeta$ (curves 1–3) and along the axis $O'\zeta$ (curves 1′ and 3′), on the parameter ϱ. As in the case $\chi > 1$, the difference of the values of $\tau_1(\zeta)$ at a fixed depth decreases as the parameter ϱ increases.

The results also show that there is the jump in the stresses at the interface for the soft coatings, but the sign of this jump may be different, depending on the layer thickness.

For soft coatings, the stress distribution inside the layer tends to uniformity with decreasing of the layer thickness or increasing of the radius of the loaded regions.

Thus, the features of internal stress and contact pressure distribution depend essentially on the relative mechanical and geometrical characteristics of the coating and also on the density of the contact spots. The discreteness of the loading plays a major role for relatively thin and hard coatings. So coating classification (relatively thin ($h/a < 1$) and thick ($h/a > 1$)) commonly used for stress evaluation is not acceptable for discrete contact; the additional geometrical parameter ϱ, which characterizes the relative size of loaded region, has to be used for contact characteristics and internal stress analysis.

Results from the internal stress analysis together with fracture criteria make

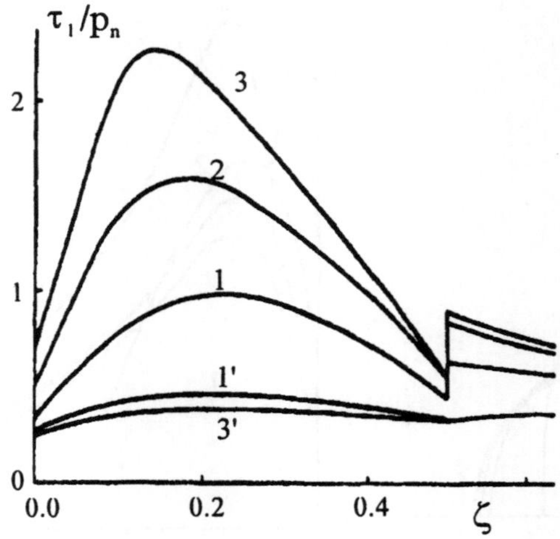

Figure 4.11: The principal shear stress along the axis $O\zeta$ (curves 1–3) and $O'z$ (curves $1'$ and $3'$) for soft coating ($\chi = 0.1$), for $p_n/E_2 = 0.005$, $\tilde{h} = 0.5$ and $\varrho = 10$ (curves 1 and $1'$), $\varrho = 2$ (curve 2), $\varrho = 1$ (curves 3 and $3'$).

it possible to predict different coatings fracture types such as wear, subsurface fracture and delamination.

In sliding contact of coated bodies with rough surfaces a cyclic stress field arises within the coating, within the substrate and at their interface due to discreteness of the contact. We can calculate the amplitude values of the principal shear stress at a fixed depth, and determine the rate of damage accumulation inside the contacting bodies. Then, using the method described in Chapter 5, we can model the fatigue fracture process of coated bodies.

On the basis of the results presented here and in Goryacheva and Torskaya (1994 and 1995) it is also possible to formulate and solve some problems of increasing of the lifetime of the coating determining its optimal mechanical and geometrical characteristics under given loading conditions.

4.3 Viscoelastic layered elastic bodies

When we investigate a sliding or rolling contact, it is important to take into account imperfect elasticity of the surface layer; this can be a coating, boundary lubricant, etc. The accurate representation of the contact normal and tangential stresses, and of the deflections and friction between coated bodies in motion, is of significant importance in tribological applications such as positioning of precision drives, bearings operating in the boundary lubrication regime, etc.

Batra and Ling (1967) investigated the deformation, friction and shear stresses in a viscoelastic layered system under the action of a moving load. Ling and

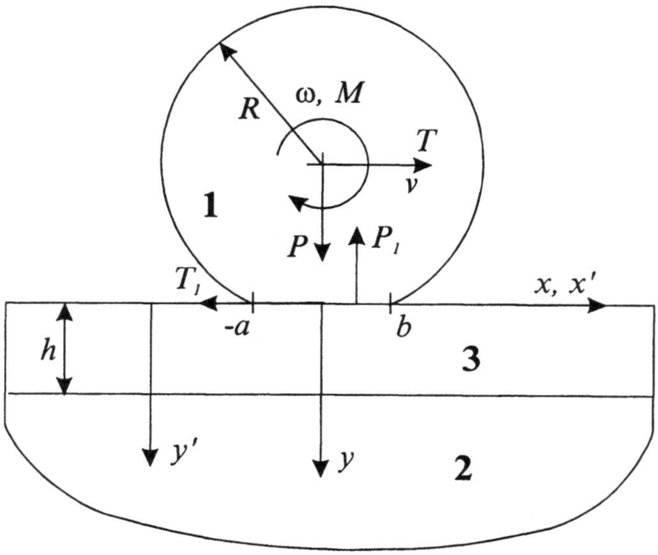

Figure 4.12: Scheme of contact of the cylinder and the layered semi-infinite plane.

Lai (1980) used Fourier transforms to consider the problem of a moving load on the viscoelastic layer bonded to a semi-infinite plane. Recently, Kalker (1991) and Braat and Kalker (1993) theoretically and experimentally analyzed the rolling contact between two cylinders coated with viscoelastic layers. They developed a numerical model for the analysis of stresses in a subsurface layer and used laser Doppler anemometry to verify their results.

In what follows we develop an analytical model to study the effects of a viscoelastic surface layer in rolling and sliding contacts.

4.3.1 Model of the contact

We consider a contact of an elastic cylinder and a foundation which consists of a viscoelastic layer of thickness h bonded to an elastic half-space (Fig. 4.12). The problem is investigated in the 2-D formulation. The cylinder rolls or slides along the base with a constant angular velocity ω and linear velocity V.

The (x', y') coordinate system is fixed on the layered semi-infinite plane while the (x, y) coordinate system moves with the cylindrical indenter. The shape function for the cylindrical indenter is $f(x) = -x^2/2R$. The relationships between the fixed (x', y') and moving (x, y) coordinate systems are

$$\begin{aligned} x' &= x + Vt, \\ y' &= y. \end{aligned} \tag{4.38}$$

In this study the quasi-stationary state is investigated. Therefore, the displacements and stresses are independent of time t in the (x, y) system.

Boundary conditions

Following Reynolds (1875), we subdivide the contact area $(-a, b)$ into slip (S) and no-slip (A) zones.

In the slip zones, the sliding friction is modelled using the Coulomb's law

$$|\tau(x)| = \mu p(x), \text{ when } y = 0,\ x \in S, \tag{4.39}$$

where $\tau(x)$ and $p(x)$ are the tangential and normal stresses in the contact zone, respectively.

For the no-slip zones, the tangential velocity of the contacting points of the cylinder and viscoelastic layer are equal. Hence, in the (x', y') coordinate system the tangential displacements u_1 and u of the cylinder and the layered semi-infinite plane, respectively, satisfy the following:

$$\frac{du}{dt} = V - \omega R + \frac{du_1}{dt}, \text{ when } y = 0,\ x \in A. \tag{4.40}$$

Eq. (4.40) in the (x, y) coordinate system can be written as

$$-\frac{du_1}{dx} + \frac{du}{dx} = \delta, \text{ when } y = 0,\ x \in A, \tag{4.41}$$

where δ is known as the apparent velocity or creep ratio

$$\delta = \frac{\omega R - V}{V}. \tag{4.42}$$

Furthermore, in the no-slip zones A, the normal and tangential stresses are related by the inequality

$$|\tau(x)| \leq \mu p(x). \tag{4.43}$$

Note that the relation (4.39) holds over the whole of the contact region $(-a, b)$ in the case of complete sliding.

It follows from the contact condition that the relation

$$v_1(x) + v_2(x) + v_3(x) = D - \frac{x^2}{2R} \tag{4.44}$$

is satisfied within the contact region $(-a, b)$. In Eq. (4.44) v_1, v_2 and v_3 are the normal displacements of the boundary of the cylinder, of the half-plane and of the layer (strip), respectively (measured positive into each body), and D is the penetration of the cylinder into the layered semi-infinite plane.

It is assumed that the viscoelastic layer is bonded to the elastic half-plane and the following boundary conditions hold at the interface $(y = h)$

$$\begin{aligned} u(x, h^-) &= u(x, h^+), & v(x, h^-) &= v(x, h^+), \\ p(x, h^-) &= p(x, h^+), & \tau(x, h^-) &= \tau(x, h^+). \end{aligned} \tag{4.45}$$

Mechanical models for the contacting bodies

Assuming that the thickness h of the viscoelastic layer is much less than the width of the contact region, we simulate its tangential and normal compliance using the one-dimensional Maxwell body, namely

$$\dot{u}_3 = \frac{h}{E_\tau}\left(\frac{\tau}{T_\tau} + \dot{\tau}\right), \qquad \dot{v}_3 = \frac{h}{E_n}\left(\frac{p}{T_n} + \dot{p}\right), \tag{4.46}$$

where u_3 and v_3 are the tangential and normal displacements of the boundary of the layer ($y = 0$) respectively, and $(\cdot)$ denotes the time derivative. As is known, the Maxwell model can be represented by a spring of modulus E_n (E_τ) in series with a dashpot of viscosity $E_n T_n$ ($E_\tau T_\tau$). For this model E_n (E_τ) and T_n (T_τ) are the elasticity modulus and the relaxation time in normal and tangential directions, respectively.

In the (x, y) system of coordinates relations (4.46) have the form

$$\frac{du_3}{dx} = \frac{h}{E_\tau}\left(-\frac{\tau(x)}{VT_\tau} + \frac{d\tau(x)}{dx}\right), \tag{4.47}$$

$$\frac{dv_3}{dx} = \frac{h}{E_n}\left(-\frac{p(x)}{VT_n} + \frac{dp(x)}{dx}\right). \tag{4.48}$$

In the model under consideration it is assumed that the same normal and tangential stresses occurring at the upper boundary of the layer ($y = 0$) occur at the layer-substrate interface ($y = h$). The displacement gradients for the elastic bodies (cylinder ($i = 1$) and substrate ($i = 2$) of the layered semi-infinite plane) can be found in Gladwell (1980) as

$$\frac{du_i}{dx} = -\frac{(1-2\nu_i)(1+\nu_i)}{E_i}p(x) - \frac{2(1-\nu_i^2)}{\pi E_i}\int_{-a}^{b}\frac{\tau(x')}{x-x'}dx', \tag{4.49}$$

$$\frac{dv_i}{dx} = \frac{(1-2\nu_i)(1+\nu_i)}{E_i}\tau(x) - \frac{2(1-\nu_i^2)}{\pi E_i}\int_{-a}^{b}\frac{p(x')}{x-x'}dx'. \tag{4.50}$$

Eqs. (4.47)–(4.50) and the boundary conditions (4.39), (4.41) and (4.44) are used to find the normal and tangential stresses in the contact region $(-a, b)$.

4.3.2 Normal stress analysis

In order to simplify the calculations, we shall neglect the effect of the tangential contact stresses on the normal contact stresses. Then, from Eqs. (4.48) and (4.50) (the latter is considered for $\tau(x) = 0$) and using the boundary condition (4.44), we obtain the following integral equation

$$\frac{h}{E_n}\frac{dp(x)}{dx} - \frac{h}{VE_nT_n}p(x) - \frac{2}{\pi E^*}\int_{-a}^{b}\frac{p(x')\,dx'}{x-x'} = -\frac{x}{R}, \tag{4.51}$$

where

$$E^* = \left(\frac{(1-\nu_1^2)}{E_1} + \frac{(1-\nu_2^2)}{E_2}\right)^{-1}. \tag{4.52}$$

Introducing the new variable ξ as

$$x = \frac{b-a}{2} + \frac{a+b}{2}\xi \tag{4.53}$$

and the dimensionless function

$$\bar{p}(\xi) = \frac{2}{\pi E^*} p\left(\frac{b-a}{2} + \frac{a+b}{2}\xi\right), \tag{4.54}$$

Eq. (4.51) can be rewritten as

$$\int_{-1}^{1} \frac{\bar{p}(\xi')\,d\xi'}{\xi - \xi'} + \alpha_n \bar{p}(\xi) - \frac{\beta_n}{L}\bar{p}'(\xi) = L(\varepsilon + \xi), \tag{4.55}$$

where

$$\alpha_n = \frac{h\pi E^*}{2VE_nT_n}, \quad \beta_n = \frac{h\pi E^*}{2RE_n}, \quad L = \frac{a+b}{2R}, \quad \varepsilon = \frac{b-a}{a+b}. \tag{4.56}$$

Bearing in mind the condition that the pressure at the ends of the contact region ($x = -a$ and $x = b$) is equal to zero, that is, $\bar{p}(-1) = \bar{p}(1) = 0$, and using the following relationships

$$\int_{-1}^{1} \frac{\bar{p}(\xi')\,d\xi'}{\xi - \xi'} = \int_{-1}^{1} \bar{p}'(\xi') \ln|\xi - \xi'|\,d\xi',$$

$$\bar{p}(\xi) = \frac{1}{2}\int_{-1}^{1} \bar{p}'(\xi')\,\mathrm{sgn}(\xi - \xi')\,d\xi',$$

we transform Eq. (4.55) to the Fredholm equation of the second kind

$$\int_{-1}^{1} F(\xi')\left[\ln|\xi - \xi'| + \frac{\alpha_n}{2}\mathrm{sgn}(\xi - \xi')\right] d\xi' - \frac{\beta_n}{L}F(\xi) = L(\varepsilon + \xi), \tag{4.57}$$

where

$$F(\xi) = \bar{p}'(\xi). \tag{4.58}$$

Integrating Eq. (4.57) on the segment $[-1, 1]$, we obtain

$$\int_{-1}^{1} F(\xi')\left[(1+\xi')\ln(1+\xi') + (1-\xi')\ln(1-\xi') - \alpha_n\xi' - 2\right] d\xi' = 2\varepsilon L. \tag{4.59}$$

From Eqs. (4.57) and (4.59) we reduce the following equation

$$\int_{-1}^{1} F(\xi') \left[\ln|\xi - \xi'| + \frac{\alpha_n}{2}\operatorname{sgn}(\xi - \xi') - \frac{1+\xi'}{2}\ln(1+\xi') - \frac{1-\xi'}{2}\ln(1-\xi') + \frac{\alpha_n \xi'}{2}\right] d\xi' - \frac{\beta_n}{L}F(\xi) = \xi L, \tag{4.60}$$

It follows from the condition of equilibrium of the normal forces applied to the cylinder that the function $F(\xi)$ also satisfies the relation

$$\bar{P} = -L\int_{-1}^{1} \xi F(\xi)\, d\xi, \tag{4.61}$$

where $\bar{P} = \dfrac{2P}{\pi R E^*}$ is the dimensionless normal load applied to the cylinder.

Eqs. (4.60) and (4.61) provide the necessary system of equations for the normal contact stress analysis. They have been simultaneously solved to determine the influence of the dimensionless parameters $\bar{P}$, α_n and β_n on the dimensionless contact characteristics, i.e. contact pressure $\bar{p}(\xi)$, the contact width L, the shift ε of the contact region, and the maximum indentation $\Delta_{\max}$ of the cylinder into the viscoelastic layer:

$$\bar{p}(\xi) = \int_{-1}^{\xi} F(\xi')\, d\xi', \quad L = -\frac{\bar{P}}{\int_{-1}^{1} \xi' F(\xi')\, d\xi'},$$
$$\varepsilon = \frac{1}{2L}\int_{-1}^{1} F(\xi')\left[(1+\xi')\ln(1+\xi') + (1-\xi')\ln(1-\xi') - \alpha_n \xi'\right] d\xi', \tag{4.62}$$
$$\Delta_{\max} = \max\frac{v_3(x)}{R} = L \cdot \max_{-1<\xi<1} \Phi(\xi),$$

where

$$\Phi(\xi) = \alpha_n\left[\int_{\xi}^{1} \xi' F(\xi')\, d\xi' + \xi\int_{-1}^{\xi} F(\xi')\, d\xi'\right] + \frac{\beta_n}{L}\int_{\xi}^{1} F(\xi')\, d\xi'.$$

Note that if we neglect the elastic properties of the substrate and the cylinder and solve Eq. (4.48) with boundary condition (4.44), we obtain

$$\bar{p}(\xi) = \frac{L}{\alpha_n}\left[\coth\frac{1}{\tilde{\zeta}} + \xi - \frac{\exp\left(\frac{\xi}{\tilde{\zeta}}\right)}{\sinh\frac{1}{\tilde{\zeta}}}\right], \tag{4.63}$$

where $\tilde{\zeta}$ is the Deborah number which represents the ratio of the relaxation time T_n of the layer material to the time taken for an element to travel through the semi-contact width $(a+b)/2$ (see Johnson, 1987)

$$\tilde{\zeta} = \frac{\beta_n}{\alpha_n L} = \frac{2T_n V}{a+b}.$$

Eq. (4.63) provides the contact pressure distribution within the contact region for the case when the normal compliance of the layer is much more than the normal compliance of the elastic substrate and cylinder (i.e. $E_n/E^* \ll 1$).

4.3.3 Tangential stress analysis

If the normal contact pressure is known, the tangential stress within the contact region can be obtained from Eqs. (4.39), (4.41), (4.45), (4.47) and (4.49). The following integral equation for determining the function $\tau(x)$ holds in the no-slip zones (A)

$$-\frac{h}{VE_\tau T_\tau}\tau(x) + \frac{h}{E_\tau}\cdot\frac{d\tau(x)}{dx} - \frac{2}{\pi E^*}\int\limits_{-a}^{b}\frac{\tau(x')}{x-x'}\,dx' = \delta - \frac{2\vartheta}{\pi E^*}p(x), \quad x\in A, \tag{4.64}$$

where

$$\vartheta = \frac{\pi E^*}{2}\left[\frac{(1-2\nu_1)(1+\nu_1)}{E_1} - \frac{(1-2\nu_2)(1+\nu_2)}{E_2}\right].$$

By introducing the following dimensionless function of variable ξ (see Eq.(4.53)) and parameters

$$\bar{\tau}(\xi) = \frac{2}{\pi E^*}\tau\left(\frac{b-a}{2} + \frac{a+b}{2}\xi\right), \quad \alpha_\tau = \frac{h\pi E^*}{2VE_\tau T_\tau}, \quad \beta_\tau = \frac{h\pi E^*}{2RE_\tau} \tag{4.65}$$

and using the method described in § 4.3.2, we reduce Eq. (4.64) to the form

$$\phi(\xi) = 0, \quad \xi\in A_\xi, \tag{4.66}$$

where

$$\phi(\xi) = -\delta + \beta\bar{p}(\xi) + \frac{\beta_\tau}{L}q(\xi) - \int\limits_{-1}^{1}\left[\ln|\xi-\xi'| + \frac{\alpha_\tau}{2}\operatorname{sgn}(\xi-\xi')\right]q(\xi')\,d\xi', \tag{4.67}$$

$$q(\xi) = \bar{\tau}'(\xi),$$

Moreover, in the no-slip zones (A_ξ), the tangential stresses satisfy the inequality

$$|\bar{\tau}(\xi)| < \mu\bar{p}(\xi),$$

which follows from Eq. (4.43).

Eq. (4.39) serves to determine the tangential stresses in the zones (S) where sliding occurs. Furthermore, in these zones the tangential stresses are opposite to the sliding direction, which leads to the relation

$$\tau(x) = \mu p(x)\,\mathrm{sgn}\left(\frac{du_2}{dx} - \frac{du_1}{dx} + \frac{du_3}{dx} - \delta\right). \tag{4.68}$$

Substituting Eq. (4.47) and (4.49) into Eq. (4.68) and using notations (4.65) and (4.67), we obtain

$$\bar{\tau}(\xi) = \mu\bar{p}(\xi)\,\mathrm{sgn}\,\phi(\xi), \quad \xi \in S_\xi. \tag{4.69}$$

The continuity equation

$$\bar{\tau}(\xi_i) = \mu\bar{p}(\xi_i), \quad i = 1, 2, \ldots, k \tag{4.70}$$

holds at the points ξ_i where one zone changes into another ($(k+1)$ is the number of the slip and no-slip zones).

Eqs. (4.66), (4.69) and (4.70) are used to determine the tangential stresses within the contact region and, also, the position and size of the slip and no-slip zones. An iterative process was used for the numerical analysis of the equations obtained.

The problem of finding the tangential stresses is simplified considerably by assuming that the cylinder and the substrate have the same elastic properties ($\vartheta = 0$) and that the tangential compliance of the layer is much greater than the normal compliance of the elastic cylinder and the substrate of the semi-infinite plane (i.e. $E_\tau/E^* \ll 1$). In this case, Eqs. (4.66) and (4.69) reduce to the following equations

$$\bar{\tau}(\xi) - \tilde{\zeta}_\tau \frac{d\bar{\tau}}{d\xi} = -B, \quad \xi \in A_\xi, \tag{4.71}$$

$$\bar{\tau}(\xi) = \mu\bar{p}(\xi)\,\mathrm{sgn}\left[-\bar{\tau}(\xi) + \tilde{\zeta}_\tau \frac{d\bar{\tau}}{d\xi} - B\right], \quad \xi \in S_\xi, \tag{4.72}$$

where

$$\tilde{\zeta}_\tau = \frac{2VT_\tau}{(a+b)}, \quad B = \frac{2\delta V E_\tau T_\tau}{h\pi E^*}.$$

The solution of the ordinary differential equation (4.71) is

$$\bar{\tau}(\xi) = -B - C\exp\left(\frac{\xi - 1}{\tilde{\zeta}_\tau}\right), \quad \xi \in A_\xi. \tag{4.73}$$

Here C is an unknown integration constant. In the no-slip zone A_ξ, the function $\bar{\tau}(\xi)$ satisfies the inequality

$$\bar{\tau}(\xi) < \mu\bar{p}(\xi). \tag{4.74}$$

The following describes the procedure used to determine the slip and no-slip zones within the contact region. We suppose that the no-slip zone begins at the leading edge ($x = b$) of the contact region. Then from Eqs. (4.72) and (4.73) and the continuous stress condition, i.e. $\bar{\tau}(1) = 0$, we obtain

$$\bar{\tau}(\xi) = \begin{cases} \mu\bar{p}(\xi), & \xi \in (-1, \xi_1)\,, \\ -B\left[1 - \exp\left(\dfrac{\xi - 1}{\tilde{\zeta}_\tau}\right)\right], & \xi \in (\xi_1, 1)\,, \end{cases} \tag{4.75}$$

where ξ_1 is the transition point between slip and no-slip zones. This point can be found from the relation

$$-B\left(1 - \exp\left(\frac{\xi_1 - 1}{\tilde{\zeta}_\tau}\right)\right) = \mu\bar{p}\,(\xi_1)\,. \tag{4.76}$$

The tangential stress $\bar{\tau}(\xi)$ given by Eq.(4.75) satisfies the relationships described in Eqs.(4.72) and (4.74) if

$$\tilde{\zeta}_\tau \mu\bar{p}'(1 - 0) - B \leq 0. \tag{4.77}$$

Eq. (4.77) is the necessary condition for a two zone contact analysis described above. If Eq. (4.77) is not satisfied, the slip zone $(\xi_2, 1)$ occurs in the leading edge of the contact region where

$$\bar{\tau}(\xi) = \mu\bar{p}(\xi), \quad \xi_2 < \xi < 1 \tag{4.78}$$

and

$$-\mu\bar{p}(\xi) + \tilde{\zeta}_\tau \mu\bar{p}'(\xi) - B \geq 0. \tag{4.79}$$

Note that when

$$\bar{\tau}(\xi) = -\mu\bar{p}(\xi), \quad \xi_2 < \xi < 1 \tag{4.80}$$

the bracket in Eq. (4.72) becomes

$$\mu\bar{p}(\xi) - \tilde{\zeta}_\tau \mu\bar{p}'(\xi) - B \leq 0, \quad \xi_2 < \xi < 1 \tag{4.81}$$

Eq. (4.81) is not satisfied near the end of the contact zone ($\xi \to 1 - 0$). Therefore, the condition of Eq. (4.80) cannot occur.

At the transition point ξ_2, the slip zone changes to the no-slip zone. In the no-slip zone Eq. (4.73) holds; therefore

$$\bar{\tau}(\xi) = -B + (\mu\bar{p}(\xi_2) + B)\exp\left(\frac{\xi - \xi_2}{\tilde{\zeta}_\tau}\right), \quad \xi_1 \leq \xi \leq \xi_2. \tag{4.82}$$

In order for Eq. (4.74) to be satisfied at the transition point ξ_2, the following condition must hold:

$$\bar{\tau}'(\xi_2 - 0) \geq \bar{\tau}'(\xi_2 + 0). \tag{4.83}$$

Substituting from Eqs. (4.78) and (4.82) into Eq. (4.83) we obtain

$$\mu\bar{p}(\xi_2) + B \geq \tilde{\zeta}_\tau \mu\bar{p}'(\xi_2). \tag{4.84}$$

Taking into account the inequalities (4.79) and (4.84), we obtain the following relation to find the point ξ_2

$$\mu\bar{p}(\xi_2) + B - \mu\tilde{\zeta}_\tau\bar{p}'(\xi_2) = 0. \tag{4.85}$$

A simple analysis of Eq. (4.82) shows that there is also the slip zone $(-1, \xi_1)$, and the following conditions are satisfied

$$-B + [\mu\bar{p}(\xi_2) + B]\exp\left(\frac{\xi_1 - \xi_2}{\tilde{\zeta}_\tau}\right) = \mu\bar{p}(\xi_1), \tag{4.86}$$

$$-\mu\bar{p}(\xi) + \tilde{\zeta}_\tau\mu\bar{p}'(\xi) - B > 0, \tag{4.87}$$

$$\tilde{\zeta}_\tau\mu\bar{p}'(\xi_1) \geq [\mu\bar{p}(\xi_2) + B]\exp\left(\frac{\xi_1 - \xi_2}{\tilde{\zeta}_\tau}\right). \tag{4.88}$$

Note that Eqs. (4.86)–(4.88) satisfy Eq. (4.72) and the continuous stress condition. Thus, when there are three zones, we have the following expression for determining the tangential stresses within the contact region $(-1, 1)$

$$\bar{\tau}(\xi) = \begin{cases} \mu\bar{p}(\xi), & \xi \in (-1, \xi_1) \cup (\xi_2, 1), \\ -B + [\mu\bar{p}(\xi_2) + B]\exp\left(\dfrac{\xi - \xi_2}{\tilde{\zeta}_\tau}\right), & \xi \in (\xi_1, \xi_2), \end{cases} \tag{4.89}$$

where ξ_1 and ξ_2 are the solution of Eqs. (4.85) and (4.86).

Therefore, the contact can have slip and no-slip zones (two zones) or slip, no-slip and slip zones (three zones). When there is no viscoelastic layer, only two zones (no-slip and slip) exist within the contact region in rolling contact of the cylinder and substrate with similar properties ($\vartheta = 0$).

4.3.4 Rolling friction analysis

A rolling cylinder is acted upon by a normal active load P and a tangential active load T, a moment M and, also, the reactions of the base P_1 and T_1 which arise as the result of the action of the normal and tangential stresses within the contact region $(-a, b)$ (see Fig. 4.12). The equations

$$M - \int_{-a}^{b} xp(x)\,dx + T_1 R = 0,$$

$$T_1 = \frac{\pi E^*(a+b)}{4}\int_{-1}^{1} \bar{\tau}(\xi)d\xi,$$

follow from the condition of equilibrium of the moments and forces.

Using the notations introduced in § 4.3.2 and § 4.3.3, we obtain the following expressions for the dimensionless magnitudes of the resistive force $\bar{T}$ and moment of rolling friction $\bar{M}$

$$\bar{T} = \frac{2T}{\pi R E^*} = -L \int_{-1}^{1} \xi q(\xi)\, d\xi,$$
$$\bar{M} = \frac{2M}{\pi E^* R^2} = -\frac{L^2}{2} \int_{-1}^{1} (\xi + L)^2 F(\xi)\, d\xi - \bar{T}. \tag{4.90}$$

The first (or second) equation of (4.90) can also be used to find the magnitude of creep ratio δ (4.40), if the value of the tangential force T (or the moment M) is known.

The coefficient of rolling friction is found from the relation

$$\mu_r = \frac{\bar{M}}{\bar{P}}, \tag{4.91}$$

where the values of $\bar{M}$ and $\bar{P}$ are determined using the second formula in (4.90) and Eq. (4.61), respectively. The case $T = 0$ corresponds to pure rolling.

When $T = \mu P$, sliding occurs over the entire contact.

4.3.5 The effect of viscoelastic layer in sliding and rolling contact

The equations for the contact normal and tangential stresses obtained in § 4.3.2 and § 4.3.3 have been used to calculate the contact characteristics and to analyze their dependence on the parameters characterizing the mechanical and geometrical properties of the surface layer for various magnitudes of the rolling (sliding) velocity.

Fig. 4.13 depicts the pressure distribution within the contact region for different values of α_n at constant $\beta_n = 0.1$ and $L = 0.1$. The contact pressure $p(\xi)$ relates to the Hertz maximum contact pressure p_0, ($p_0 = E^* L/2$), so $p(\xi)/p_0 = \pi \bar{p}(\xi)/L$. The solid curves correspond to the general case of the contact interaction of elastic bodies when there is a viscoelastic layer between them. The dashed curves have been constructed using formula (4.63) in the case when the elastic properties of the cylinder and the substrate of the semi-infinite plane are neglected. In calculations, the contact width was held constant and the load was varied. The results show that, as the velocity V decreases (the parameter α_n (Eq. (4.55) increases), the contact pressure $p(\xi)$ becomes non-symmetric. This is mainly due to the cylindrical indenter having time to affect the viscoelastic properties of the surface layer. The figure demonstrates also that for specified viscoelastic characteristics of the surface layer, the contact pressure and its maximum value essentially depend on

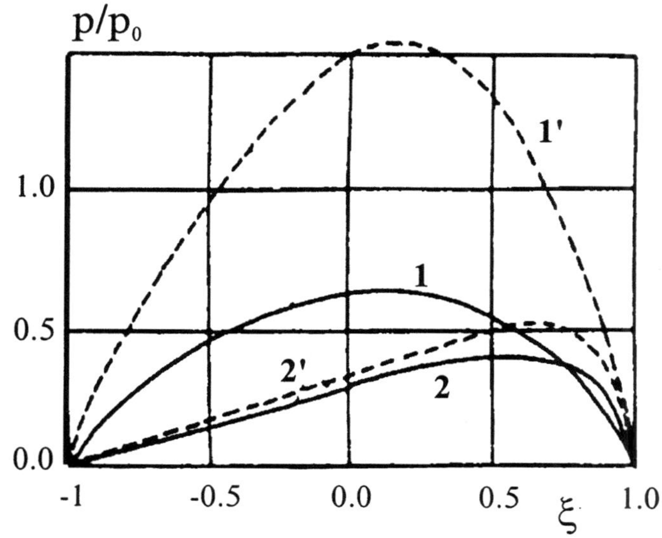

Figure 4.13: Contact pressure distribution calculated from Eqs. (4.60) and (4.61) – solid lines, and from Eq. (4.63) – dashed lines, for $\beta_n = 0.1$, $L = 0.1$ and $\alpha_n = 1$ (curves 1 and 1′), $\alpha_n = 10$ (curves 2 and 2′).

the elastic properties of the indenter and the base for small values of α_n (for high velocities). However, when the velocity decreases ($\alpha_n = 10$), the difference between the pressure distribution in the two cases becomes negligibly small. Hence, the viscoelastic surface layer mainly influences the contact pressure distribution at low velocities of motion.

Fig. 4.14 illustrates the influence of the parameter β_n/α_n on the size and shift of the contact region, and the maximum indentation of the cylinder into the viscoelastic layer for $\beta_n = 1$ (curve 1) and $\beta_n = 0.1$ (curve 2). The parameter $\beta_n/\alpha_n = T_n V/R$ depends on the relaxation time T_n and the velocity V. The results indicate that as the parameter β_n/α_n increases, the contact semi-width L decreases and tends to a constant value ($L = 1.49L_0$ and $L = 2.71L_0$ when $\beta_n = 0.1$ and $\beta_n = 1$, respectively; L_0 is the dimensionless semi-contact width in the case of the Hertz contact, $L_0 = \sqrt{2\bar{P}}$). For small values of the parameter β_n/α_n the contact width increases considerably, especially as the parameter β_n increases (Fig. 4.14(a)). We note that the parameter β_n depends on the thickness of the layer and the relative elastic properties of the layer, substrate and the cylinder. As the parameter β_n/α_n decreases there is an increase in the shift ε of the contact region (Fig. 4.14(b)) and the maximum penetration $\Delta_{\max}$ of the cylinder into the viscoelastic layer (Fig. 4.14(c)). This is because the viscoelastic properties of the surface layer are dominant for small values of the parameter β_n/α_n. As the

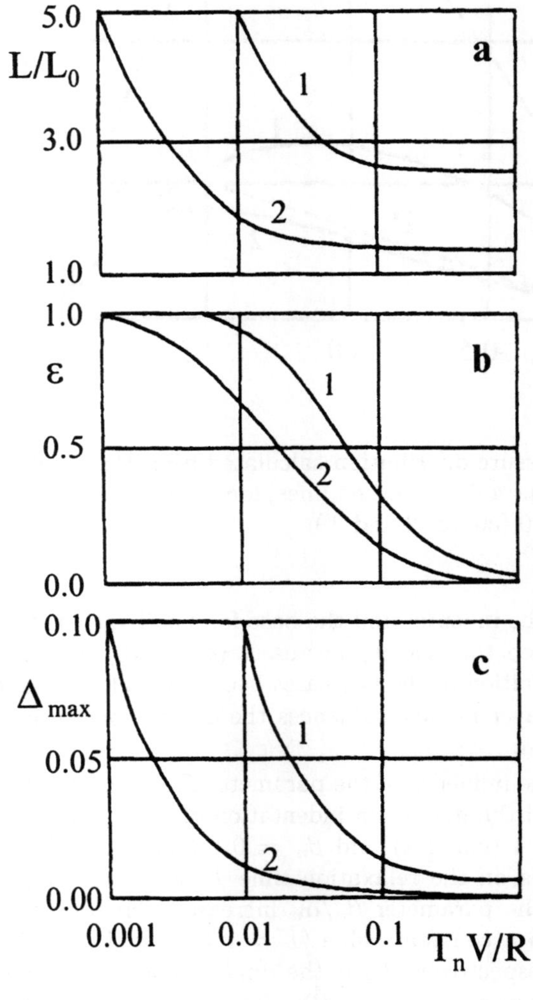

Figure 4.14: Size (a) and shift (b) of the contact region, and the maximum indentation of the cylinder into the viscoelastic layer (c) vs. parameter $T_n V/R$ for $\bar{P} = 0.001$ and $\beta_n = 1$ (curve 1), $\beta_n = 0.1$ (curve 2).

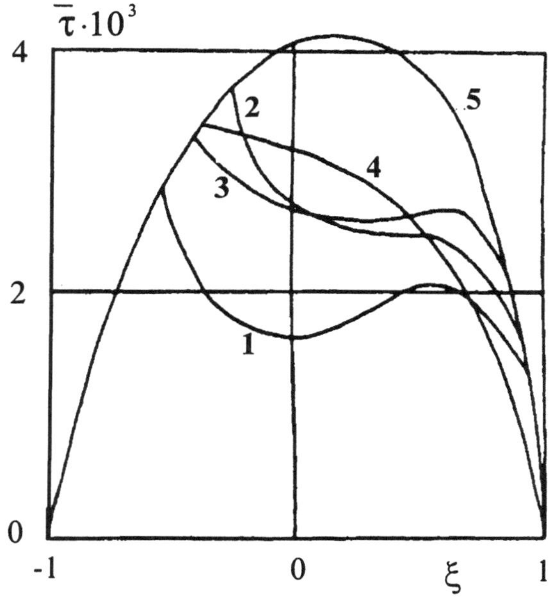

Figure 4.15: Tangential contact stresses for $\beta_n = 0.1$, $\alpha_n = 1$, $\bar{P} = 0.01$, $\mu = 0.1$, $\beta_\tau = 0.1$ and $\bar{T} = 0.6\mu\bar{P}$, $\theta = 0.1$, $\vartheta = -0.4$ (curve 1); $\bar{T} = 0.8\mu\bar{P}$, $\theta = 1$, $\vartheta = -0.4$ (curve 2); $\bar{T} = 0.8\mu\bar{P}$, $\theta = 0.1$, $\vartheta = -0.4$ (curve 3), $\bar{T} = 0.8\mu\bar{P}$, $\theta = 0.1$, $\vartheta = 0.4$ (curve 4) and $\bar{T} = \mu\bar{P}$ (curve 5).

relaxation time or the velocity of the indenter increases, the contact shift becomes negligibly small for all values of the parameter β_n.

The results of the calculations of the tangential stresses within the contact region from Eqs. (4.66), (4.69) and (4.70) are shown in Fig. 4.15. The properties of the surface viscoelastic layer in this analysis are described by the parameter $\theta = T_\tau/T_n$ which is the ratio of the relaxation times in the tangential and normal directions ($\theta = (\beta_\tau\alpha_n)/(\beta_n\alpha_\tau)$) and, also, by the parameter β_τ (Eq. (4.65)), which depends on the relative thickness of the layer and the relative elastic properties of the layer, substrate and the cylinder.

The results show that, as the parameter θ increases, there is an increase in the values of the maximum tangential stresses within the contact region and a decrease in the size of the no-slip zone. With the same layer characteristics ($\beta_\tau = 0.1$ and $\theta = 0.1$), a change in the relative elastic characteristics of the cylinder and the substrate from $\vartheta = -0.4$ (curve 3) to $\vartheta = 0.4$ (curve 4) leads to a transition from a three-zone contact to a two-zone contact. Furthermore, it was established that, as the value of the tangential force $\bar{T}$ becomes smaller, the contact passes from a completely sliding contact (curve 5) to a three-zone and, then, to a two-zone case.

The same results were obtained in calculations using Eqs. (4.75), (4.89) in the

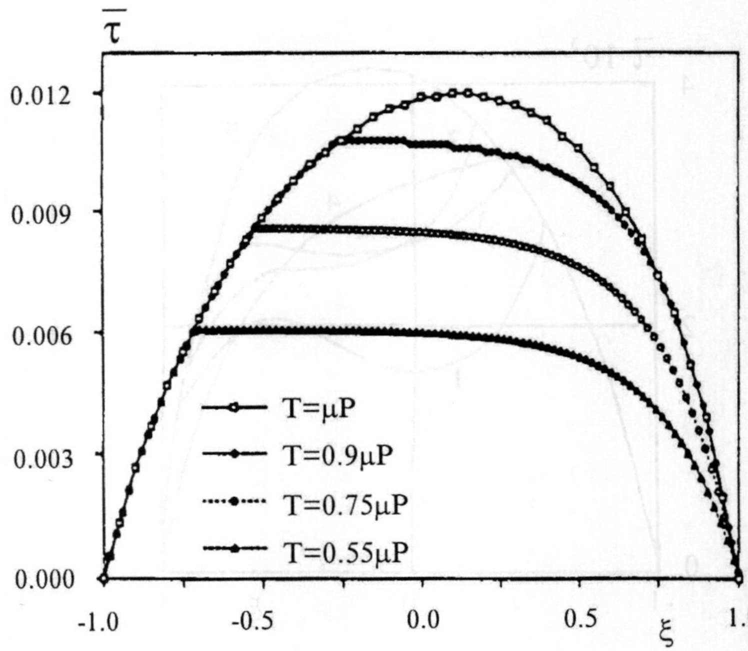

Figure 4.16: Tangential contact stresses in the case $\vartheta = 0$ and $E_\tau/E^* \ll 1$, $\beta_n = 0.1$, $\alpha_n = 1$, $\bar{P} = 0.01$, $\mu = 0.3$ for various values of $\bar{T}$.

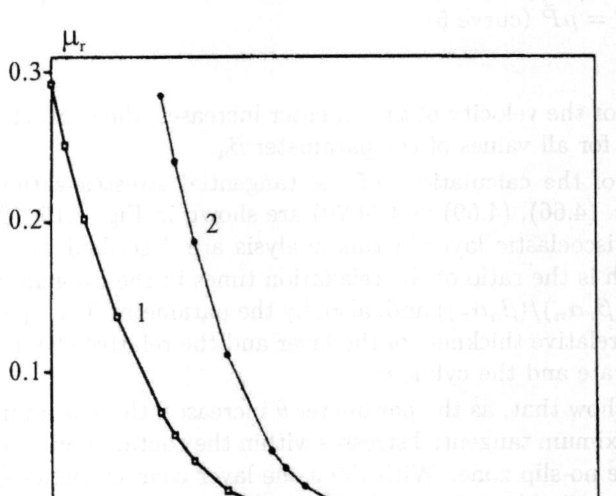

Figure 4.17: Rolling friction coefficient vs. parameter T_nV/R for $\bar{P} = 0.001$, $\bar{T} = 0$ and $\beta_n = 0.1$ (curve 1), $\beta_n = 1$ (curve 2).

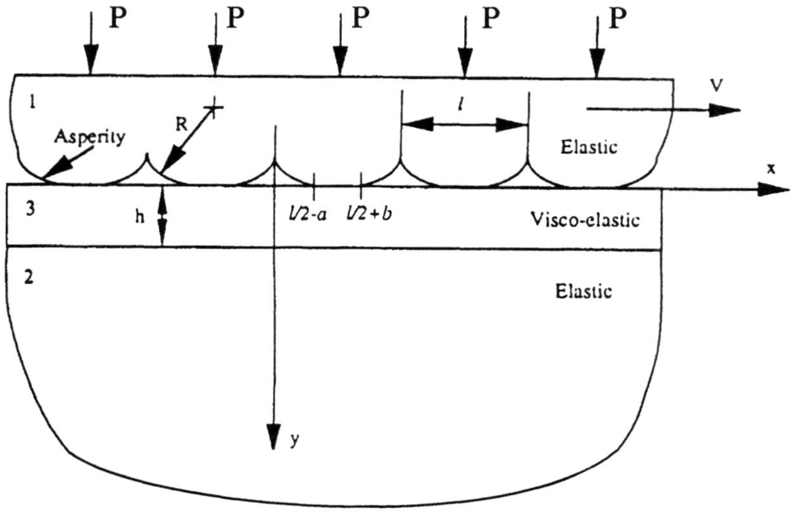

Figure 4.18: Scheme of contact of the periodic indenter and the layered semi-infinite plane.

particular case of identical elastic characteristics of the cylinder and the substrate ($\vartheta = 0$) and $E_\tau/E^* \ll 1$. Fig. 4.16 illustrates the tangential stress distribution within the contact region calculated in this particular case for the various values of $\bar{T}$. The results indicate that the size of the no-slip zone increases for decreasing values of the tangential force.

Graphs of the coefficient of rolling friction μ_r, calculated from Eq. (4.91), vs. the parameter $\beta_n/\alpha_n = T_n V/R$ for $\bar{P} = 0.001$ and $\bar{T} = 0$ are shown in Fig. 4.17. The coefficient of rolling friction for the model of a viscoelastic layer under consideration (the Maxwell body) decreases monotonically as the parameter $T_n V/R$ increases and $\mu_r \to 0$ as $T_n V/R \to +\infty$.

Thus, this analysis shows that the inelastic properties of the surface layer are significant in rolling and sliding contact, especially for small values of the parameter $T_n V/R$.

4.4 The effect of roughness and viscoelastic layer

The results given in § 4.2 make it possible to analyze the combined effect of both surface roughness and surface layer properties in normal contact of coated elastic bodies.

As was pointed out in the previous section, in sliding contact the imperfect elasticity of the surface layer has a marked influence on contact characteristics and the friction coefficient. The more complicated dependence of the contact characteristics on the mechanical properties of surface layer and velocity of motion occurs in sliding contact of rough bodies. We will analyze the combined effect of surface layer properties and surface roughness in sliding contact for the 2-D contact problem of a periodic system of cylindrical asperities sliding over a viscoelastic layer bonded to an elastic half-space.

4.4.1 Model of the contact and its analysis

Fig. 4.18 illustrates a schematic of a rough elastic indenter in contact with a layered semi-infinite plane. The roughness of the indenter is taken into account by a periodic function $f(x) = f(x+l)$, where l is the period. The indenter slides on the base with a constant velocity V, and it is acted upon by an l-periodic external force. The total normal force (load) per one period of the indenter is P.

We introduce the (x', y') coordinate system fixed on the layered semi-infinite plane, and the (x, y) coordinate system which moves with the indenter. The relationships between the systems are given in Eq. (4.38). In this study the steady state is investigated. Therefore, all stresses and strains are independent of time in the (x, y) coordinate system.

As in § 4.3, we will use a one-dimensional model of the layer. For this model the same normal stress $p(x)$ occurring on the surface ($y = 0$) occurs at the layer and the semi-infinite plane interface ($y = h$). To simplify the problem the effect of tangential stress on normal stress is neglected. As is shown by Staierman (1949), the following relationship holds for periodic contact of elastic bodies ($0 \leq x \leq l$)

$$v_1(x) + v_2(x) = -\frac{2}{\pi E^*} \int_{\frac{l}{2}-a}^{\frac{l}{2}+b} p(x') \ln\left|2\sin\left[\frac{\pi(x'-x)}{l}\right]\right| dx', \tag{4.92}$$

where v_1 and v_2 are the normal displacements of the elastic indenter and the elastic semi-infinite plane respectively, and E^* is the equivalent modulus (see Eq. (4.52)). Eq. (4.92) was derived by summing the displacements at any point x due to the contact pressure at all intervals

$$\frac{2n+1}{2}l - a \leq x \leq \frac{2n+1}{2}l + b \ (n = 0, \pm1, \pm2, \ldots).$$

In this analysis it is assumed that the function $f(x)$ is smooth and has the following form for $x \in (0, l)$

$$f(x) = \frac{\left(x - \frac{l}{2}\right)^2}{2R}. \tag{4.93}$$

So the normal stress $p(x)$ at the ends of the contact zones satisfies the condition

$$p\left(\frac{l}{2} - a\right) = p\left(\frac{l}{2} + b\right) = 0. \tag{4.94}$$

The function $p(x)$ also satisfies the periodicity condition

$$p(x) = p(x + l). \tag{4.95}$$

The following displacement condition occurs within the contact region

$$v_1(x) + v_2(x) + v_3(x) = D - f(x), \tag{4.96}$$

where D is the penetration of the indenter into the layered semi-infinite plane and v_3 is the displacement of the layer due to its deformation. We consider below two different models of the layer deformation.

The Maxwell model

As in § 4.3 we use the Maxwell model to describe the viscoelastic layer normal compliance. So the relation between the derivative of the normal displacement v_3 of the layer and the contact pressure $p(x)$ has the form of Eq. (4.48) in the (x, y) coordinate system.

By using Eq. (4.94) and differentiating Eqs. (4.92) and (4.96) and then substituting from Eq. (4.48), (4.92) in the differential form of Eq. (4.96), we obtain the following integral equation in the range $x \in [(l/2) - a, (l/2) + b]$:

$$-\frac{h}{VE_nT_n}p(x) + \frac{h}{E_n} \cdot \frac{dp}{dx} -$$
$$-\frac{2}{\pi E^*} \int\limits_{\frac{l}{2}-a}^{\frac{l}{2}+b} p'(x') \ln \left| 2 \sin \left[\frac{\pi(x' - x)}{l} \right] \right| dx' = -f'(x). \tag{4.97}$$

Introducing the new variables ξ and η as

$$x = \frac{b - a + l}{2} + \frac{a + b}{2}\xi, \quad y = \frac{a + b}{2}\eta + h \tag{4.98}$$

and the dimensionless function

$$\bar{p}(\xi) = \frac{2}{\pi E^*} p \left(\frac{b - a + l}{2} + \frac{a + b}{2}\xi \right) \tag{4.99}$$

and substituting Eqs. (4.93), (4.98) and (4.99) into Eq. (4.97), we obtain

$$\int\limits_{-1}^{1} F(\xi') \left[\ln \left| \sin \left[\frac{\pi}{2} \cdot \frac{L}{\bar{l}}(\xi' - \xi) \right] \right| + \frac{\alpha_n}{2} \operatorname{sgn}(\xi' - \xi) \right] d\xi' -$$
$$\frac{\beta_n}{L} F(\xi) = L(\varepsilon + \xi), \tag{4.100}$$

where $F(\xi) = \bar{p}'(\xi)$, $\bar{l} = l/(2R)$ and parameters α_n, β_n, L and ε are determined in Eq. (4.56).

Eq. (4.100) is a Fredholm equation of the second kind. It follows from Eq. (4.94) that

$$\int_{-1}^{1} F(\xi)\, d\xi = 0. \tag{4.101}$$

The equilibrium condition takes the form

$$-L \int_{-1}^{1} \xi F(\xi)\, d\xi = \bar{P}. \tag{4.102}$$

where $\bar{P}$ is the dimensionless load applied to the period, i.e. $\bar{P} = 2P/(\pi E^* R)$.

Eq. (4.100), (4.101) and (4.102) provide the necessary system of equations for contact stress analysis. The solution of the system of equations is used to determine the normal contact pressure $\bar{p}(\xi)$, the size L and the shift ε of the contact zone

$$\bar{p}(\xi) = \int_{-1}^{\xi} F(\xi')\, d\xi', \quad -1 \le \xi \le 1, \tag{4.103}$$

$$L = -\bar{P} \left[\int_{-1}^{1} \xi F(\xi)\, d\xi \right]^{-1}, \tag{4.104}$$

$$\varepsilon = \frac{1}{2L} \left[\int_{-1}^{1}\int_{-1}^{1} F(\xi') \ln \left| \sin \frac{\pi}{2} \cdot \frac{L}{\bar{l}} (\xi' - \xi) \right| d\xi' d\xi - \alpha_n \int_{-1}^{1} \xi' F(\xi')\, d\xi' \right]. \tag{4.105}$$

It must be noted that when $L/\bar{l} \ll 1$, Eq. (4.100) becomes

$$\int_{-1}^{1} F(\xi') \left[\ln|\xi' - \xi| + \frac{\alpha_n}{2} \operatorname{sgn}(\xi' - \xi) \right] d\xi' - \frac{\beta_n}{L} F(\xi) = L(\varepsilon + \xi),$$

which is the case of a single indenter with the shape function described by Eq. (4.93) (see § 4.3).

The Kelvin solid

Since the Maxwell model is valid only during time intervals when the strains remain small, the Kelvin model has been also included in the analysis. For this model the relation between normal pressure $p(x)$ and displacement v_3 is given by

$$v_3 + T_\epsilon \frac{\partial v_3}{\partial t} = \frac{h}{E_L} \left(p + T_\sigma \frac{\partial p}{\partial t} \right), \tag{4.106}$$

where T_ϵ and T_σ characterize the retardation and relaxation times, E_L is the longitudinal modulus of elasticity. The model exhibits delayed elasticity.

Eq. (4.106) has the following form in the moving system of coordinates (x, y)

$$v_3 - T_\epsilon V \frac{\partial v_3}{\partial x} = \frac{h}{E_L}\left(p - T_\sigma V \frac{\partial p}{\partial x}\right). \tag{4.107}$$

Note, that the normal displacement of the viscoelastic layer also satisfies the periodicity condition

$$v_3(x) = v_3(x + l). \tag{4.108}$$

By differentiating Eq. (4.92), we exclude the functions v_i $(i = 1, 2, 3)$ from Eqs. (4.92), (4.96), (4.107). Then taking into account the shape function (4.93), we obtain for $x \in \left[\frac{l}{2} - a, \frac{l}{2} + b\right]$

$$-\frac{2}{\pi E^*} \int\limits_{\frac{l}{2}-a}^{\frac{l}{2}+b} p(x')\left\{\ln\left|2\sin\frac{\pi(x'-x)}{l}\right| + \frac{\pi T_\epsilon V}{l}\cot\frac{\pi(x'-x)}{l}\right\} dx' +$$
$$\frac{h}{E_L}p(x) - \frac{hT_\sigma V}{E_L}p'(x) = D - \frac{\left(x - \frac{l}{2}\right)^2}{2R} + \frac{T_\epsilon V}{R}\left(x - \frac{l}{2}\right). \tag{4.109}$$

The solution of Eq. (4.107) in the unloaded zone $\left(-\frac{l}{2} + b \le x \le \frac{l}{2} - a\right)$ where $p(x) = 0$ is the function

$$v_3 = v_0 \exp\left(x/T_\epsilon V\right). \tag{4.110}$$

This function satisfies the contact condition (4.96) at $a_1 = \frac{l}{2} - a$ and $b_1 = \frac{l}{2} + b$

and also the periodicity condition (4.108). From these conditions it follows that

$$v_0 \exp\left(\frac{b_1 - l}{T_\epsilon V}\right) = D - \frac{b^2}{2R} + \frac{2}{\pi E^*}\int\limits_{a_1}^{b_1} p(x')\ln\left|2\sin\frac{\pi(x' + l - b_1)}{l}\right| dx',$$
$$v_0 \exp\left(\frac{a_1}{T_\epsilon V}\right) = D - \frac{a^2}{2R} + \frac{2}{\pi E^*}\int\limits_{a_1}^{b_1} p(x')\ln\left|2\sin\frac{\pi(x' - a_1)}{l}\right| dx'. \tag{4.111}$$

Eliminating the constants D and v_0 from Eqs. (4.109) and (4.111), we obtain

$$-\frac{2}{\pi E^*}\int_{a_1}^{b_1} p(x')\left[\ln\left|2\sin\frac{\pi(x'-x)}{l}\right| + \frac{\pi T_\epsilon V}{l}\cot\frac{\pi(x'-x)}{l} + \right.$$

$$\left.\frac{\ln\left|2\sin\frac{\pi(x'+l-b_1)}{l}\right|\exp\left(\frac{l+a_1-b_1}{T_\epsilon V}\right) - \ln\left|2\sin\frac{\pi(x'-a_1)}{l}\right|}{1-\exp\left(\frac{l+a_1-b_1}{T_\epsilon V}\right)}\right]dx' + \tag{4.112}$$

$$\frac{h}{E_L}p(x) - \frac{hT_\sigma V}{E_L}p'(x) = \frac{T_\epsilon V}{R}\left(x-\frac{l}{2}\right) +$$

$$\frac{(b_1-x)(b_1+x-l)}{2R} + \frac{(b_1-a_1)(l-b_1-a_1)}{2R\left(1-\exp\left(\frac{l+a_1-b_1}{T_\epsilon V}\right)\right)}.$$

Then introducing the dimensionless parameters

$$\beta_\epsilon = \frac{h\pi E^*}{2RE_L}, \quad \alpha_T = \frac{T_\epsilon}{T_\sigma}, \quad \zeta_0 = \frac{a+b}{2VT_\epsilon}, \quad \bar{l} = \frac{l}{2R} \tag{4.113}$$

and using the variables, functions and parameters introduced in Eqs. (4.56), (4.98) and (4.99), we obtain

$$\int_{-1}^{1} \bar{p}(\xi')H(\xi,\xi')\,d\xi' - \frac{\beta_\epsilon}{L}\bar{p}(\xi) + \frac{\beta_\epsilon}{L\zeta_0\alpha_T}\bar{p}'(\xi) = G(\xi), \tag{4.114}$$

where

$$H(\xi,\xi') = \ln\left|2\sin\frac{\pi L}{2\bar{l}}(\xi'-\xi)\right| + \frac{\pi L}{2\zeta_0\bar{l}}\cot\frac{\pi L}{2\bar{l}}(\xi'-\xi) +$$

$$\frac{\ln\left|2\sin\frac{\pi L}{2\bar{l}}(1-\xi')\right|\exp\left(2\zeta_0\left(\frac{\bar{l}}{L}-1\right)\right) - \ln\left|2\sin\frac{\pi L}{2\bar{l}}(1+\xi')\right|}{1-\exp\left(2\zeta_0\left(\frac{\bar{l}}{L}-1\right)\right)},$$

$$G(\xi) = \frac{L}{2}\xi^2 + L\xi\left(\varepsilon - \frac{1}{\zeta_0}\right) - \frac{L\varepsilon}{\zeta_0} - \frac{L}{2}(1+2\varepsilon) + \frac{2\varepsilon L}{1-\exp\left(2\zeta_0\left(\frac{\bar{l}}{L}-1\right)\right)}.$$

The condition (4.94) and the equilibrium equation take the forms

$$\bar{p}(-1) = \bar{p}(1) = 0,$$

$$\int_{-1}^{1} \bar{p}(\xi)\,d\xi = \bar{P}, \tag{4.115}$$

where $\bar{P} = \dfrac{2P}{\pi R E^*}$.

The linear integro-differential equation (4.114) and Eqs. (4.115) are used to find the dimensionless contact pressure $\bar{p}(\xi)$ $(-1 \le \xi \le 1)$ and the dimensionless width L and shift ε of the contact zone, if the layer properties are described by Eq. (4.106).

The tangential stress distribution within the contact zone $(-1 \le \xi \le 1)$ is given by

$$q(\xi) = \mu \bar{p}(\xi), \tag{4.116}$$

where μ is the coefficient of friction.

4.4.2 The method of determination of internal stresses

The subsurface stresses at any point (ξ, η) inside the solid (indenter or substrate of the semi-infinite plane) can be calculated according to the following relationships (Johnson, 1987):

$$\sigma_\xi(\xi,\eta) = -\frac{2}{\pi} \sum_{n=-\infty}^{+\infty} \left[\int_{\bar{a}_n^-}^{\bar{a}_n^+} \frac{\eta(\xi-\xi')^2}{\Lambda^4} \bar{p}(\xi_1)\, d\xi' + \int_{\bar{a}_n^-}^{\bar{a}_n^+} \frac{(\xi-\xi')^3}{\Lambda^4} q(\xi_1)\, d\xi' \right], \tag{4.117}$$

$$\sigma_\eta(\xi,\eta) = -\frac{2}{\pi} \sum_{n=-\infty}^{+\infty} \left[\int_{\bar{a}_n^-}^{\bar{a}_n^+} \frac{\eta^3}{\Lambda^4} \bar{p}(\xi_1)\, d\xi' + \int_{\bar{a}_n^-}^{\bar{a}_n^+} \frac{\eta^2(\xi-\xi')}{\Lambda^4} q(\xi_1)\, d\xi' \right], \tag{4.118}$$

$$\tau_{\xi\eta}(\xi,\eta) = -\frac{2}{\pi} \sum_{n=-\infty}^{+\infty} \left[\int_{\bar{a}_n^-}^{\bar{a}_n^+} \frac{\eta^2(\xi-\xi')}{\Lambda^4} \bar{p}(\xi_1)\, d\xi' + \int_{\bar{a}_n^-}^{\bar{a}_n^+} \frac{\eta(\xi-\xi')^2}{\Lambda^4} q(\xi_1)\, d\xi' \right], \tag{4.119}$$

where

$$\Lambda = \sqrt{(\xi-\xi')^2 + \eta^2}, \quad \bar{a} = \frac{L}{\bar{l}} = \frac{a+b}{l}, \quad \bar{a}_n^\pm = \frac{2n}{\bar{a}} \pm 1, \quad \xi_1 = \xi' - \frac{2n}{\bar{a}}.$$

Taking ξ_1 as a new variable, we reduce Eqs. (4.117), (4.118) and (4.119) to

$$\sigma_\xi(\xi,\eta) = -\frac{2}{\pi} \sum_{n=-\infty}^{+\infty} \chi_n^1(\xi,\eta), \tag{4.120}$$

$$\sigma_\eta(\xi,\eta) = -\frac{2}{\pi} \sum_{n=-\infty}^{+\infty} \chi_n^2(\xi,\eta), \tag{4.121}$$

$$\tau_{\xi\eta}(\xi,\eta) = -\frac{2}{\pi} \sum_{n=-\infty}^{+\infty} \chi_n^3(\xi,\eta), \tag{4.122}$$

where

$$\chi_n^1(\xi,\eta) = \int_{-1}^{1} \frac{\eta\zeta^2}{\Lambda_1^4}\bar{p}(\xi_1)\,d\xi_1 + \int_{-1}^{1} \frac{\zeta^3}{\Lambda_1^4}q(\xi_1)\,d\xi_1,$$

$$\chi_n^2(\xi,\eta) = \int_{-1}^{1} \frac{\eta^3}{\Lambda_1^4}\bar{p}(\xi_1)\,d\xi_1 + \int_{-1}^{1} \frac{\eta^2\zeta}{\Lambda_1^4}q(\xi_1)\,d\xi_1,$$

$$\chi_n^3(\xi,\eta) = \int_{-1}^{1} \frac{\eta^2\zeta}{\Lambda_1^4}\bar{p}(\xi_1)\,d\xi_1 + \int_{-1}^{1} \frac{\eta\zeta^2}{\Lambda_1^4}q(\xi_1)\,d\xi_1,$$

$$\zeta = \xi - \xi_1 - \frac{2n}{\bar{a}}, \quad \Lambda_1 = \sqrt{\zeta^2 + \eta^2}.$$

Due to the periodic nature of the solution, we need to consider the results only in the region Ω_0 ($\Omega_0 : \xi, \eta| - \varepsilon - (1/\bar{a}) \le \xi \le (1/\bar{a}) - \varepsilon, \eta \ge 0$). In this case, $|\xi_1| \ll |\xi - (2n/\bar{a})|$ when $n > N$, therefore the infinite series of Eqs. (4.120) through (4.122) can be approximated by

$$\sigma_\xi(\xi,\eta) = -\frac{2}{\pi}\sum_{n=-N}^{N} \chi_n^1(\xi,\eta) + \sigma_\xi^N(\xi,\eta), \tag{4.123}$$

$$\sigma_\eta(\xi,\eta) = -\frac{2}{\pi}\sum_{n=-N}^{N} \chi_n^2(\xi,\eta) + \sigma_\eta^N(\xi,\eta), \tag{4.124}$$

$$\tau_{\xi\eta}(\xi,\eta) = -\frac{2}{\pi}\sum_{n=-N}^{N} \chi_n^3(\xi,\eta) + \tau_{\xi\eta}^N(\xi,\eta), \tag{4.125}$$

where $\sigma_\xi^N(\xi,\eta)$, $\sigma_\eta^N(\xi,\eta)$, $\tau_{\xi\eta}^N(\xi,\eta)$ are the stresses produced in $(\xi,\eta) \in \Omega_0$ due to the normal and tangential stresses

$$\bar{p}_n = \frac{\bar{a}}{2}\int_{-1}^{1} \bar{p}(\xi)\,d\xi, \tag{4.126}$$

$$\bar{\tau}_n = \frac{\bar{a}}{2}\int_{-1}^{1} q(\xi)\,d\xi, \tag{4.127}$$

applied within the area Ω_∞:

$$\left(\xi,\eta : -\infty < \xi < -\frac{2N+1}{\bar{a}} - \varepsilon, \quad \frac{2N+1}{\bar{a}} - \varepsilon < \xi < +\infty, \quad \eta = 0\right).$$

For numerical calculation the functions $\sigma_\xi(\xi,\eta)$, $\sigma_\eta(\xi,\eta)$, $\tau_{\xi\eta}(\xi,\eta)$ are divided into two parts

$$\sigma_\xi(\xi,\eta) = \tilde{\sigma}_\xi(\xi,\eta) + \bar{p}_n, \tag{4.128}$$

$$\sigma_\eta(\xi,\eta) = \tilde{\sigma}_\eta(\xi,\eta) + \bar{p}_n, \tag{4.129}$$

$$\tau_{\xi\eta}(\xi,\eta) = \tilde{\tau}_{\xi\eta}(\xi,\eta) + \bar{\tau}_n, \tag{4.130}$$

where $\tilde{\sigma}_\xi(\xi,\eta)$, $\tilde{\sigma}_\eta(\xi,\eta)$, $\tilde{\tau}_{\xi,\eta}(\xi,\eta)$ are the internal stresses due to the normal pressure $\tilde{p}(\xi)$ and tangential stress $\tilde{\tau}(\xi)$ occurring on the interface ($\eta = 0$) given by

$$\tilde{p}(\xi) = \begin{cases} p\left(\xi - \dfrac{2n}{\bar{a}}\right) - \bar{p}_n, & \xi \in \omega_i, \\ -\bar{p}_n, & \xi \in \Omega \setminus \omega_i, \\ 0, & \xi \notin \Omega, \end{cases} \tag{4.131}$$

$$\tilde{\tau}(\xi) = \mu\tilde{p}(\xi), \tag{4.132}$$

where ω_i denotes $(2i/\bar{a}) - 1 < \xi < (2i/\bar{a}) + 1$, $i = 0, \pm1, \ldots, \pm N$ and Ω denotes $-(2N+1)/\bar{a} - \varepsilon < \xi < (2N+1)/\bar{a} - \varepsilon$. The second parts in Eqs. (4.128) through (4.130) are the uniform stress field resulted from the constant normal pressure $\bar{p}_n$ and tangential stress $\bar{\tau}_n$ on the interface ($\xi, \eta :\ -\infty < \xi < +\infty, \quad \eta = 0$).

4.4.3 Contact characteristics

The equations developed in § 4.4.2 for two different models of the viscoelastic layer were used to investigate the effect of viscoelastic layer properties and roughness parameters on the normal pressure distribution, size and shift of contact zones.

For the Maxwell model the viscoelastic layer properties are characterized by parameters β_n and β_n/α_n. The parameter $\beta_n = h\pi E^*/(2RE_n)$ relates the thickness of the layer h to the radius of the asperity R, and the relative modulus of the viscoelastic layer E_n to the equivalent modulus of the substrate and the indenter E^*. Therefore, when β_n is varied, it can be thought of as either the thickness or the relative modulus of the layer being varied. The parameter $\beta_n/\alpha_n = T_nV/R$ depends on the velocity V of the indenter and the relaxation time T_n of the viscoelastic layer.

If the viscoelastic layer is modeled by the Kelvin solid, its relative properties are described by the parameters α_T, ζ_0, β_ε (see Eq. (4.113). The parameter α_T is the ratio of the retardation time T_ϵ to the relaxation time T_σ of the material of the layer. The parameter ζ_0 represents the ratio of the time taken for an element to travel through the semi-contact width $(a+b)/2$ to the retardation time T_ϵ. The parameter β_ε characterizes the relative thickness and the relative elastic modulus of the layer and has the same sense as the parameter β_n in the Maxwell model.

The surface roughness of the indenter is characterized by the dimensionless contact density $\bar{a}$ or dimensionless distance $\bar{l}$ between centers of asperities. When $\bar{a}$ is small the contact density is low and the asperities are far apart (parameter $\bar{l}$ is large).

The results presented in Fig. 4.19–4.21 have been calculated based on the Maxwell model of the surface layer.

Figs. 4.19 and 4.20 demonstrate that, due to the viscoelastic layer, the contact pressure distribution becomes nonsymmetrical compared to the symmetrical

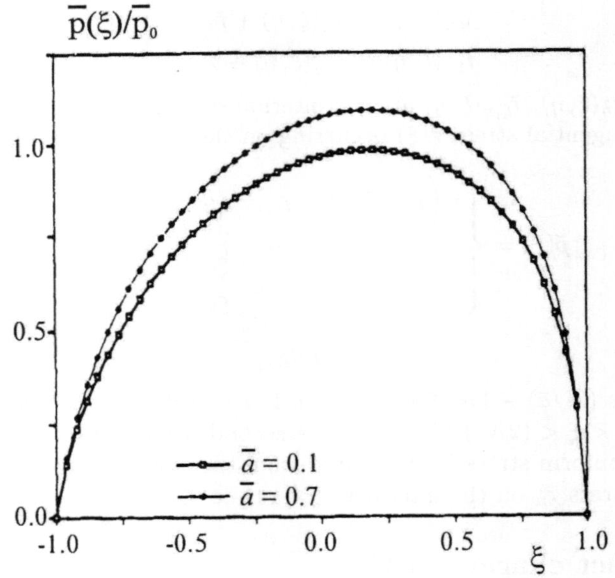

Figure 4.19: Contact pressure for various contact densities at $\beta_n = 0.01$, $\beta_n/\alpha_n = 0.01$ and $\bar{P} = 0.01$; layer is modelled by the Maxwell body.

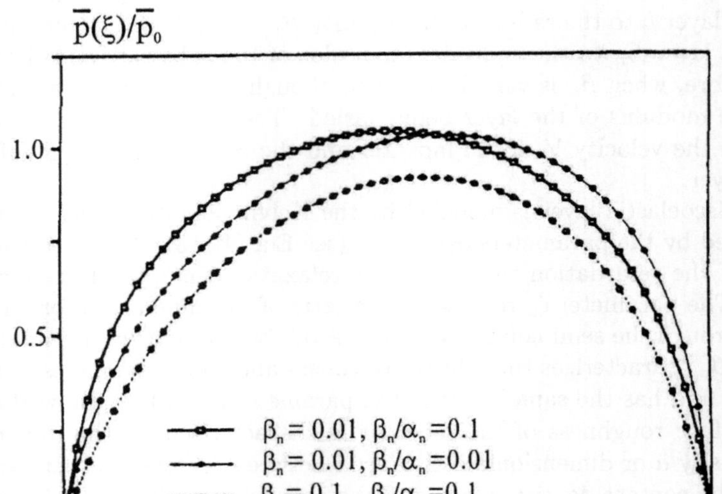

Figure 4.20: Contact pressure for various of values β_n and β_n/α_n at $\bar{a} = 0.5$ and $\bar{P} = 0.01$; layer is modelled by the Maxwell body.

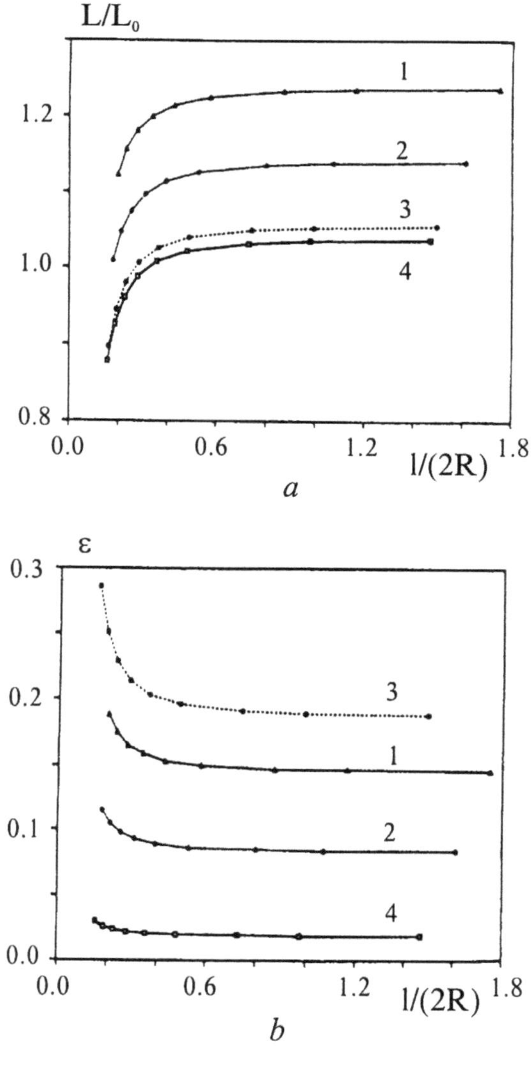

Figure 4.21: Size (a) and shift (b) of the contact zone vs. the dimensionless distance between asperities centers for $\bar{P} = 0.01$ and $\beta_n = 0.1$, $\beta_n/\alpha_n = 0.1$ (curve 1), $\beta_n = 0.05$, $\beta_n/\alpha_n = 0.1$ (curve 2), $\beta_n = 0.01$, $\beta_n/\alpha_n = 0.01$ (curve 3), $\beta_n = 0.01$, $\beta_n/\alpha_n = 0.1$ (curve 4); layer is modelled by the Maxwell body.

Hertzian pressure distribution. Fig. 4.19 depicts the influence of the contact density parameter $\bar{a}$ on the contact pressure $\bar{p}(\xi)$ related to the dimensionless Hertz pressure, $\bar{p}_0 = \sqrt{2\bar{P}}/\pi$. The results show that the maximum contact pressure increases when contact density increases. The parameter β_n also influences the magnitude of the maximum contact pressure; however, the character of the pressure distribution (its nonsymmetry) depends more on the parameter β_n/α_n (Fig. 4.20). For small values of β_n/α_n, the nonsymmetry increases.

Fig. 4.21 illustrates the variation of the dimensionless contact width L/L_0 (L_0 is the dimensionless Hertz width of the contact zone, $L_0 = \sqrt{2\bar{P}}$) and the shift of the contact zone ε as a function of dimensionless distance $\bar{l} = l/(2R)$ between the centers of asperities for various values of β_n and β_n/α_n. The results indicate that the parameter $\bar{l}$ significantly affects the contact characteristics if it is small ($\bar{l} \leq 0.6$). In this range of the parameter $\bar{l}$ the contact width decreases and the shift of the contact zone increases as the parameter $\bar{l}$ decreases. For higher values of this parameter the results coincide with those obtained in § 4.3 for a single asperity in sliding contact with the layered foundation. In this case it is possible to neglect the interaction between asperities. The results also indicate that when parameter β_n/α_n increases, the size of each contact and its shift decrease. Contact size and the shift also decrease when the parameter β_n decreases.

We have obtained similar results using the Kelvin solid as the model of the surface layer. The contact pressure is also distributed nonsymmetrically in this case, and its nonsymmetry increases as the parameter α_T increases. However, differences occur for small values of the parameter $l/2R$. The results calculated from the Kelvin model show that a decrease of the distance between asperities causes a decrease of the width and the shift of the contact zones. Fig. 4.22 demonstrates the dependence of the contact shift ε on the dimensionless distance $l/2R$ between asperities for different values of the parameter β_ε and VT_ϵ/R. The case $\beta_\varepsilon \to +\infty$ corresponds to the model of the viscoelastic layer bounded to the rigid substrate. The decrease of the contact shift for small values of $l/2R$ can be explained by the mutual influence of contact zones taken into account in the framework of the Kelvin model. This model describes the restoration of the layer in unloaded zones. It follows from Eq. (4.110), that the displacements of the ends of the unloaded zone satisfy the relationship

$$\frac{v_3\left(\dfrac{l}{2}-a\right)}{v_3\left(-\dfrac{l}{2}+b\right)} = \exp\frac{l-a-b}{T_\epsilon V}.$$

If the distance $(l-a-b)$ between contact zones is small, the layer does not recover its original shape until the time $(l-a-b)/V$ ($(l-a-b)/V \ll T_\epsilon$), and there is a decrease of the width and the shift of the contact zone. The effect of decreasing contact width for small values of $l/2R$ also arises due to elasticity of the substrate, and it is taken into account by both models.

The contact width and the maximum penetration of the indenter into the

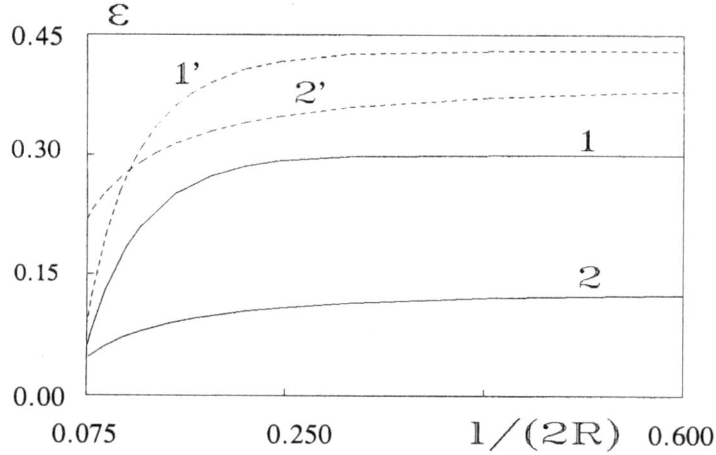

Figure 4.22: Shift of the contact zone vs. the distance between asperities centers for $\alpha_T = 50$, $\bar{P} = 1/15$ and $T_\epsilon V/R = 1$ (curves 1, 1′), $T_\epsilon V/R = 10$ (curves 2, 2′); $\beta_\varepsilon = 10$ (solid lines) and $\beta_\varepsilon \to +\infty$ (dashed lines); layer is modeled by the Kelvin solid.

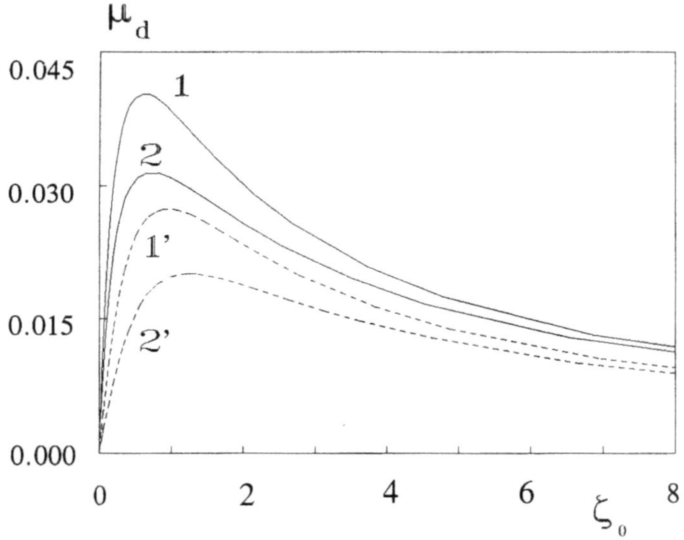

Figure 4.23: Friction coefficient vs. the parameter ζ_0 for $\beta_\varepsilon = 10$, $\bar{P} = 1/15$ and $l/2R = 5$ (curves 1, 1′), $l/2R = 1$ (curves 2, 2′); $\alpha_T = 50$ (solid lines), $\alpha_T = 5$ (dashed lines); layer is modelled by the Kelvin solid.

β_n	β_n/α_n	μ	$(\tau_1)_{max}$	ξ_m	η_m
0.01	0.01	0.00	0.292	0.32	0.72
		0.10	0.302	0.60	0.60
		0.26	0.342	0.88	0.34
		0.28	0.349	0.92	0.28
0.01	0.1	0.00	0.293	0.04	0.76
		0.10	0.299	0.36	0.70
		0.26	0.330	0.76	0.46
		0.28	0.335	0.80	0.42
0.1	0.1	0.00	0.261	0.20	0.70
		0.10	0.268	0.44	0.62
		0.26	0.296	0.76	0.42
		0.28	0.301	0.32	0.00
0.1	1	0.00	0.262	0.04	0.70
		0.10	0.266	0.28	0.68
		0.26	0.290	0.20	0.02
		0.28	0.300	0.20	0.00

Table 4.1: The magnitude and location of the principal shear stress for $\bar{a} = 0.1$, $\bar{P} = 0.01$ in the case of the Maxwell model of the layer.

viscoelastic layer calculated from the Kelvin model for a given load is limited by the values following from the solution of the corresponding contact problem for the elastic indenter and the elastic layer bonded to the elastic half-plane. The elastic layer is characterized by the modulus E_L if $\zeta_0 \to +\infty$, and by the modulus $\alpha_T E_L$ (the instantaneous modulus of elasticity) if $\zeta_0 \to 0$. The contact shift tends to zero as $\zeta_0 \to 0$ and $\zeta_0 \to +\infty$.

The mechanical friction coefficient μ_d calculated from this model also tends to zero as $\zeta_0 \to 0$ and $\zeta_0 \to \infty$. Fig. 4.23 illustrates the dependence of μ_d calculated from Eq. (3.78) on the parameter ζ_0 for different values of $l/2R$ and α_T. The results show that the mechanical friction coefficient has a maximum value for magnitudes of the parameter ζ_0 commensurable with unity. The decrease of the parameter $l/2R$ leads to the decrease of μ_d. So the increasing contact density (decreasing $l/2R$) is equivalent to decreasing effective layer viscosity.

4.4.4 Internal stresses

The results from the contact stress analysis were used to investigate the influence of the viscoelastic layer, surface roughness and friction on the internal stresses within the semi-infinite plane. The results of the principal shear stress $\tau_1(\xi, \eta)$ calculation for different values of parameters $\bar{a}$, β_n, β_n/α_n and μ are presented in Table 4.1 and in Fig. 4.24. The principal shear stress is nondimensionalized with respect to the maximum Hertzian pressure $\bar{p}_0 = \sqrt{2\bar{P}}/\pi$. The table contains the magnitude $(\tau_1)_{max}$ and location (ξ_m, η_m) of the maximum value of the principal

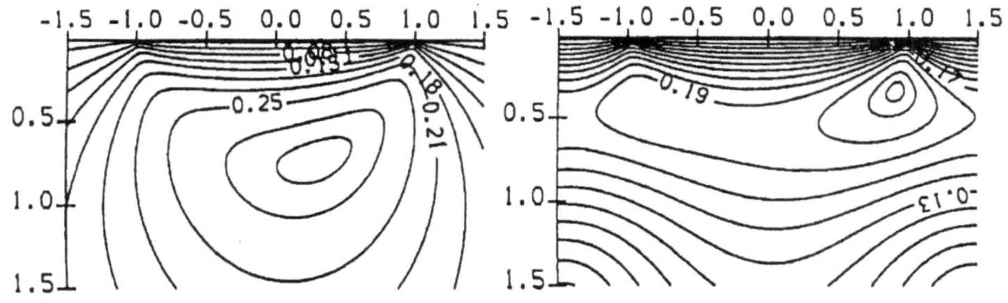

a b

Figure 4.24: Contours of the function $\tau_1(\xi, \eta)$ for $\beta_n = 0.01$, $\beta_n/\alpha_n = 0.01$, $\mu = 0$ and different contact density: $(a+b)/l = 0.1$ (a), $(a+b)/l = 0.7$ (b) in the case of the Maxwell model of the layer.

shear stress within the substrate of the semi-infinite plane. Fig. 4.24 illustrates the contours of the function $\tau_1(\xi, \eta)$ for two different densities of asperity arrangement. The contact pressure at any period is applied within the interval $(-1, 1)$ on the ξ-axis.

Comparing the results for the case $\mu = 0$ and low contact density parameter ($(a+b)/l = 0.1$) with Hertz internal stress distribution, we conclude that due to the viscoelastic layer the principal shear stresses $\tau_1(\xi, \eta)$ are distributed non-symmetrically with respect to the axis of symmetry of the contact zone. For the same value of β_n/α_n, when β_n increases, the location (ξ_m, η_m) of the maximum value of $\tau_1(\xi, \eta)$ approaches the interface (η_m decreases) and the magnitude of $(\tau_1)_{\max}$ decreases (Table 4.1).

In the presence of the viscoelastic surface layer, the maximum value of the principal shear stress occurs on the interface at a higher coefficient of friction, compared to the case when the two elastic bodies are in contact. Note that for this case, the viscoelastic properties of the layer have a significant effect on the contact characteristics (Figs.4.19–4.21) and consequently on the internal stresses.

When the contact density parameter $(a+b)/l$ is high, the amplitude of the principal shear stress at the fixed depth below the surface is small. This conclusion is similar to that obtained in Chapter 2 for the high contact density of elastic bodies.

The results of this analysis indicate that the viscoelastic properties of the surface layer, and the surface roughness parameter significantly affect the pressure and internal stresses in sliding contact of coated elastic bodies. These surface nonhomogeneity parameters must be taken into account in prediction of wear of tribo-contacts. The principal shear stress distribution for different values of the contact density parameter obtained in this section can be used for calculation of the surface fatigue fracture in sliding contact of rough bodies (Chapter 5).

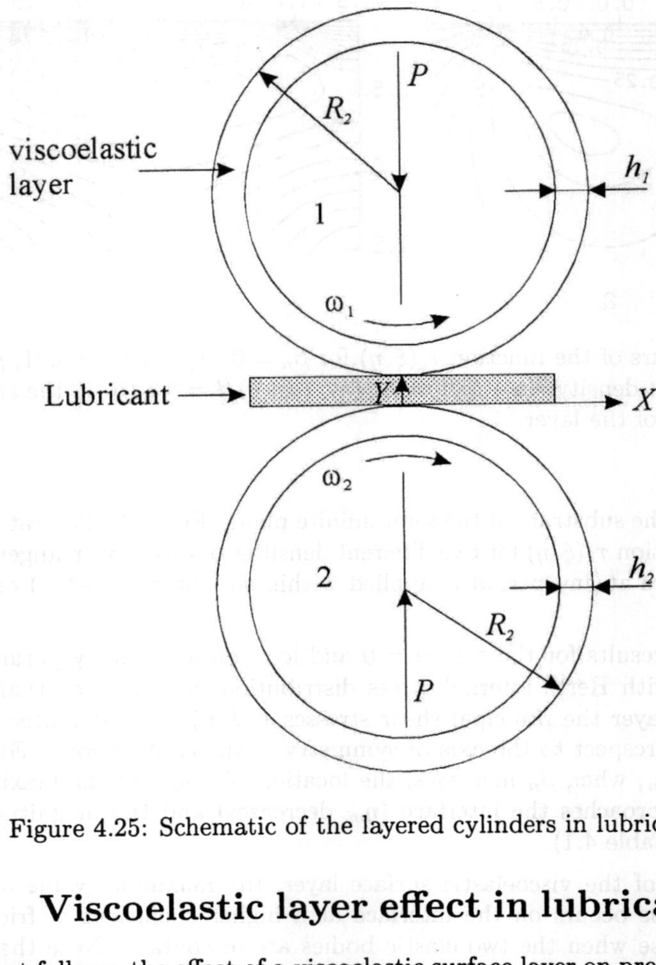

Figure 4.25: Schematic of the layered cylinders in lubricated contact.

4.5 Viscoelastic layer effect in lubricated contact

In what follows, the effect of a viscoelastic surface layer on pressure, film thickness and coefficient of friction in lubricated contact is investigated. Hydrodynamic and elasto-hydrodynamic lubrication has been studied extensively by a number of investigators (see the monographs by Dowson and Higginson, 1966, and Hamrock, 1994). These studies in general concentrated on pressure and film thickness and demonstrated that the Newtonian fluid model predicts satisfactory film thickness between contacting bodies. However, the Newtonian fluid model fails to predict friction and power loss similar to experimental results at high loads and low velocities. In order to predict friction results correlating with experiments, Hirst and Moore (1974), Johnson and Tevaarwerk (1977), Conry, Wang and Cusano (1987), Sadeghi and Dow (1987), Sui and Sadeghi (1991) have introduced thermal effects or non-Newtonian fluid behavior, or both, in their studies. The Eyring fluid model, the non-linear Maxwell model and others have been used to describe the lubricant shear behavior at high loads. Recently, Elsharkawy and

Hamrock (1994) investigated the effect of elastic coating on the pressure and the film thickness on elasto-hydrodynamic lubrication of multi-layered elastic bodies.

In this study, the combined effect of viscoelastic layer bonded to elastic cylinders and a thin film of lubricant is investigated. The influence of geometry and mechanical properties of a thin viscoelastic layer on contact stresses, film thickness and friction coefficient is analyzed for various operating conditions.

4.5.1 Problem formulation

Fig. 4.25 illustrates a schematic of the contact between two layered rotating cylinders, separated by a thin film of lubricant. The (r_i, θ_i) coordinate system is fixed on each cylinder and rotates with angular velocity ω_i ($i = 1, 2$ for upper and lower cylinder, respectively). The (x,y) coordinate system is fixed in the plane such that the y-axes coincides with the line between the centers of cylinders. The shape functions of the cylinders are $f_i(x) = \kappa_i \pm x^2/2R_i$ (κ_1 and κ_2 are constants).

The relationships between the moving (r_i, θ_i) and the fixed (x, y) coordinate systems are

$$\begin{aligned} x &= r_i \cos(\theta_i + \omega_i t), \\ y &= r_i \sin(\theta_i + \omega_i t) + y_{0i}, \end{aligned} \qquad i = 1, 2, \tag{4.133}$$

where $y_{0,1}$, $y_{0,2}$ are the coordinates of the centers of the cylinders.

The surface layers are modelled as a one-dimensional Maxwell (viscoelastic) body. The one dimensional time-dependent relationship between the normal pressure p, which is assumed to be uniform across the viscoelastic layer thickness h_i, and the normal displacement v_i^L is used to describe the layer compliance in the normal direction

$$\frac{dv_i^L}{dt} = \frac{h_i}{E_n}\left(\frac{p}{T_n} + \frac{dp}{dt}\right). \tag{4.134}$$

Here E_n and T_n are the elastic modulus and the relaxation time of the layer material in the normal direction.

To simplify the analysis, we assume that the mechanical properties of the layers at the upper and lower cylinders are the same. However, the method developed here can be used to consider the general case of different mechanical properties of layers.

Using Eq. (4.133), we can write Eq. (4.134) in the system of coordinates (x, y) as

$$\frac{dv_i^L}{dx} = \frac{h_i}{E_n}\left(-\frac{p}{V_i T_n} + \frac{dp}{dx}\right), \tag{4.135}$$

where V_1 and V_2 are the linear velocities of the cylinder surfaces.

The Reynolds equation is used to describe the two-dimensional flow of a thin lubricating film between two surfaces moving with velocities V_1 and V_2

$$\frac{d}{dx}\left(\frac{H^3}{\eta} \cdot \frac{dp}{dx}\right) = 6\left(V_1 + V_2\right)\frac{dH}{dx}, \tag{4.136}$$

where $H(x)$ is the film thickness and $p(x)$ is the pressure. The variation of viscosity η with pressure-viscosity effect is taken into account by the relationship proposed by Barus (1893):

$$\eta = \eta_0 \exp(\alpha p), \tag{4.137}$$

η_0 is an absolute viscosity at ambient pressure and temperature, α is the pressure-viscosity coefficient of the lubricant.

The boundary conditions for the Reynolds equation (4.136) are

$$p(-\infty) = p(b) = 0, \quad \left.\frac{dp}{dx}\right|_{x=b} = 0. \tag{4.138}$$

where b is the exit point.

The substrate of the layered cylindrical rollers is considered to be two-dimensional, isotropic, homogeneous and linearly elastic. The displacement gradient for the substrate is

$$\frac{dv_i}{dx} = -\frac{2\left(1-\nu_i^2\right)}{\pi E_i} \int\limits_{-\infty}^{b} \frac{p(x')\,dx'}{x-x'} \tag{4.139}$$

Within the contact region, the thickness of the film can be expressed by

$$\begin{aligned} H(x) = H(b) + \frac{x^2-b^2}{2R} + [v_1(x)+v_2(x)] - \\ [v_1(b)+v_2(b)] + \left[v_1^L(x)+v_2^L(x)\right] - \left[v_1^L(b)+v_2^L(b)\right], \end{aligned} \tag{4.140}$$

where $\dfrac{1}{R} = \dfrac{1}{R_1} + \dfrac{1}{R_2}$. The displacements $v_i^L(x)$, $v_i(x)$ are given by Eqs. (4.135) and (4.139).

The force equilibrium condition within the contact region is given by

$$\int\limits_{-\infty}^{b} p(x)\,dx = P. \tag{4.141}$$

Eqs. (4.139), (4.135)–(4.141) are used to determine the pressure $p(x)$, the film profile $H(x)$, the elastic $v_i(x)$ and viscoelastic $v_i^L(x)$ displacements of the contacting surfaces.

4.5.2 Method of solution of the main system of equations

We will analyze pressure and film thickness for the cases of a constant viscosity and a variable viscosity relationship according to Eq. (4.137).

For constant viscosity η_0, the system of Eqs. (4.139)–(4.141) can be reduced to one equation to determine the film thickness. Integrating Eq. (4.139) and

Eq. (4.135) and using boundary condition Eq. (4.138), we obtain

$$v^L(x) - v^L(b) = -\frac{1}{2}\int_{-\infty}^{b}\left\{\frac{1}{E_n T_n}\left(\frac{h_1}{V_1}+\frac{h_2}{V_2}\right)[(x-x') - |x-x'|)] - \frac{h_1+h_2}{E_n}[\text{sgn}(x-x') - 1]\right\}\frac{dp}{dx'}\,dx', \tag{4.142}$$

$$v(x) - v(b) = \frac{2}{\pi E^*}\int_{-\infty}^{b}[(b-x')\ln(b-x') - (x-x')\ln|x-x'|]\frac{dp}{dx'}\,dx', \tag{4.143}$$

where

$$v^L(x) = v_1^L(x) + v_2^L(x), \tag{4.144}$$

$$v(x) = v_1(x) + v_2(x). \tag{4.145}$$

Substituting Eqs. (4.142) and (4.143) in Eq. (4.140) and using the integral form of the Reynolds equation (4.136), we obtain

$$H(x) = H(b) + \frac{x^2-b^2}{2R} + \int_{-\infty}^{b} k(x,x',b)\frac{H(x')-H^*}{H^3(x')}\,dx', \tag{4.146}$$

where H^* is the film thickness at which the pressure gradient is zero, $dp/dx = 0$. The kernel in Eq. (4.146) is

$$k(x,x',b) = 6\eta_0(V_1+V_2)\left\{\frac{2}{\pi E^*}[(b-x')\ln(b-x') - (x-x')\ln|x-x'|] - \frac{1}{2E_nT_n}\left(\frac{h_1}{V_1}+\frac{h_2}{V_2}\right)[(x-x') - |x-x'|] + \frac{h_1+h_2}{2E_n}[\text{sgn}(x-x') - 1]\right\}. \tag{4.147}$$

In the dimensionless form Eqs. (4.146) and (4.147) are given by

$$\bar{H}(\xi) = 1 + \frac{\bar{b}^2}{2\bar{H}_*}(\xi^2-1) + \frac{3\bar{b}^2}{\bar{H}_*^3}\int_{-\infty}^{1}\bar{K}(\xi,\xi',\bar{b})\frac{\bar{H}-1}{\bar{H}^3}\,d\xi', \tag{4.148}$$

$$\bar{K}(\xi,\xi',\bar{b}) = \frac{4\bar{P}S}{\pi}[(1-\xi')\ln(1-\xi') - (\xi-\xi')\ln|\xi-\xi'|] - \bar{\eta}\frac{1-\Lambda\gamma}{1-\gamma^2}[(\xi-\xi') - |\xi-\xi'|] + \frac{\bar{P}S}{\beta\bar{b}}[\text{sgn}(\xi-\xi') - 1]. \tag{4.149}$$

where

$$\xi = \frac{x}{b}, \quad \bar{H}(\xi) = \frac{H(b\xi)}{H(b)}, \quad \bar{b} = \frac{b}{R}, \quad \bar{H}_* = \frac{H(b)}{R},$$
$$\bar{P} = \frac{P}{E^* R}, \quad S = \frac{\eta_0 (V_1 + V_2)}{P}, \quad h = \frac{h_1 + h_2}{2}, \tag{4.150}$$
$$\bar{\eta} = \frac{4\eta_0 h}{E_n T_n R}, \quad \Lambda = \frac{h_1 - h_2}{2h}, \quad \beta = \frac{E_n R}{2E^* h}, \quad \gamma = \frac{V_1 - V_2}{V_1 + V_2}.$$

Combining Eqs. (4.136), (4.138) and (4.141) and using the dimensionless function and parameters (4.150), we obtain

$$\frac{1}{S} = -\frac{6\bar{b}^2}{\bar{H}_*^2} \int_{-\infty}^{1} \xi \frac{\bar{H}(\xi) - 1}{\bar{H}^3(\xi)} \, d\xi, \tag{4.151}$$

$$\int_{-\infty}^{1} \frac{\bar{H}(\xi) - 1}{\bar{H}^3(\xi)} \, d\xi = 0. \tag{4.152}$$

The system of Eqs. (4.148), (4.151) and (4.152) is used to determine the function $\bar{H}(x)$ and the two parameters $\bar{b}$ and $\bar{H}_*$. Dimensionless contact pressure $\bar{p}(x) = p(x)R/P$ is obtained by integrating Eq. (4.136),

$$\bar{p}(\xi) = -\frac{6\bar{b}S}{\bar{H}_*^2} \int_{\xi}^{1} \frac{\bar{H}(\xi') - 1}{\bar{H}^3(\xi')} \, d\xi'. \tag{4.153}$$

The Newton–Kantorovich method was used to solve the system of Eqs. (4.148), (4.151) and (4.152). In the numerical analysis, the infinite interval $(-\infty, 1)$ was divided into two parts $(-\infty, \bar{d})$ and $(\bar{d}, 1)$, and $\bar{H}_1(\xi)$ was approximated by

$$\bar{H}_1(\xi) = \begin{cases} \bar{H}(\xi), & \bar{d} < \xi < 1, \\ \dfrac{\bar{b}^2 \xi^2}{2\bar{H}_*} + \bar{\delta}, & -\infty < \xi < \bar{d}. \end{cases} \tag{4.154}$$

Parameters $\bar{\delta}$ and $\bar{d}$ are found to satisfy the conditions

$$\bar{H}(\bar{d} - 0) = \bar{H}(\bar{d} + 0),$$
$$\bar{H}'(\bar{d} - 0) = \bar{H}'(\bar{d} + 0).$$

The system of equations was reduced to linear algebraic equations which were solved by the Gauss elimination.

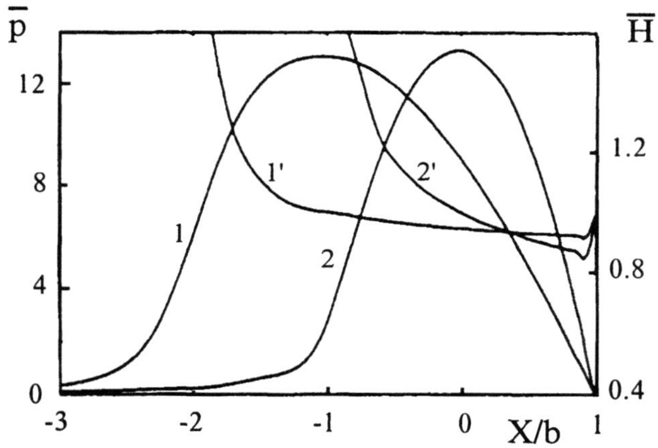

Figure 4.26: Pressure (curves 1 and 2) and film thickness (curves 1′ and 2′) for $\bar{P} = 10^{-4}$, $\beta = 1$, $S = 2 \cdot 10^{-5}$ and $\bar{\eta}/(1-\gamma^2) = 5 \cdot 10^{-8}$ (1, 1′), $\bar{\eta} = 0$ (2, 2′).

4.5.3 Film profile and contact pressure analysis

The results of calculations are presented as functions of five dimensionless parameters. The parameter $\bar{\eta} = 4\eta h/E_n T_n R$ characterizes the relative fluid film and layer viscosities; h is the layer thickness which has been assumed to be the same on the upper and lower cylinders. The relative elastic modulus of the layer and substrate is characterized by $\beta = E_n R/2E^* h$. The Sommerfeld number, $S = \eta_0(V_1 + V_2)/P$, describes the average velocity effect, and γ denotes the difference in velocities, $\gamma = (V_1 - V_2)/(V_1 + V_2)$. The dimensionless load is given by $\bar{P} = P/(E^* R)$.

Fig. 4.26 demonstrates the influence of the relative viscosity parameter $\bar{\eta}$ on pressure and film thickness. When $\bar{\eta} = 0$ it is the condition of elastic layer on elastic substrate. The results indicate that the maximum pressure is nearly the same for both curves; however, when viscoelastic effects are included (curves 1 and 1′), the pressure distribution becomes more asymmetrical. Fig. 4.26 also shows that the film thickness is nearly constant in the contact when $\bar{\eta} = 5 \cdot 10^{-8}$.

Fig. 4.27 illustrates contact pressure and film thickness for various Sommerfeld numbers. The results show that for low velocity conditions $(S = 10^{-5})$, the film thickness within the contact zone is nearly constant, and the pressure distribution is similar to the case without the lubricating film (compare with the results presented in § 4.3). For increasing values of velocity $(S = 10^{-3})$, the pressure is distributed over a large area, and the film shape develops features corresponding to the hydrodynamic regime. The maximum contact pressure also decreases, and the point where it occurs moves toward the exit of the contact as velocity increases. The contact exit location $\bar{b}$ strongly depends on velocity for low values of velocity. Under low velocity conditions, when $\bar{\eta}$ increases the exit location approaches the axis of symmetry of the cylinder. For high velocity conditions, the exit location is nearly the same for all values of $\bar{\eta}$.

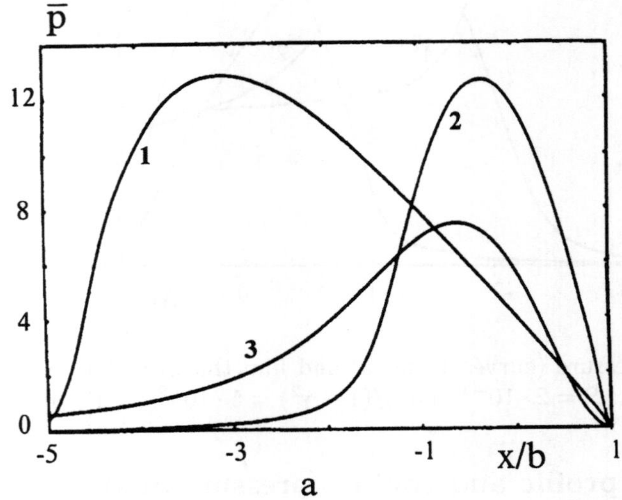

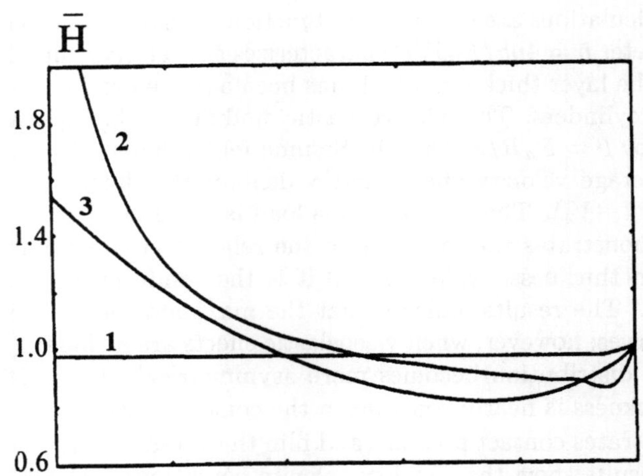

Figure 4.27: Contact pressure (a) and film thickness (b) for $\bar{P} = 10^{-4}$, $\beta = 1$, $\bar{\eta}/(1-\gamma^2) = 5 \cdot 10^{-8}$ and $S = 10^{-5}$ (curve 1), $S = 5 \cdot 10^{-5}$ (curve 2), $S = 10^{-3}$ (curve 3).

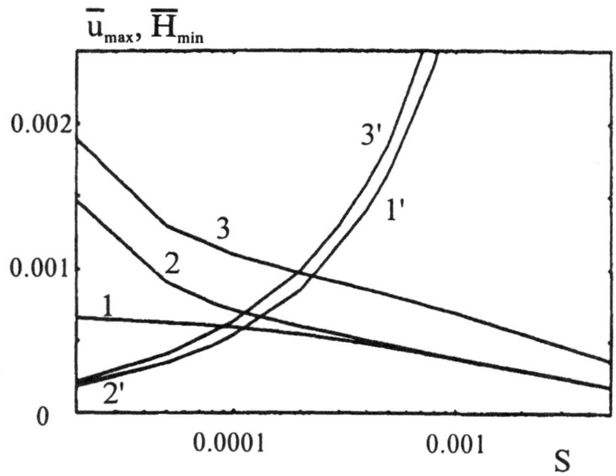

Figure 4.28: Maximum displacement of viscoelastic layer (curves 1,2,3) and minimum film thickness (curves 1′, 2′, 3′) as a function of Sommerfeld number for $\bar{P} = 10^{-4}$ and $\beta = 1$, $\bar{\eta} = 0$ (curves 1, 1′), $\beta = 1$, $\bar{\eta} = 5 \cdot 10^{-8}$ (curves 2, 2′), $\beta = 0.5$, $\bar{\eta} = 5 \cdot 10^{-8}$ (curves 3, 3′).

The comparison of the minimum film thickness $\bar{H}_{\min}$ and maximum indentation $\bar{u}_{\max} = u_{\max}/R$ of the cylinder in the viscoelastic layer is shown in Fig. 4.28. The figure indicates that under low velocity conditions, $\bar{H}_{\min}$ is much lower than the maximum indentation. However, as the Sommerfeld number increases, the minimum film thickness is much larger than the indentation. This shows that, at low Sommerfeld values, the viscoelastic layer properties largely determine the contact characteristics. For high Sommerfeld numbers, the minimum film thickness depends hardly at all on the viscosity $E_n T_n$ of the layer. Minimum film thickness and maximum indentation increase as parameter β decreases.

Fig. 4.29 illustrates the effect of variable viscosity (see Eq. (4.137)) on pressure and film thickness profile. The results indicate that when we take into account the dependence of viscosity on pressure, the film thickness is larger than when viscosity is constant, and the film thickness exhibits a small reduction near the exit. The pressure profile is negligibly higher.

This analysis shows that, for low velocities, the contact characteristics are dominated by the properties of the viscoelastic layer and the elastic substrate, and as the velocity increases the viscoelastic layer effect becomes negligible.

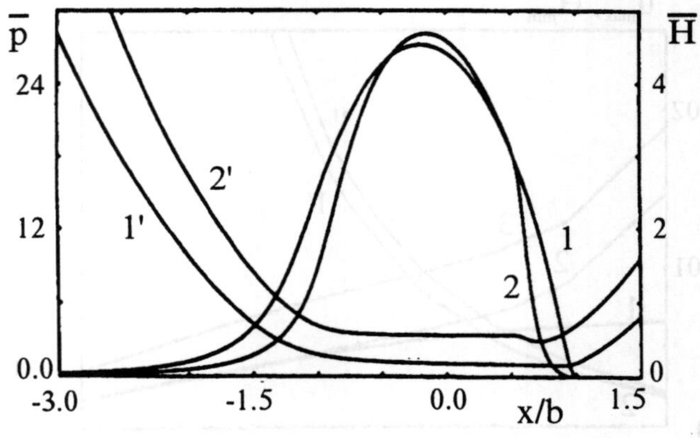

Figure 4.29: Variable viscosity effect on pressure (1,2) and film thickness (1′, 2′) for $\bar{\eta} = 10^{-9}$, $\alpha P/R = 2 \cdot 10^{-11}$ and $S = 7 \cdot 10^{-6}$: viscosity constant (curves 1, 1′), viscosity changes with pressure (curves 2, 2′).

4.5.4 Rolling friction and traction analysis

The asymmetry of the pressure distribution within the contact zone generates the rolling resistance couple M which is determined by the formula

$$M = \int_{-\infty}^{b} xp(x)\, dx. \tag{4.155}$$

In lubricated contacts the traction resistance due to the film of lubricant for the lower ($i = 1$) and upper ($i = 2$) cylinders is given by

$$T_1 = -\int_{-\infty}^{b} \frac{H}{2} \frac{dp}{dx}\, dx + \int_{-\infty}^{b} \frac{V_1 - V_2}{H} \eta\, dx. \tag{4.156}$$

$$T_2 = \int_{-\infty}^{b} \frac{H}{2} \cdot \frac{dp}{dx}\, dx + \int_{-\infty}^{b} \frac{V_1 - V_2}{H} \eta\, dx, \tag{4.157}$$

Then the rolling resistance coefficient and the traction coefficients are defined as

$$\mu_r = \frac{M}{PR}, \quad \mu_t = \frac{T}{P}.$$

Using Eqs. (4.155)–(4.156), we write the rolling and traction coefficients in the dimensionless form

$$\mu_r = \frac{3\bar{b}^3 S}{\bar{H}_*^2} \int_{-\infty}^{1} \xi^2 \frac{\bar{H} - 1}{\bar{H}^3}\, d\xi,$$

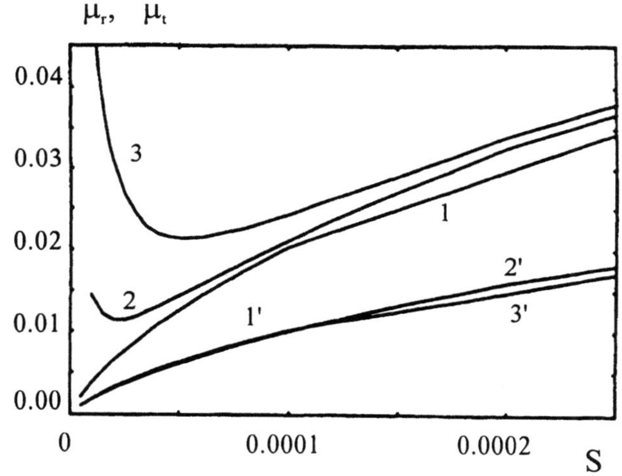

Figure 4.30: Rolling friction coefficient μ_r (curves 1, 2, 3) and traction coefficient μ_t (curves $1'$, $2'$, $3'$) as a function of Sommerfeld number for $\bar{P} = 10^{-4}$ and $\beta = 1$, $\bar{\eta} = 0$ (1, $1'$), $\beta = 1$, $\bar{\eta} = 5 \cdot 10^{-8}$ (2, $2'$), $\beta = 0.5$, $\bar{\eta} = 5 \cdot 10^{-8}$ (3, $3'$).

$$\mu_t^{\pm} = \pm \frac{3\bar{b}S}{\bar{H}_*} \int_{-\infty}^{1} \frac{\bar{H} - 1}{\bar{H}^2} d\xi + \frac{\gamma \bar{b} S}{\bar{H}_*} \int_{-\infty}^{1} \frac{d\xi}{\bar{H}},$$

where μ_t^+ (μ_t^-) is the traction coefficient for upper (lower) cylinder. The results of calculations of the rolling and traction coefficients for various values of dimensionless parameters are presented in Figs. 4.30 and 4.31.

Fig. 4.30 depicts the coefficient of rolling friction μ_r and traction coefficient μ_t^+ as a function of the Sommerfeld number for different values of layer viscosity, and thus various values of the parameter $\bar{\eta}$. The results indicate that for high values of $\bar{\eta}$ ($\bar{\eta} = 5 \cdot 10^{-8}$) the rolling friction coefficient monotonically reduces as the Sommerfeld number increases. At the definite value $S = S_*$ which depends on the parameters $\bar{\eta}$ and β the friction coefficient reaches its minimum and then increases as the velocity increases. The plots of pressure distribution for various Sommerfeld numbers and $\bar{\eta}/(1 - \gamma^2) = 5 \cdot 10^{-8}$ (see Fig. 4.27) conform to the non-monotonic dependence of μ_r on S illustrated by curves 2 and 3 in Fig. 4.30. When $\bar{\eta} = 0$, which is the case for the elastic coated elastic body, the coefficient of rolling friction monotonically increases as the Sommerfeld number increases. The traction coefficient is nearly the same for all values of $\bar{\eta}$ and increases as the Sommerfeld number increases. However, its magnitude in general is lower than the rolling friction coefficient.

Fig. 4.31 illustrates the dependence of rolling friction and traction coefficients on the difference in sliding velocities γ of cylinders. The results show that in general the friction coefficients hardly depend at all on the parameter γ for $\gamma \leq 0.1$.

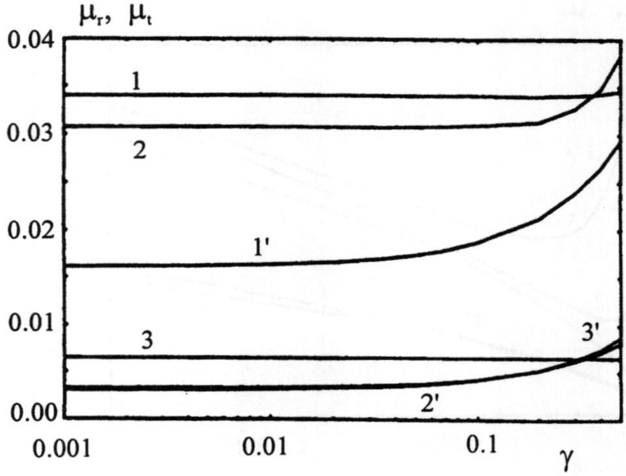

Figure 4.31: Rolling friction coefficient μ_r (curves 1, 2, 3) and traction coefficient μ_t (curves 1′, 2′, 3′) as a function of sliding velocity for $\bar{P} = 10^{-4}$, $\beta = 1$ and $\bar{\eta} = 5 \cdot 10^{-8}$ (1, 1′), $\bar{\eta} = 10^{-8}$ (2, 2′), $\bar{\eta} = 0$ (3, 3′).

However, for larger values of γ, they monotonically increase as the parameter γ increases.

Thus, the results indicate that due to a viscoelastic layer the rolling friction coefficient is a non-monotonic function of the Sommerfeld number. The results are in good qualitative agreement with the well-known experimental results of Stribeck (see, for example, Moore, 1975).

Chapter 5

Wear Models

5.1 Mechanisms of surface fracture

5.1.1 Wear and its causes

The study of wear is one of the main targets of tribology. Let us remember, that a brief definition of tribology is as follows: "Tribology is friction, wear and lubrication".

Wear is defined as a process of progressive loss of material from the operating surfaces of solids arising from their contact interaction. The dimensions of body and its mass are diminished by wear.

There can be many causes for wear. First of all, it is caused by material fracture under stresses in the process of friction. This widespread type of wear is classified as mechanical wear and is often taken to be a synonym of the word "wear".

Among other wear causes, chemical reactions and electrochemical processes can be mentioned. Corrosive wear is an example of this type of surface fracture. It is the main wear mechanism in moving components operating in a chemically aggressive environment.

Some physical processes can also cause wear. For example, it is known that almost all the energy dissipated in friction is converted into heat. An increase of the surface layer temperature can change the aggregate state of the material. In such a case the wear is provided because of melting and flowing of the melt out of the interface (ablation wear) or because of evaporation (breaks, high speed guides, plane wheels, etc.). High temperature accelerates diffusion processes which can influence wear in some cases (cutting tools). For these cases, wear occurs at atomic and molecular levels.

It should be mentioned, that in operation of moving contact wear can be conditioned by several causes simultaneously. That is why a description of wear as a result of one of the causes, mentioned above, is basically an idealization of this sophisticated phenomenon.

Since the mechanical wear of two bodies in contact can be studied by the

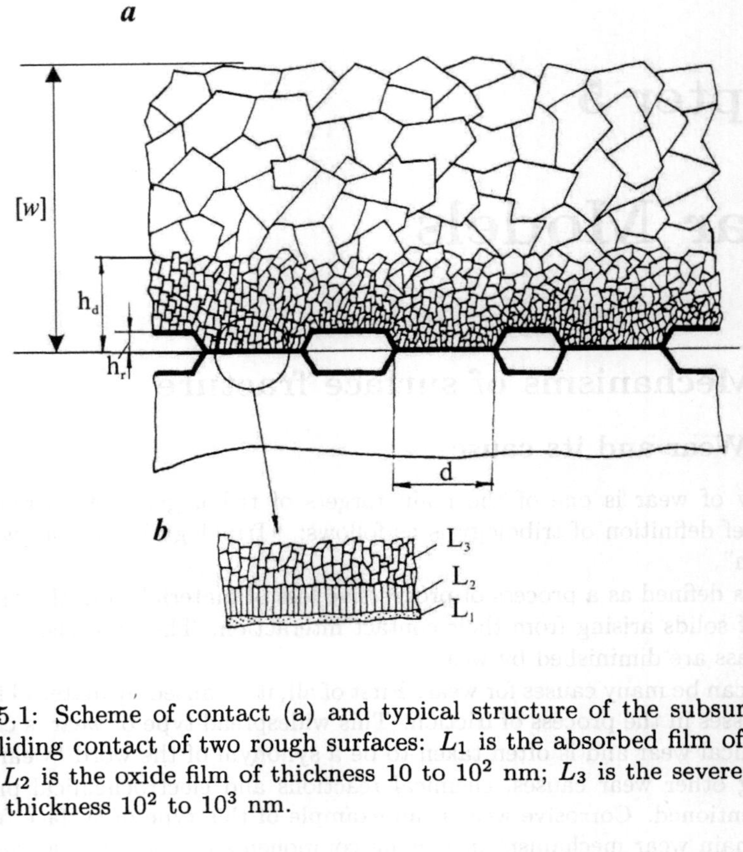

Figure 5.1: Scheme of contact (a) and typical structure of the subsurface layer (b) in sliding contact of two rough surfaces: L_1 is the absorbed film of thickness 10 nm; L_2 is the oxide film of thickness 10 to 10^2 nm; L_3 is the severe deformed layer of thickness 10^2 to 10^3 nm.

methods and approaches used in contact mechanics and fracture mechanics, it will be the subject of our investigation in this book.

5.1.2 Active layer

The first thing to do, when starting the analysis of the wear mechanism, is to identify the area where fracture takes place. This is usually done in fracture mechanics, when the most dangerous pieces of the structure or the specimen are identified. Unfortunately, it is not always possible to identify such pieces. Tribology is more *fortunate*, from this point of view. Numerous studies of wear particles (they are also called wear debris) including their shape, size distributions and composition, and wear scars on rubbing surfaces (Rabinowicz, 1965, Tsuchiya and Tamai, 1970, Seifert and Westcott, 1972, Sasada and Kando, 1973, Sasada and Norose, 1975) witness that fracture occurs within a thin subsurface layer.

In order to visualize the area, consider two solids in contact (Fig. 5.1a). In describing this area, we shall use the scale of the contact spot diameter d which is typically within the interval from 1 to 10 μm. Note, that the height of the surface roughness h_r is (0.1 to 1)d. The wear particle dimensions vary within wide limits, but seldom are they greater than the value of one or two diameters of the contact spot. This allows us to estimate the thickness of the layer h_d near the surface, where fracture takes place; this is often called *the active layer*.

Tribology is *unfortunate* because of extreme complexity of this layer. It is usually an inhomogeneous complex structure. Practically all engineering surfaces are contaminated. Naturally occurring contaminant films range from a single layer of molecules adsorbed from the atmosphere, to much thicker oxide and other films formed by chemical reactions between the surface and environment (Fig. 5.1b).

Besides, each surface is the product of a manufacturing process which changes the properties of the substrate material. Defects of different scale and nature are produced, residual stresses appear etc. within the subsurface layer.

The properties of the active layer have not been studied in as much detail as the properties of the bulk material. Special tools and facilities are required, because of the small thickness, and large depth variation of properties of the active layer. Such methods as electron microprobe analysis (EMA), X-ray photoelectron spectroscopy (XRPS), and sliding beam X-ray diffraction allow us to study the elemental and chemical composition as well as the structure of the active layer. The method of nano-indentation, in combination with the analysis of the contact problem can also be used to determine the mechanical properties such as hardness and Young's modulus of thin surface layers.

In contact interaction, the active layer is highly and nonuniformly loaded due to the roughness of contacting bodies. This can be supported by the following estimates. The contact occurs within spots, the total area A_r of which is only a small part of the nominal (apparent) area of contact A_a. For contacting surfaces described by the various micro- and macro- geometry parameters, the following estimate is valid: $\dfrac{A_r}{A_a} \sim 10^{-3}$ to 10^{-1}. So the mean real contact pressure, which is the load divided by the real contact area, is 10 or 1000 times greater than the nominal contact pressure. Furthermore, the maximum pressure within a contact spot can be several times greater than the mean one.

It is worth noting that, unlike the nominal pressure which can be controlled by the load applied to the contacting bodies, the mean real contact pressure does not change essentially when the load varies. Many experimental and theoretical investigations of contact characteristics of rough solids (Kragelsky, Bessonov and Shvetsova, 1953, Demkin, 1963, Hisakado, 1969, Gupta, 1972) give conclusive proof that the mean value p_r is practically independent of the compressive load P, and is determined mainly by the roughness parameters and mechanical properties of the contacting bodies. The estimate of the pressure p_r shows that the contact is accompanied by heavy loading conditions. For very smooth surfaces $\dfrac{p_r}{E} \sim 10^{-3}$; for rough metallic surfaces it is approximately 10^{-2}. The latter value indicates that plastic deformations play a significant role in contact of rough surfaces.

Because of the high and nonuniform loading of deformable bodies in contact, the internal stresses are distributed nonuniformly within the active layer. Examples of internal stress distribution in the subsurface layer for different microgeometry parameters were presented in Chapter 2.

Sliding contact of rough bodies has a further peculiarity: the cyclic character of loading caused by migration of contact spots due to the relative motion. In this case there is a cyclic stress field in the subsurface layer.

5.1.3 Types of wear in sliding contact

Fracture is usually realized near contact spots characterized by high normal and tangential stresses. It can occur due to single or repeated loading of contact spots.

Fracture under single loading occurs if internal stresses caused by this loading are so high that the fracture criterion is satisfied at some point of the contacting bodies. This type of fracture is observed in *adhesive wear* characterized by transport of the material from one contacting surface to another. The high adhesion of contacting bodies is a necessary condition for realization of this type of wear. As a rule, surface contaminations such as adsorbed molecules of oxygen, water vapour, films of metal oxides, and other chemical constitution decrease adhesion. However, high contact stresses can cause plastic flow of contacting surfaces, and rupture of these films. This kind of film removal from the surfaces becomes more effective if the plastic flow occurs within both contacting bodies (if their hardnesses are not too different).

In frictional contact of bodies with essentially different hardnesses, the other surface fracture mechanism, *abrasive wear*, is realized. In abrasive wear, the asperities of the hard body push the soft material of the other body out of the way due to material plastic deformation. As we mentioned earlier, plastic deformations arise in contact of very rough surfaces (two-body abrasion) or in the presence of hard wear or abrasive particles in a frictional zone (three-body abrasion).

Abrasive wear can also occur in single loading of contact spots under high stresses in the active layer. This kind of abrasive wear is known as *micro-cutting.* In micro-cutting, the hard asperity plays the part of the cutter which removes thin chips from the surface of the soft body. Micro-cutting is similar to some technological operations such as treatment by file or abrasive paper (two-body abrasion) and lapping or polishing (three-body abrasion). It is usually characterized by a high wear rate.

If stresses near contact spots are not so high (for example, the contact pressure does not exceed the yield stress) and there is no strong adhesion between the contacting bodies, the fracture does not occur in a single loading. However, the cyclic character of loading in combination with high level of stresses in the active layer (p_r is always more than the fatigue limit) creates preconditions for intensive accumulation of defects in the material and its failure as the result of *fatigue.* It is known that we cannot prevent frictional fatigue failure, as we cannot decrease the frictional contact stresses below the fatigue limit. It was established experimentally that, in fatigue wear, particles are detached at discrete instants of time, and the size of each particle is comparable with the contact spot diameter.

Fatigue wear usually occurs in predominantly elastic contact. However, this mechanism of wear occurs elsewhere, and can be of considerable importance in adhesive and abrasive wear.

5.1.4 Specific features of surface fracture

The process of wear has some peculiarities, which suggest that we consider it as a special form of fracture.

Usually, admissible limit wear $[w]$ (see Fig. 5.1) of moving parts which have been designed well from a tribological point of view, is much more than the typical size of wear debris. Thus, repeated particle detachment can occur during the life of the parts. *Repeated fracture* of the material in wear is the distinctive feature of the process, as opposed to the bulk material fracture. In classical fracture mechanics, we ask, *How long will the material or structure operate before failure?* In tribology we may ask a similar question, *How long can the material be detached and removed before it will be finally worn?*

After removal of surface material due to wear, subsurface layer enters the contact. The characteristics of this layer, including ones that determine wear intensity, depend on the entire history of the frictional interaction. Thus, wear can be considered as the process of *hereditary type.*

In many cases, wear is a *feedback process.* One of the characteristics that control the process is surface roughness; this influences the stress field and fracture of the surfaces and, on the other hand, is formed due to this fracture. Self-organization and equilibrium structure formation in wear occur as the result of the feedback action. Equilibrium roughness observed by a set of researchers (see Khrushchov, 1946, Shchedrov, 1950, Kragelsky, Dobychin and Kombalov, 1982) is a typical example of the structure formed in such a self-organization process.

5.1.5 Detached and loose particles

The material particles detached from the surface in fracture process are not yet the wear particles, but the mostly probable candidates for this role. These particles may have various possible futures: they can be reduced, adhere to the mother surface again or to the counter-body (adhesive wear); they can charge into the more soft surface and then play the part of abrasive grains with respect to the counter-body (abrasive wear); finally, they can leave the contact zone forever, in this case the loose particles are called wear particles.

Note, that the problems of transportation and behavior of detached and foreign particles in the contact zone are still not clearly understood, but yet are important for the description of wear process and for the analysis of such types of wear as fretting and three-body abrasion, and for the construction of a model of the third body (the interface layer consisting of the particles, lubricant, etc.).

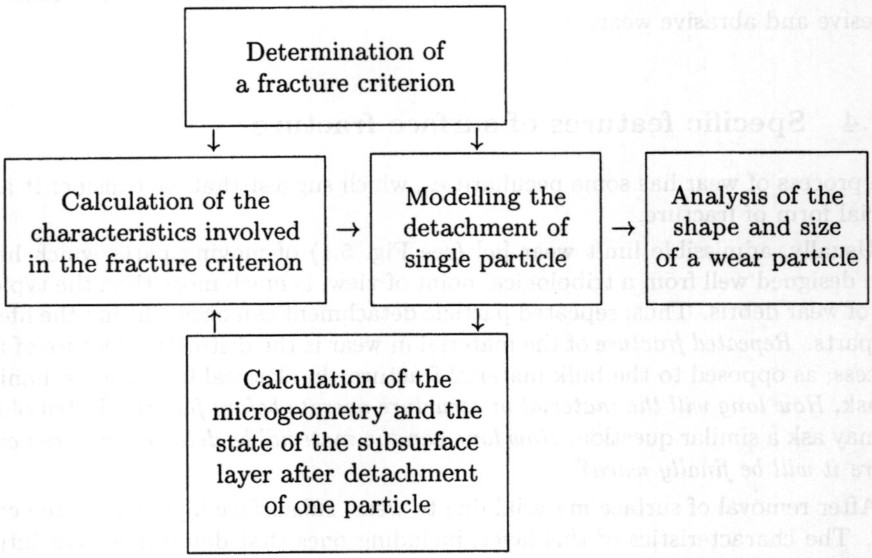

Figure 5.2: The main stages in wear modelling, and their mutual relations.

5.2 Approaches to wear modelling

5.2.1 The main stages in wear modelling

The phenomenon of wear described above includes the following: modification of material in the active layer, surface fracture and, finally, removal of the wear particles away from the contact zone.

Modelling this phenomenon is a very complicated problem. Hence, quite simple models are usually proposed in tribology, which describe in detail only a limited number of features of the wear process.

The subject of the investigation is a thin surface layer whose thickness is comparable to the contact spot size.

There are several main stages in wear modelling which are shown in Fig. 5.2. This figure also indicates the mutual relations between the different stages.

The first stage consists of the analysis of the wear mechanism, and determining the fracture criterion corresponding to this mechanism. As a rule, the fracture criterion depends on the absolute or the amplitude value of stresses, on the temperature, mechanical characteristics of the materials and so on.

The next stage is determination of the stresses and strains, temperature and other functions involved in the fracture criterion, and characterizing the state of thin surface layers of the contacting bodies.

The problem of calculation of a stress distribution near the surface of a deformable body which is in contact with the rough surface can be investigated by the methods of contact mechanics described in Chapters 2 and 3, and in the monographs by Gladwell (1980), Galin (1980, 1982), Johnson (1987), etc.

The methods of fracture mechanics are used to determine the onset of failure, and to model the particle detachment, based on the fracture criterion and on the state of the subsurface layer of the body in contact. The shape and the size of the particle detached from the surface can be evaluated as well.

As a rule, the wear is not characterized as a catastrophic state of the moving part; this contrasts with construction failure where crack propagation is tantamount to catastrophe. The specific feature of wear is its repeated character. To describe the succeeding particle detachment from the surface we need to calculate the state of the surface layer (stress and temperature distributions, etc.) after the particle detachment at the previous stage of the wear process. The change of the microgeometry caused by the surface fracture leads to redistribution of the contact pressure and internal stresses which control the wear process.

As discussed in § 5.1, some particles detached from the surface remain in the friction zone and influence the contact characteristics and wear process. Modelling of their motion is a complicated problem of substance transportation in the third body; this is beyond the scope of the present book.

So the modelling of wear must involve contact mechanics problems, and take into account the macro- and microgeometry of the contacting bodies, the inhomogeneity of the mechanical properties of the subsurface layer, and also the fracture mechanics problems used to describe the particle detachment from the surface.

In our opinion, the choice of the fracture criterion is the most difficult problem in modelling, because the processes that cause the wear particle detachment can be of different kinds. This explains the large variety of wear mechanisms.

5.2.2 Fatigue wear

The results of many experimental researches prove that surface fracture can be explained very often by the concept of fatigue, i.e. by the damage accumulation process in cyclic loading. When two rough surfaces move along each other, an inhomogeneous cyclic stress field with high amplitude values of stresses occurs in the subsurface layer, and causes damage accumulation near the surface.

Below we investigate the surface fatigue wear, and use the fatigue damage model developed by Ionov and Ogibalov (1972) and Collins (1981) based on the macroscopic approach. It involves the construction of the positive function $Q(M,t)$, non-decreasing in time, characterizing the measure of material damage at the point M. Failure occurs when this function reaches a threshold level. This concept of fatigue is applied to the investigation of surface failure as well as bulk failure of materials. Moreover, there are experimental data which demonstrate quantitative coincidence of surface and bulk fatigue failure parameters for some materials. For example, Nepomnyashchy showed (see Kragelsky, Dobychin and Kombalov, 1982) that it occurs for some types of rubbers.

However, unlike the catastrophic character of fatigue failure in a mechanical structure, fracture in a wear process may occur again and again. After a fracture event at the instant of time t_*, and removal of the wear debris, the remaining part of the material, characterized by the known damage distribution function $Q(M, t_*)$, comes into contact again, i.e. the material has in itself traces of the process history. This circumstance leads to several specific features of the fatigue wear process which will be investigated in § 5.3 and § 5.4.

There are many different physical approaches to the damage concept in which the damage accumulation rate $\dfrac{\partial Q(M,t)}{\partial t}$ is considered as a function of the stress field and other parameters, depending on the fracture mechanism, the kind of the material and so on. In what follows we will use two different functions: the power dependence of $\dfrac{\partial Q(M,t)}{\partial t}$ on the amplitude stresses at the given point, which is based on Wöhler's curve; and the dependence which is based on the thermokinetics strength theory developed by Regel, Slutsker and Tomashevsky (1974). The latter approach also allows us to analyze the effect of temperature on the wear process.

5.3 Delamination in fatigue wear

5.3.1 The model formulation

We consider the wear of a half-space which is acted upon by a cyclic surface loading. The oscillating undersurface stress field causes a damage accumulation process. We assume that the rate of damage accumulation $q = \dfrac{\partial Q}{\partial t} \geq 0$ is a function of the amplitude value of the load $P(t)$, and the distance Δz from the surface of the half-space to a given point. Since the stress field vanishes at infinity,

$$\lim_{\Delta z \to +\infty} q(\Delta z, P) = 0$$

We introduce a stationary coordinate system $0xyz$, with its origin at the half-space surface at the initial time $t = 0$, the z-axis directed into the half-space, and the x- and y-axis along the half-space surface.

It will be shown below that in the wear process under consideration the z-coordinate of the surface changes due to wear, and it is a monotonically increasing piecewise continuous function of time $Z(t)$, where $Z(0) = 0$. For each time interval $[t_n, t_{n+1}]$ $(n = 0, 1, 2, \ldots)$ $Z(t)$ is continuous, and we can determine the damage accumulation function by the equation $(z > Z(t))$

$$Q(z,t) = \int_{t_n}^{t} q(z - Z(t'), P(t'))\, dt' + Q_n(z), \tag{5.1}$$

where $Q_n(z) = Q(z, t_n)$, $0 \leq Q_n(z) < 1$. Failure occurs at the point z^* at the instant of time t^*, $\left(z^* \in [Z(t^*), +\infty)\right)$ if the following condition is satisfied

$$Q(z^*, t^*) = 1. \tag{5.2}$$

We investigate the wear process from the initial time $t_0 = 0$. It follows from Eqs. (5.1) and (5.2) that the failure process is determined by the functions $q(\Delta z, P)$ and $Q_0(z)$ which are assumed to be continuous. If $q(z - Z(t), P)$ and $Q_0(z)$ are monotonically decreasing function with respect to z, i.e. $\dfrac{\partial q}{\partial z} < 0$ and $\dfrac{dQ_0}{dz} \leq 0$, the condition (5.2) is satisfied at the surface $z = Z(t)$ beginning from the time $t = t_1$, which is determined on the basis of the condition $\int\limits_0^{t_1} q(0, P(t'))\, dt' + Q_0(0) = 1$, and $Z(t) = 0$ for $t \leq t_1$. We shall term the resulting continuous change of the body linear dimension $z = Z(t)$ *the surface wear.*

If one of the functions $q(z - Z(t), P)$ and $Q_0(z)$ (or both of them) is not monotone with respect to z but rather has, for example, a maximum at some distance from the half-space surface, the condition (5.2) may be satisfied at the internal point $z = Z_1$ of the half-space at the time instant t_1. In this case *subsurface fracture* which is a separation of a layer of thickness $\Delta Z_1 = Z_1$ occurs. At subsequent instants of time continuous change of the linear dimensions $Z(t)$ $(t > t_1)$ will occur as a result of surface wear. For determination of the further course of the process, $t > t_1$, we examine the function $Q(z, t)$ (5.1) for $z > Z_1$ as in the previous step, etc. We may obtain the next subsurface failure at the time instant t_n at the point $Z_n = Z(t_n + 0)$, $(n = 2, 3, \ldots)$. The thickness of the layer which is separated is determined by the relation

$$\Delta Z_n = Z(t_n + 0) - Z(t_n - 0).$$

Hence, $Z(t)$ is a piecewise continuous function in this case.

5.3.2 Surface wear rate

We can determine the surface wear rate $\dfrac{dZ(t)}{dt}$ in each interval (t_n, t_{n+1}) where the function $Z(t)$ is continuous. To this end, we obtain the equation for determination of the function $\dfrac{dt}{dZ}$. Since $Q(Z(t), t) \equiv 1$ then

$$\frac{dt}{dZ} = -\frac{\partial Q}{\partial Z} \Big/ \frac{\partial Q}{\partial t} \tag{5.3}$$

or, considering the values of the derivatives $\dfrac{\partial Q}{\partial Z}$ and $\dfrac{\partial Q}{\partial t}$ along the line $t = t(Z)$ based on Eq. (5.1)

$$\frac{\partial Q}{\partial Z} = \int\limits_{Z_n}^{Z} \frac{\partial q(Z - Z', P(Z'))}{\partial Z} \cdot \frac{dt}{dZ'} dZ' + \frac{\partial Q_n(Z)}{\partial Z},$$
$$\frac{\partial Q}{\partial t} = q\Big(0, P\big(t(Z)\big)\Big) \tag{5.4}$$

we obtain the following integral equation for determination of the function $\dfrac{dt}{dZ}$ in interval $[t_n, t_{n+1}]$

$$\frac{dt}{dZ} = -\frac{1}{q(0, P(t(Z)))} \left[\int\limits_{Z_n}^{Z} \frac{\partial q(Z - Z', P(Z'))}{\partial Z} \cdot \frac{dt}{dZ'} dZ' + \frac{\partial Q_n(Z)}{\partial Z} \right] \tag{5.5}$$

Eq. (5.5) for $P(t) = P_0$ (P_0 is a constant) is a Volterra integral equation of the second kind which can be solved using the Laplace transformation. Detailed discussion of this question is in the paper by Goryacheva and Checkina (1990).

Thus, if we know the functions $q(z, P)$, $Q_n(z)$, it is possible to describe the kinetics of the surface wear.

As an example, we consider here the wear process described by the monotonically decreasing function $q(z, P)$

$$q(z, P) = \left(\frac{P}{P_*}\right)^N \frac{1}{T_*} \exp\left(-\frac{z}{a(P)} N\right),$$

where P_* and T_* are the characteristic load and time, $a(P) > 0$ is a quantity having the dimension of length and depending on the load P, N is a constant ($N > 0$). We assume also that $Q_0(z) = 0$, and that the load $P(t)$ is the step function

$$P(t) = \begin{cases} P_0, & t < t_1, \\ P_1, & t \geq t_1, \end{cases}$$

where $t_1 = T_* \left(\dfrac{P_*}{P_0}\right)^N$. For $t = t_1$ the function $Q_1(z) = Q(z, t_1)$ can be obtained from Eq. (5.1)

$$Q_1(z) = \exp\left(-\frac{z}{a\,(P_0)} N\right)$$

We use the Laplace transformation method to determine the surface wear rate for $t > t_1$. The function $q(z, P)$ has the Laplace transform with respect to z

$$\mathcal{L}\left[q(z, P); p\right] = \left(\frac{P}{P_*}\right)^N \frac{1}{T_*} \cdot \frac{p}{p + \dfrac{N}{a(P)}}, \tag{5.6}$$

where

$$\mathcal{L}\left[f(z);p\right] \equiv \int_0^{+\infty} \exp(-pz) f(z)\,dz.$$

Using Eqs. (5.5) and (5.6), we obtain

$$\mathcal{L}\left[\frac{dt}{dZ};p\right] = \frac{NT_*}{a(P_0)}\left(\frac{P_*}{P_1}\right)^N \frac{p+\dfrac{N}{a(P_1))}}{p\left(p+\dfrac{N}{a(P_0)}\right)}$$

Using the inverse transformation, we have

$$\frac{dt}{dZ} = \frac{NT_*}{a(P_0)}\left(\frac{P_*}{P_1}\right)^N \left[\frac{a(P_0)}{a(P_1)} - \left(\frac{a(P_0)}{a(P_1)} - 1\right)\exp\left(-\frac{N}{a(P_0)}Z\right)\right].$$

Integrating this relation, we find the dependence of the surface coordinate Z on the time t

$$t = \frac{NT_*}{a(P_1)}\left(\frac{P_*}{P_1}\right)^N Z - \left(\frac{a(P_0)}{a(P_1)} - 1\right)\left(\frac{P_*}{P_1}\right)^N T_*\left(1 - \exp\left(-\frac{N}{a(P_0)}Z\right)\right),$$

from which it follows that the surface wear rate in this case has the limit

$$\lim_{t\to+\infty}\frac{dZ}{dt} = \frac{a(P_1)}{NT_*}\left(\frac{P_1}{P_*}\right)^N,$$

i.e., the steady-state wear rate is independent of the initial load P_0.

This result shows that for monotonically decreasing (with respect to z) function $q(z,P)$ and $Q_0(z) = 0$ $\left(\text{or } \dfrac{\partial Q_0}{\partial z} \le 0\right)$ only surface wear occurs.

5.3.3 Wear kinetics in the case $q(z,P) \sim \tau_{\max}^N$, $\quad P = \text{const}$

In case of a complex stress state, the fatigue damage accumulation rate is usually associated with the values of the equivalent stresses (for example, principal shear or tensile stresses), that are responsible for the damage mode under examination (Pisarenko and Lebedev, 1976, Collins, 1981).

We consider here the following relation for the function $q(z,P)$

$$q(z,P) = \left(\frac{\tau_{\max}(z,P)}{\tau_*}\right)^N \frac{1}{T_*}. \tag{5.7}$$

where $\tau_{\max}(z,P)$ is the amplitude value of the principal shear stress at the given depth z. Values of τ_*, T_* and N can be determined in special frictional fatigue tests, for example, by the method described below.

We suppose further that the oscillating stress field in an elastic half-space is caused by sliding of a periodic system of indenters. The analysis of internal stresses in periodic contact problems for an elastic half-space for different values of friction coefficient and density of contact spots presented by Kuznetsov and Gorokhovsky (1978a and 1978b) (in two-dimensional formulation), and also in Chapter 2 shows that the cases of monotone and nonmonotone function $\tau_{\max}(z)$ actually take place and, consequently, the fatigue wear features which follow from analysis of Eqs. (5.1) for different functions $q(z, P)$ given in § 5.3.1 are realistic.

In what follows we consider a system of spherical indenters sliding along the surface of the half-space. We assume that the distance between indenters is sufficiently large that they do not influence each other. The model can be applied to analysis of the fatigue wear of an elastic half-space which is in frictionless contact with a moving wavy surface.

Using the relationship for the amplitude value $\tau_{\max}$ at the fixed depth z (Hamilton and Goodman, 1966) and Eq. (5.7) we can write

$$q(z,P) = \left(\frac{P}{P_*}\right)^{N/3} \frac{1}{T_*} q_1\left(\frac{z}{a(P)}\right),$$

$$q_1(\zeta) = \left| C - \frac{3}{2\,(1+\zeta^2)} - C\zeta \arctan\left(\frac{1}{\zeta}\right)\right|^N. \tag{5.8}$$

$$C = 1+\nu, \quad P_* = \frac{4\pi^3\tau_*^3 R^2}{3E_*^2}, \quad a(P) = \sqrt[3]{\frac{3}{4}\cdot\frac{P}{E_*}R}, \quad \frac{1}{E_*} = \frac{1-\nu_1^2}{E_1} + \frac{1-\nu_2^2}{E_2}.$$

The specific feature of the function $q(z, P)$ which determines the wear process is its nonmonotone character (the presence of maximum at the depth $\zeta = 0.48$). This function satisfies also the condition $\lim_{z\to+\infty} q(z,P) = 0$.

It is supposed for the contact under consideration that the damage $Q(z, P)$ at each instant t is the same at all half-space points at the fixed depth z. Thus fracture of the half-space has a delamination character, and the contact geometry does not change during the wear process.

If $P(t) = \text{const}$, in the dimensionless coordinates

$$\zeta = \frac{z}{a(P)}, \quad \theta = \frac{t}{T_*}\left(\frac{P}{P_*}\right)^N,$$

the function $Q(\zeta, \theta)$ does not depend on the load. Consequently, the influence of the load magnitude on the damage process shows up only in the choice of time and distance scale (in accordance with coordinate transformation above).

The kinetics of the process described by Eqs. (5.1), (5.2) and (5.8) were studied numerically. The function $Q(\zeta, \theta)$ is shown in Fig. 5.3 at various instants of time for $N = 5$, $Q_0(z) = 0$. Before the first fracture at the instant θ_1, the curve $Q(\zeta, \theta)$ has the characteristic form (I) with a subsurface maximum point. After the first subsurface fracture at the point ζ_1, $Q(\zeta, \theta)$ has the form of the monotone function

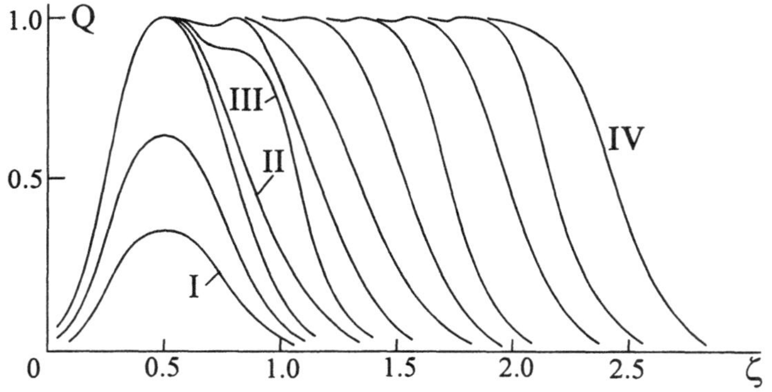

Figure 5.3: Damage accumulation function Q in wear process under a constant load ($N = 5$).

(II) with its maximum at the surface. The surface wear process occurs at this stage. In the course of damage accumulation, an inflection point appears at some depth (III), as the function $q(z, P)$ is nonmonotone with respect to z. When, at the instant θ_2, the subsurface maximum value is equal to unity, $Q(\zeta_2, \theta_2) = 1$, the next subsurface fracture occurs at the point ζ_2 and so on. Then the subsurface fracture terminates, and the surface wear rate approaches a constant value. The curve $Q(\zeta, \theta)$ now takes on the form that is characteristic for the steady-state surface wear (IV).

Fig. 5.4 illustrates the wear process for this case; we show the dependence of the dimensionless surface coordinate $\tilde{\zeta} = Z/a(P)$ on the dimensionless time θ. The instants of subsurface fracture are marked with stars, numbers near the stars show the dimensionless depth $\Delta\tilde{\zeta}_n$ of the detached layer.

Calculations reveal the influence of the exponent N on the process. For $N = 3$, only a single subsurface fracture event occurs. For $N = 5$ six events occur, while for $N = 5.5$ twenty-eight subsurface fracture events occur. However, provided that $P = \text{const}$, there are common features of the fracture processes: monotone diminishing of the detached layer thickness, cessation of subsurface fracture, and transition to the steady-state surface wear with a constant rate.

5.3.4 Influence of the load variations $P(t)$ on wear kinetics

In real contacts, the function $P(t)$ has typical features as a result of the discrete contact area, waviness, periodic character of the loading, etc. We simulate it in a simple manner by a periodic function $P(t)$, and study its influence on the fracture process.

The numerical analysis was made for the function $q(z, P)$ determined by Eq. (5.8), and the damage accumulation law (5.1). In calculations we took $N = 5$ and $C = 1.25$ and introduced the dimensionless functions and variable

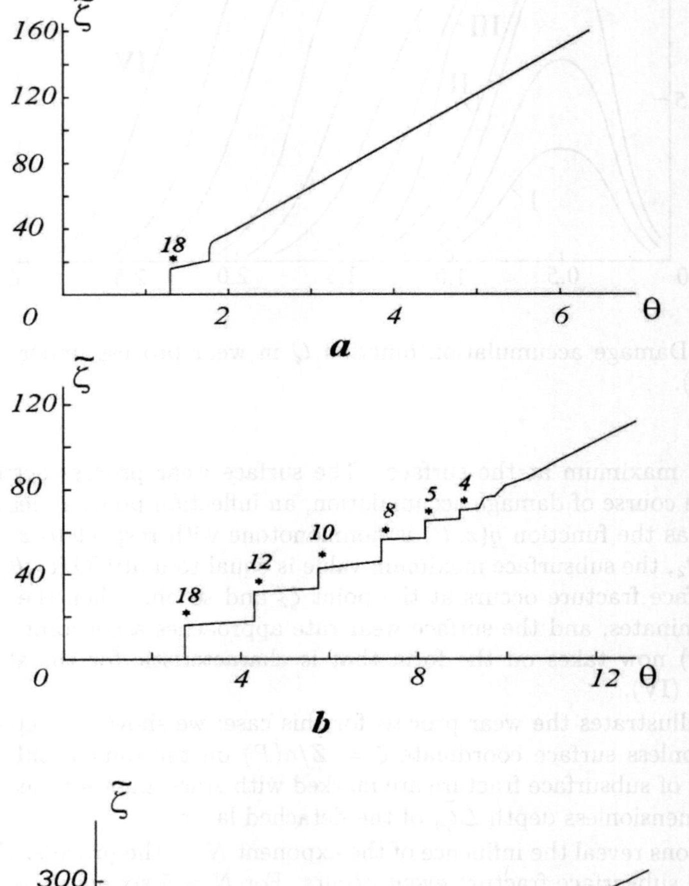

Figure 5.4: Kinetics of wear process $\tilde{\zeta}(\theta)$ under a constant load for $N = 3$ (a), $N = 5$ (b) and $N = 5.5$ (c).

$$\bar{P}(t) = \frac{P(t) \cdot 3^6 E_*^2}{R^2 \tau_*^3}, \quad \bar{w}(t) = \frac{w(t) \cdot 150\sqrt{15} E_*}{R \tau_*}, \quad \bar{t} = \frac{t}{10^9 T_*},$$

where the wear $w(t)$ coincides in magnitude with the surface coordinate $Z(t)$.

At first we consider $\bar{P}(\bar{t})$ as a periodic step function with period $\bar{t}_0$:

$$\bar{P}(\bar{t}) = \begin{cases} 1 - \delta, & \left(k - \dfrac{1}{2}\right) \bar{t}_0 \le \bar{t} < k\bar{t}_0, \\ 1 + \delta, & k\bar{t}_0 \le t < \left(k + \dfrac{1}{2}\right) \bar{t}_0, \end{cases} \qquad k = 0, \pm 1, \pm 2, \ldots \tag{5.9}$$

We have analyzed the influence of δ on the process. The results of calculation are depicted in Fig. 5.5 a, b, c for $\delta = 0.2$, $\delta = 0.5$, $\delta = 0.8$, respectively, $\left(\bar{t}_0 = \dfrac{2}{5}\right)$. From here on we shall show the function $\bar{P}(\bar{t})$ in the upper part of the graphs for clarity. In spite of the fact the average value of $\bar{P}(\bar{t})$ is the same for the three processes , there are qualitative differences between them. For small δ, the process is similar to that for the case $P(t) = \text{const}$, i.e. after several events of subsurface fracture only surface wear occurs with a periodically varying rate. With increase of δ, the subsurface fracture arises. In the course of subsurface fracture, we can easily identify two stages: the initial stage, when the occurrence of fracture is not directly associated with the change of $\bar{P}(\bar{t})$; and the stage when subsurface fracture occurs periodically with some lag in relation to the instant of increase of the function $\bar{P}(\bar{t})$; in the latter case the number of fracture events per period increases with increase of δ.

The calculations, made for the function $\bar{P}(\bar{t})$ (Eq. (5.9)) with fixed δ and different values of the period $\bar{t}_0$ (see Fig. 5.6), made it possible to establish the significant influence of change of the period on the nature of the process. If the period is small, the system does not *sense* the change of the magnitude of $\bar{P}(\bar{t})$ and the subsurface fracture process terminates (Fig. 5.6a). With increase of the period, the subsurface fracture does not terminate (Fig. 5.6b,c); for large periods it has a periodic nature in accordance with the nature of the function $\bar{P}(\bar{t})$ (Fig. 5.6c).

We also studied the time dependence of wear in the cases of different functions $\bar{P}(\bar{t})$, which were characterized by the same limits of variation and characteristic times $\left(\delta = 0.7, \quad \bar{t}_0 = \dfrac{1}{5}\right)$ (see Fig. 5.7). The results show that for the smooth function $\bar{P}(\bar{t}) = 1 + \delta \cos\left(\dfrac{2\pi \bar{t}}{\bar{t}_0}\right)$, the subsurface fracture terminates (Fig. 5.7a), while the step function $\bar{P}(\bar{t})$ yields steady-state subsurface fracture (Fig. 5.7b). Fig. 5.7c illustrates the wear kinetics when $\bar{P}(\bar{t})$ is a piecewise constant random function with uniform distribution within the interval $(0.3, 1.7)$. This situation is closer to the real wear conditions. The results indicate that subsurface fracture does not terminate in this case, and the instants of its arising are correlated with large jumps in $\bar{P}(\bar{t})$.

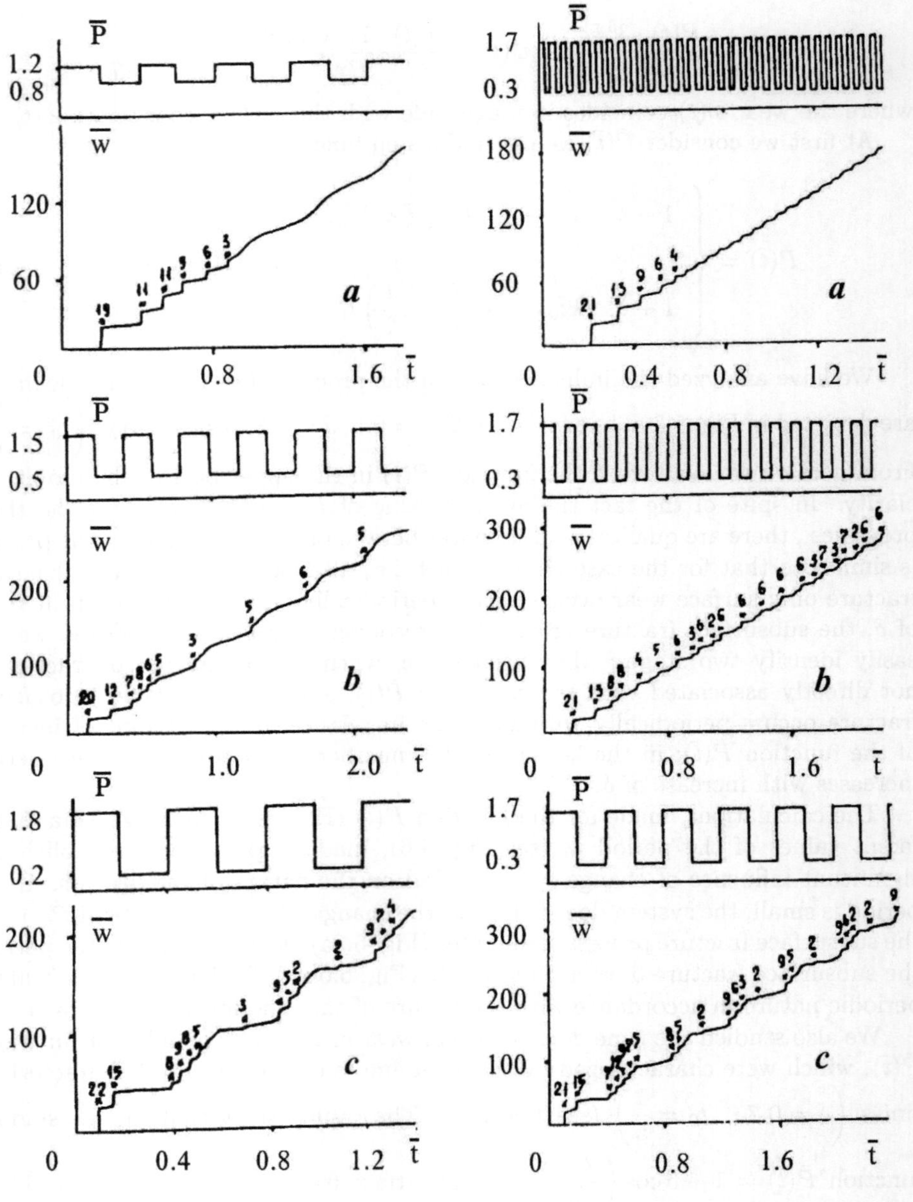

Figure 5.5: Kinetics of wear process for periodic function $\bar{P}(\bar{t})$ (Eq. (5.9)): $\delta = 0.2$ (a), $\delta = 0.5$ (b), $\delta = 0.8$ (c); $\bar{t}_0 = 0.4$, $N = 5$ (left figure).

Figure 5.6: Kinetics of wear process for periodic function $\bar{P}(\bar{t})$ (Eq. (5.9)): $\bar{t}_0 = 1/20$ (a), $\bar{t}_0 = 2/15$ (b), $\bar{t}_0 = 2/5$ (c); $\delta = 0.7$, $N = 5$ (right figure).

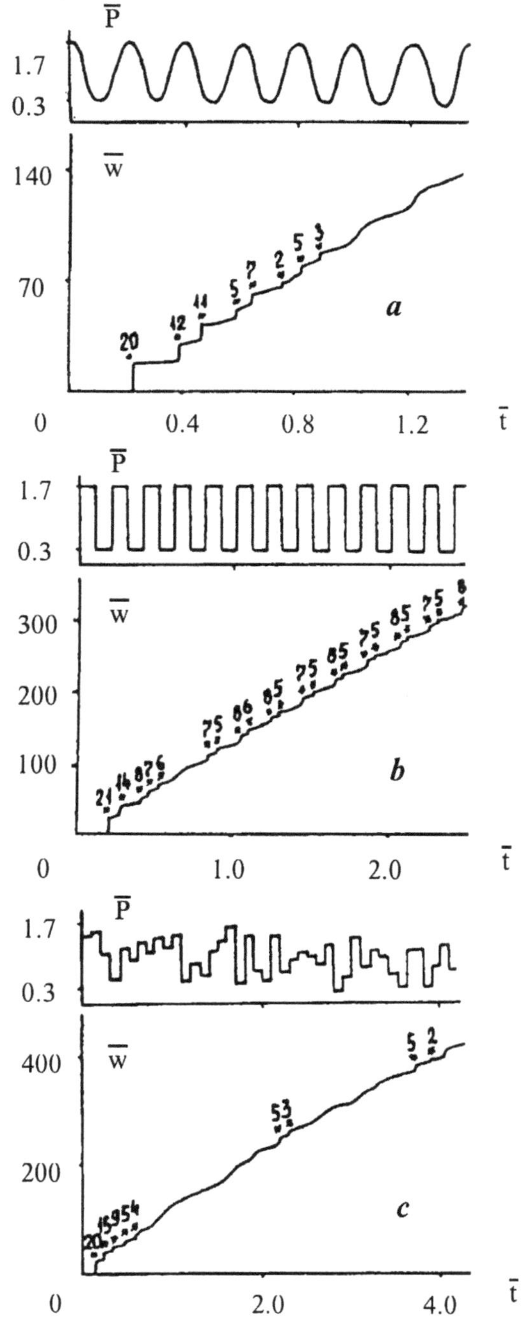

Figure 5.7: Kinetics of wear process for various types of function $\bar{P}(\bar{t})$: sinusoidal function (a), step function (b), random function, uniformly distributed within the interval $(0.3, 1.7)$ (c); $\delta = 0.7$, $N = 5$, $\bar{t}_0 = 1/5$.

$\bar{t}_0$	Wear rate modes		
	I_s	I_{ss}	I_w
1.00	100.0	30.0	130.0
0.40	92.5	45.0	137.5
0.20	70.0	62.5	132.5
0.13	76.0	59.0	135.0
0.04	128.0	0	128.0

Table 5.1: Wear rates for the periodic step function $\bar{P}(\bar{t})$ ($\delta = 0.7$).

5.3.5 Steady-state stage characteristics

Based on the examples considered above, we can conclude that the unsteady fracture process is followed by the steady-state stage in the case of a periodic function $P(t)$. The steady-state stage can be described by the practically important characteristics: the average rates of surface I_s, subsurface I_{ss}, and total I_w wear; each of these are determined as $\frac{\Delta\bar{w}}{\bar{t}_0}$, where $\Delta\bar{w}$ is the change of the body linear dimension as a result of wear of the given mode over the period $\bar{t}_0$.

Table 5.1 illustrates the wear rates calculated for the periodic step function $\bar{P}(\bar{t})$ (Eq. (5.9)) ($\delta = 0.7$). The results show that, even though the ratios of the rates I_s and I_{ss} in the examined cases are different, the quantities I_w differ only moderately (by less than 10%); the minimum total wear rate is reached when there is no subsurface fracture. The following arguments show that these results are quite legitimate.

We define the total damage $\Omega(t)$ accumulated by material at an instant t as $\Omega(t) = \int\limits_{Z(t)}^{+\infty} Q(z)\,dz$ where $Z(t)$ is the surface coordinate at the instant t.

Change of $\Omega(t)$ over the time interval Δt occurs, on the one hand, due to detachment of damaged material $\Delta\Omega_1 = \Delta Z \cdot Q_{av}$ where Q_{av} is an average damage (over the time interval Δt) of detached material, ΔZ is its thickness. On the other hand, $\Omega(t)$ increases during Δt due to the damage accumulation process

$$\Delta\Omega_2 = \int\limits_{t}^{t+\Delta t}\int\limits_{0}^{+\infty} q(z, P(t))\,dz\,dt$$

If $P(t)$ is a periodic function with period t_0, then in the steady-state stage of the wear process

$$\Omega(t+t_0) - \Omega(t) = -\Delta\Omega_1 + \Delta\Omega_2 = -\Delta Z \cdot Q_{av} + \int\limits_{t}^{t+t_0}\int\limits_{0}^{+\infty} q(z, P(t))\,dz\,dt = 0. \quad (5.10)$$

Averaging $q(z, P(t))$ over the period, $\overline{q}(z) = \frac{1}{t_0}\int\limits_t^{t+t_0} q(z, P(t))\,dt$, we obtain from Eq. (5.10)

$$\frac{\Delta Z}{t_0} = \frac{1}{Q_{av}}\int\limits_0^{+\infty} \overline{q}(z)\,dz. \tag{5.11}$$

If there is only surface wear ($Q_{av} = 1$), the average total wear rate, determined from Eq. (5.11) has the minimum value. If a subsurface fracture occurs also, then $Q_{av} < 1$, and the average total wear rate is higher than in the case of pure surface wear. As we can see from Fig. 5.3, there is no great difference between Q_{av} and 1 for the process described by the damage accumulation rate function (5.8). Hence in a steady-state stage, the total wear rates in the presence or absence of subsurface fracture do not greatly differ. All these conclusions agree with the results presented in Table 5.1.

Consequently, if the wear process tends to the steady-state one, the total wear rate can be evaluated based on the function $q(z, P(t))$ and Eq. (5.11) without examining the kinetics of the process. We note that for $P(t) =$ const, $\overline{q}(z) = q(z, P)$, and Eq. (5.11) for the steady-state surface wear takes the form

$$\frac{dZ}{dt} = \int\limits_0^{+\infty} q(z, P)\,dz \tag{5.12}$$

On the basis of this relation we can determine the steady-state surface wear rate for the example that was considered in § 5.3.2.

5.3.6 Experimental determination of the frictional fatigue parameters

Based on the model under consideration, we can propose a method for the experimental determination of frictional fatigue parameters in Eq. (5.8).

In pin-disk experiments, a dependence $w(t)$ similar to one represented in Fig. 5.4, can be obtained for given values of radius R and load P. If there is a qualitative coincidence between the wear process for the material tested and the model wear process, we can determine values N and $(P_*)^{N/3}T_*$ in Eq. (5.8) based on the analysis of the characteristics of the steady-state stage of the wear process.

For this purpose the wear rate under the constant load P must be examined. It was shown in §5.3.3 that, in the steady-state stage, only the surface wear occurs with a constant rate $Z' = \frac{dZ}{dt}$ which can be determined from Eqs. (5.8) and (5.12)

$$Z' = \left(\frac{P}{P_*}\right)^{N/3}\frac{1}{T_*}\left(\frac{3PR}{4E_*}\right)^{1/3}\int\limits_0^{+\infty} q_1(\zeta)\,d\zeta \tag{5.13}$$

The wear rates Z_1' and Z_2' for two different loads P_1 and P_2, respectively, for the steady-state stage of the process are measured experimentally. Then we obtain from Eq. (5.13)

$$N = 3\frac{\ln(Z_1'/Z_2')}{\ln(P_1/P_2)} - 1$$

The value $(P_*)^{N/3}T_*$ involved in Eq. (5.8) can be calculated from Eqs. (5.8) and (5.13).

A similar method can be applied for other functions $q(z,P)$ used to describe the damage accumulation process in the test under different conditions.

Thus, in the model described here we have identified two types of fracture process which may occur depending on the character of function $q(z,P)$: continuous surface wear (when the failure condition is satisfied only at the surface) or continuous surface wear which is accompanied by detachment of layers of some thickness at discrete instants (delamination).

Within the framework of the model it is possible to determine numerically such characteristics of the process as the rate of surface, subsurface and total wear, instants of delamination, thickness of layers detached from the surface, etc.

It is shown that for the periodic function $P(t)$ (in particular for $P(t) = \text{const}$) the process of wear may be divided into two stages: initial and steady-state. For the steady-state stage of wear an integral evaluation of total wear rate may be obtained analytically if $q(z,P)$ is known.

The conclusions relating to the discontinuous nature of the fatigue failure of surfaces is qualitatively confirmed by the results (Cooper, Dowson and Fisher, 1991) on testing polymer material used for artificial joints (Ultra-high molecular weight polyethylene) in contact with a steel pin (pin-disk tests) and by experimental results (Kragelsky and Nepomnyashchy, 1965) on the frictional fatigue for a set of different rubbers, the scheme of the test was similar to the previous one.

5.4 Fatigue wear of rough surfaces

The model of fatigue wear described in the previous section allows us to simulate some important features of this type of surface fracture. However, the assumptions adopted in the model formulation provide essential limitations on the geometry of the bodies in contact and the shape of the wear fragments. Hence, only one particular case of surface fracture – delamination – can be described by this model.

The development of fatigue wear models for the contact of rough surfaces without any *a priori* assumptions on the shape of detached particles is of great importance. One such model (Checkina, 1996) is presented in this section. Consideration of successive acts of particle detachment within the framework of the model provides a means for investigating microgeometry change in wear process.

5.4.1 The calculation of damage accumulation on the basis of a thermokinetic model

We will use an approach to fatigue wear modelling similar to one described in § 5.2.2. The first stage of the model construction is the calculation of the damage field inside the body as a function of time. Since the processes leading to fatigue wear take place in the subsurface layers, they are greatly affected by friction between the contacting surfaces. Friction influences the stress field, and also causes frictional heating of surface layers. To take into account both this factors, we will use a thermokinetic model (Regel, Slutsker and Tomashevsky, 1974) for description of the damage accumulation process in subsurface layers.

In accordance with this model, the rate of damage accumulation is given by the relation

$$\frac{dQ(x,y,z,t)}{dt} = \frac{1}{\tau_*} \exp\left(-\frac{U - \gamma\sigma(x,y,z,t)}{kT(x,y,z,t)}\right), \tag{5.14}$$

where U is the activation energy, τ_* and γ are the material characteristics, k is the Boltzman coefficient, and $\sigma(x,y,z,t)$ is a characteristic of the stress field at the point (x,y,z) within the deformable body at an instant t. Using various stress field characteristics $\sigma(x,y,z,t)$ in Eq. (5.14), we can reproduce different types of fracture. The value of the principal shear stress τ_1 is used as the stress field characteristic in what follows.

Note that the thermokinetic mechanism of damage accumulation implies consideration of the thermal effects in an explicit form. The temperature field $T(x,y,z,t)$ in subsurface layers is essentially nonhomogeneous, hence, its calculation must be carried out with high accuracy (it is not possible to use averaged temperature characteristics).

We will consider the damage accumulation process in contact interaction of two bodies for the case of a 2-D periodic problem (Fig. 5.8). The system of coordinates Oxz is fixed at one of the contacting bodies, and the z-axis is directed inside the body. The shapes of the bodies $f_1(x)$ and $f_2(x)$ are periodic functions (with the same period l) which are represented by Fourier series.

The body 1 slides with the constant velocity V along the surface of the body 2 and is acted by the vertical force P per period. To calculate the internal stress and the temperature fields we first solve the contact problem for different relative positions of the bodies. The solution of the plane periodic contact problem obtained by Staierman (1949) and Kuznetsov (1976, 1985) is used to calculate the contact pressure. The relation for the contact pressure for complete contact (all points of the surfaces are in contact) can be derived from the solution

$$p(x) = \frac{E^*}{2l} \sin\left(\frac{\pi x}{l}\right) \int_0^l \frac{f'(x')\,dx'}{\sin\left(\frac{\pi x'}{l}\right) \sin\left(\frac{\pi(x'-x)}{l}\right)} + p_0 \tag{5.15}$$

where $E^* = \left(\frac{1-\nu_1^2}{E_1} + \frac{1-\nu_2^2}{E_2}\right)^{-1}$, p_0 is a constant calculated from the equilibri-

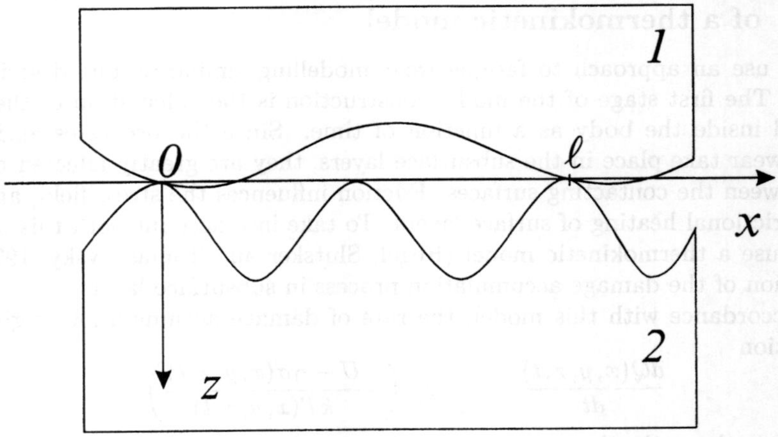

Figure 5.8: Shape of bodies in contact.

um condition

$$\int_0^l p(x)dx = P, \tag{5.16}$$

$f(x)$ is the equivalent shape of contacting surfaces at any fixed instant t: $f(x) = f_2(x) - f_1(x - Vt)$ $\left(f'(x) = \dfrac{df(x)}{dx}\right)$.

In fact, contact of rough bodies is discrete and for the calculation of the contact pressure on the basis of Eq. (5.15) we use a step by step numerical procedure. The pressure $p_i(x)$ at the i-th step is calculated by the formula

$$p_i(x) = \frac{E^*}{2l}\sin\left(\frac{\pi x}{l}\right)\int_0^l \frac{\tilde{f}_i'(x')\,dx'}{\sin\left(\dfrac{\pi x'}{l}\right)\sin\left(\dfrac{\pi(x'-x)}{l}\right)} + p_{0i}, \quad i = 0, 1, 2, \ldots$$

where

$$\tilde{f}_{i+1}(x) = f(x) \quad \text{if} \quad p_i(x) > 0,$$

$$\tilde{f}_{i+1}(x) = \tilde{f}_i(x) + \alpha p_i(x) \quad \text{if} \quad p_i(x) < 0,$$

$$\tilde{f}_0(x) = f(x),$$

and p_{0i} satisfies Eq. (5.16) at the i-th step. The calculation is completed if the minimum pressure is greater than some negative parameter of small absolute value. This procedure converges to the function which is the contact pressure within the contact zones and zero outside the contact zones. Simultaneously, the function $\tilde{f}_i(x)$ tends to the elastic displacement of the boundary of the bodies within and outside the contact zone. The convergence of the procedure is provided by appropriate choice of α.

The stress and temperature fields inside the body 2 are then calculated for the geometry of contact shown in Fig. 5.8. Since asperities of the surfaces usually are rather sloping we investigate the case when the ratio of the height of the asperity to the period is ≈ 0.03. So the Green functions for a homogeneous half-plane (Johnson, 1987) can be used for the calculation of the internal stresses.

The temperature field inside the body 2 is calculated by solving the two-dimensional problem of heat conductivity (Carslaw and Jaeger, 1960). The following boundary conditions at the surface ($z = 0$) are used

- inside the contact zones:

$$-\lambda \frac{\partial T(x,z,t)}{\partial z} = q^T(x,t),$$

- outside the contact zones:

$$-\lambda \frac{\partial T(x,z,t)}{\partial z} = 0,$$

where λ is the heat conductivity coefficient for the body 2. The temperature at some depth inside the body is supposed to be equal to the environmental temperature T_0, the heat flow at the side boundaries $x = 0$, $x = l$ is obtained from the periodicity condition. The heat flow $q^T(x,t)$ into unmovable body 2 is described by the relation

$$q^T(x,t) = K\mu p(x,t)V,$$

where K is the coefficient of distribution of thermal flux, μ is the friction coefficient, $p(x,t)$ is the contact pressure at an instant t. The heat conductivity problem was solved numerically using a step-by-step procedure.

Then the damage accumulation function $Q(x,z,t)$ is calculated from Eq. (5.14). In summary, the following parameters determine the damage accumulation process:

- the initial shape of the bodies in contact: $f_i(x)$;
- the dependence of external load on time: $P(t)$;
- the elastic characteristics of the bodies: E_i, ν_i;
- the parameters of the heat conductivity equation for the body 2: heat capacity C, heat conductivity λ, density ρ;
- the characteristics describing the damage accumulation inside the body 2: U, γ, τ_*;
- the coefficient of distribution of thermal flux: K;
- the friction coefficient: μ;

- the sliding velocity: V;
- the environmental temperature: T_0.

From a tribological standpoint, it is of interest to investigate the influence of such parameters as the friction coefficient μ and the coefficient of distribution of thermal flux K on the damage field in the subsurface layers. Fig.5.9 shows the damage field in the instant when fracture arises ($\max\limits_{x,z} Q(x,z,t) = 1$) for the cases (a) $\mu = 0$, $K = 0$; (b) $\mu = 0.2$, $K = 0$; (c) $\mu = 0.2$, $K = 1$; (d) $\mu = 0.5$, $K = 0$ and the trajectories of cracks that will be discussed in § 5.4.2. The level curves of the function $Q(x,z,t)$ are represented by solid lines. The point where fracture begins ($Q(x^*,z^*,t^*) = 1$) is indicated by the dark mark. We can see that for small values of μ and K (Fig. 5.9(a) and Fig. 5.9(b), the critical value of $Q(x,z,t)$ is reached under the surface. On an increase of the friction coefficient (Fig.5.9(d)) or the coefficient of distribution of thermal flux (Fig.5.9(c)), the maximum of damage takes place at the surface. Hence, surface wear described in § 5.3.1 occurs at first. We can see, however, that local subsurface maxima also occur for these cases, that can arise at some instant in the subsurface fracture. We can conclude that, since the subsurface maximum is small for the case (d), the probability of the subsurface fracture in this case is small. If the surface of the body 2 is flat, the damage field inside this body depends only on the z-coordinate and the fracture process has the character of delamination similar to the case considered in § 5.3.

To complete the model of the fatigue wear, we must propose some fracture mechanism for the body with a nonhomogeneous damage field.

5.4.2 Particle detachment

The process of particle detachment arises from crack propagation; the typical time of the process and the trajectory of the crack depend on the type of the material, its stress-strain state and damage field.

We consider here the simplest mechanism of fatigue fracture that allows us to model repeated particle detachment. We assume that the fracture crack propagation occurs at the instant t^* when the damage function reaches a fixed critical value at some point (x^*,z^*). Since (x^*,z^*) is the point of maximum of the function $Q(x,z,t)$, the level curves of this function near this point are ellipses. If the eccentricity of the ellipse is not equal to zero, we assume that crack propagates in the direction of its major axis. The direction θ_i of the crack at any point (x_i,z_i) of its trajectory passing through the points of the mesh is chosen from the condition (see Fig. 5.9)

$$\max_{\theta} Q(x_i + \epsilon\cos\theta, z_i + \epsilon\sin\theta) = Q(x_i + \epsilon\cos\theta_i, z_i + \epsilon\sin\theta_i)$$

where $\theta \in (-\frac{\pi}{2} + \theta_{i-1}, \frac{\pi}{2} + \theta_{i-1})$.

Fig. 5.9 shows the trajectories of cracks at the instant of first fracture calculated on the basis of this approach. The direction of the crack propagation from the

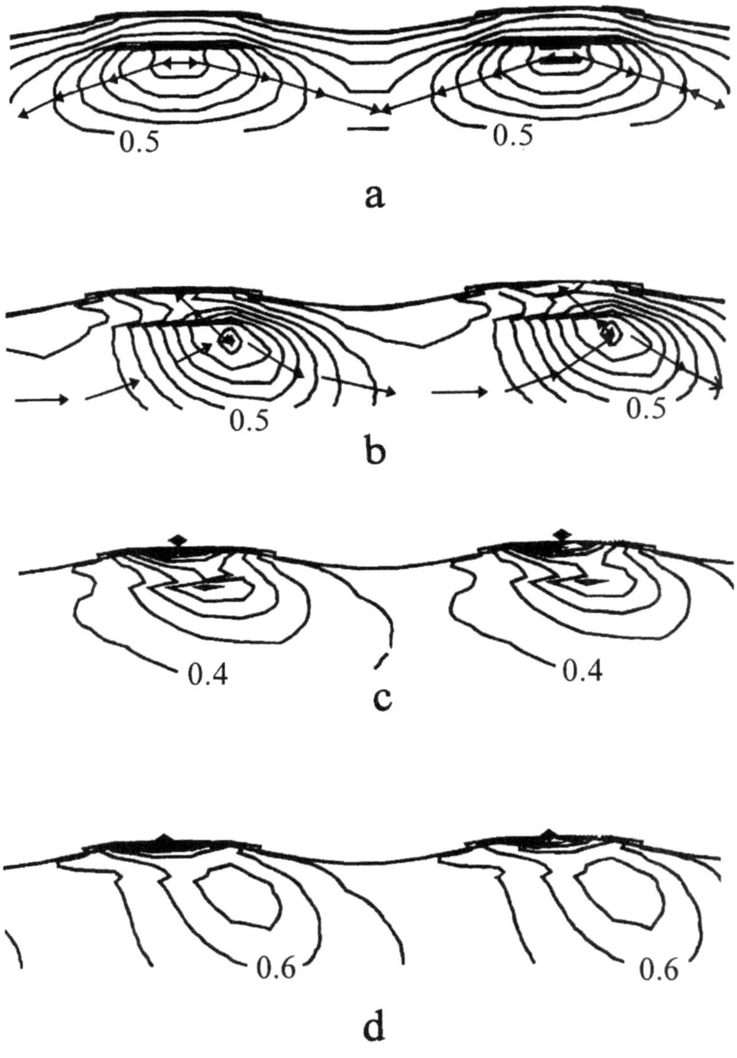

Figure 5.9: Damage field in the instant when fracture arises ($\max_{x,z} Q(x,z,t) = 1$): (a) $\mu = 0$, $K = 0$; (b) $\mu = 0.2$, $K = 0$; (c) $\mu = 0.2$, $K = 1$; (d) $\mu = 0.5$, $K = 0$.

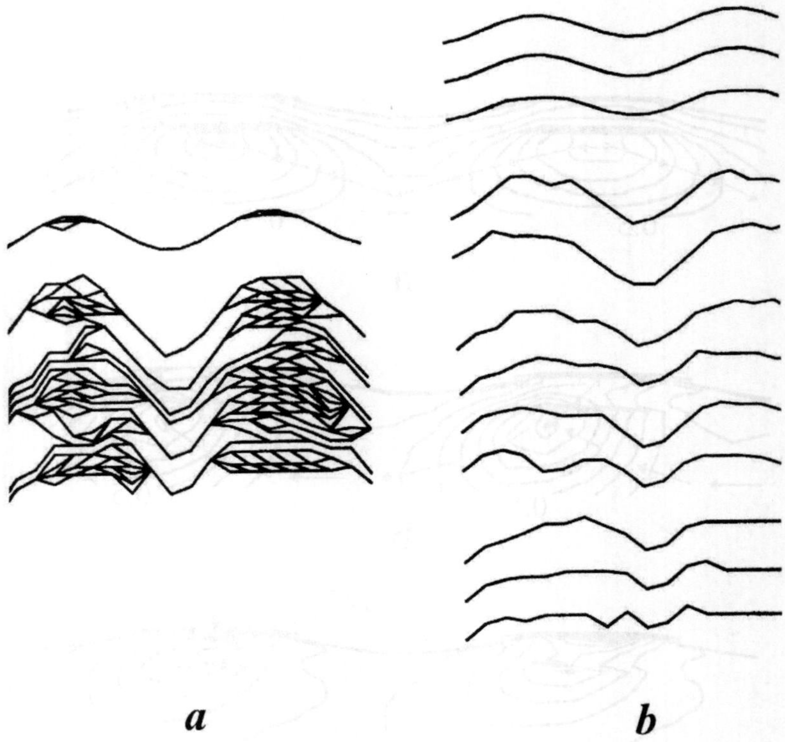

Figure 5.10: The shape of wear particles (a) and change of the surface profile (b) in wear process ($\mu = 0.2,\ \ K = 1$).

point where critical value of Q is reached is shown by arrows. Due to the periodicity of the body shape, crack propagation arises simultaneously at each period. As a result of this process, fragments of materials are detached. In case (a) in Fig. 5.9 fracture has the character of delamination.

We consider the process of crack propagation as an instantaneous one. It is known that the relation of the time of fatigue crack propagation to the time of damage accumulation is different for different materials. The consideration of the process as an instantaneous one, naturally reduces the class of materials which are described by this model. However, from our point of view, this assumption is inevitable. Really, the stress field calculation for a body containing even one crack of an arbitrary shape is an extremely laborious procedure and the allowance of *slow* crack propagation in wear modelling would make calculation impossible.

The assumption that the direction of crack propagation coincides with the direction in which a function determining the fracture criterion diminishes the most slowly was used before. For example, Sikarskie and Altiero (1973) used a

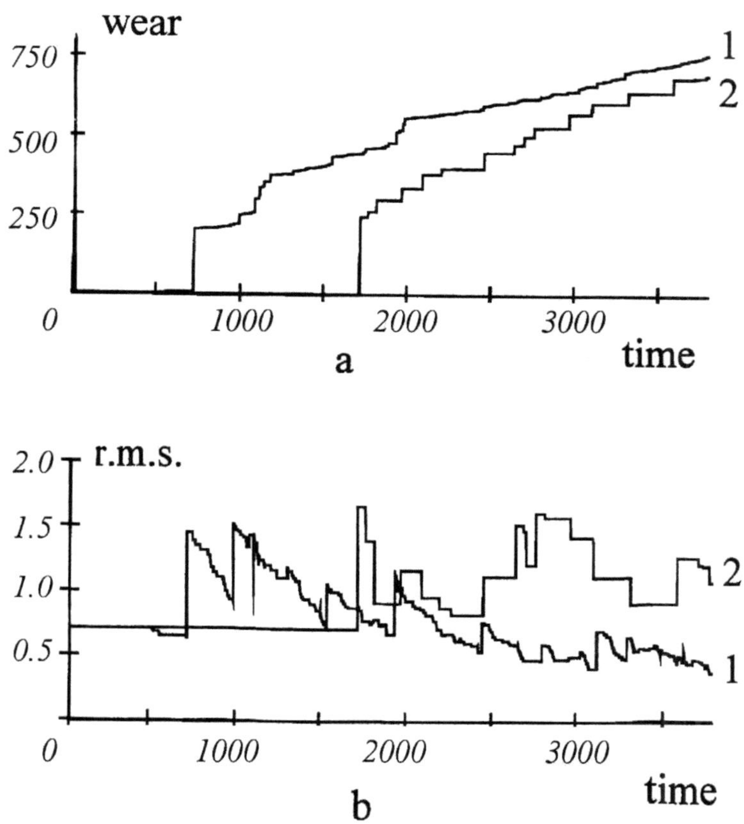

Figure 5.11: Characteristics of wear process: total wear (a) and the root-mean-square deviation of the profile (b) vs.time for $\mu = 0.2$ and $K = 1$ (curve 1), $K = 0$ (curve 2).

similar assumption in modelling brittle fracture of elastic materials.

Each act of fracture causes a change in the surface microgeometry. The modified surface shape is used for further calculations.

5.4.3 The analysis of the model

The proposed model allows us to describe the wear process, and to determine its characteristics: wear intensity, size and shape of wear particles, surface shape variation in wear process.

An example of the calculation is shown in Fig. 5.10 and Fig. 5.11. Fig. 5.10 illustrates the wear process for parameters μ=0.2, K=1 (the damage field for this case is shown in Fig.5.8(c)). We can see from this figure that both surface fracture (small particles detachment) and subsurface one (large particle detachment) occur in this case. The root-mean-square deviation of the profile, and the total wear are shown in Fig.5.11 for $\mu = 0.2$ and two values of the parameter K.

The analysis of the model shows that the size of the particles depends substantially on the friction coefficient. Its change influences both the stress field and the temperature field. The increase of the friction coefficient causes an increase in heat generation, and also in the displacement toward the surface of the point where the value of the principal shear stress is maximal. Both these factors lead to decreases in the depth of fracture and in the thickness of detached particles. Surface wear similar to one described in § 5.3 can appear as a limiting case.

The influence of the contact pressure on the size of particles is not so unidirectional. The increase of the contact pressure causes an increase in the contact spot size, and, hence, increase of the depth of fracture initiation (remember that we use the principal shear stress as the characteristic that determines damage accumulation), on the other hand, it stimulates heating of the surface layers that leads to a decrease in the depth of fracture initiation.

As the result of wear, the root-mean-square deviation of the profile can increase or decrease in comparison with the initial one (see Fig. 5.11(b)). Both these situations were observed experimentally by different authors (see, for example, Kragelsky, Dobychin and Kombalov, 1982).

The incubation period, that is the time interval between the beginning of interaction and fracture origination, is a typical feature of this type of wear. The wear intensity during the incubation period is zero. The incubation period becomes shorter if the rate of damage accumulation increases, that is if there is an increase of temperature and stresses in the subsurface layers. This can be caused e.g. by an increase of load, friction coefficient or the quantity of heat absorbed by the body under consideration. The factors that lead to a shortening of the incubation stage also cause an increase in the wear rate.

Thus the proposed model provides many possibilities for fatigue wear process analysis and for the study of the surface microgeometry changes and equilibrium roughness formation in the steady-state stage of wear.

Chapter 6

Wear Contact Problems

6.1 Wear equation

One of the principal results of wear is that there are irreversible changes in the shape of the surfaces. These changes are comparable to elastic deformations and thus should be taken into account in the estimation of the contact characteristics of the bodies in sliding contact (distribution of stresses, dimensions of contact area etc.).

In order to solve those problems it is necessary to have information about the wear laws for materials. Such laws in tribology are called wear equations: they establish a relation between some characteristics of wear and a set of parameters characterizing the properties of friction surfaces and operating conditions.

6.1.1 Characteristics of the wear process

Among a great number of characteristics of wear processes analyzed in tribology, we select two of them which are convenient for contacting body wear estimation: *wear rate* and *wear intensity*.

The selection of these characteristics can be defended by several arguments. First, they are based on continuity of the wear process in time and they are described by continuous functions; the space-time discreteness of the wear process does not need to be taken into account at this scale. Secondly, these characteristics are directly related to the changes in the surface shape.

The wear rate is defined as the volume of material that is worn from a unit area of surface per unit time.

Generally, different points of the surface have different wear rates, and thus it is reasonable to speak about the wear rate at the given surface point (x, y), which can be estimated according to the definition as follows:

$$\frac{dw_*}{dt} = \lim_{\Delta t \to 0} \lim_{\Delta A \to 0} \frac{\Delta v_w(x, y)}{\Delta A \Delta t} = \lim_{\Delta t \to 0} \frac{\Delta w_*(x, y)}{\Delta t}, \tag{6.1}$$

Friction unit, part	Wear resistance class
Cylinder-piston groups of automobiles	11–12
Gauges	10–11
Slideways of machine tools	9–10
Cutting tools	7–8
Brake shoes	6–7
Sliding bearings	5–8

Table 6.1: Wear resistance classes for some parts and joints

where ΔA is the surface element in the vicinity of point (x, y), $\Delta v_w(x, y)$ is the volume of the material worn from the surface element ΔA during the time interval Δt (wear time), $\Delta w_*(x, y)$ is the linear wear at the point (x, y), which characterizes the depth of the layer worn during the time interval Δt.

The quantity $\frac{dw_*}{dt}$ has a dimension of velocity; it gives the rate of change of the surface position due to wear.

If all points of the rubbing surface are wearing under similar conditions, the ratio $\frac{\Delta v_w}{\Delta A}$ will not depend on the coordinates on the surface, and will be equal to the ratio of the material volume worn from the rubbing surface, to the area of this surface.

In addition to the wear rate, the ***wear intensity*** factor is used. This is defined as the volume of the material worn from the rubbing surface unit, per sliding distance unit, i.e.,

$$\frac{dw_*}{dl} = \lim_{\Delta l \to 0} \lim_{\Delta A \to 0} \frac{\Delta v_w}{\Delta A \Delta l} = \lim_{\Delta l \to 0} \frac{\Delta w_*}{\Delta l}, \tag{6.2}$$

where Δl is a sliding distance.

The wear intensity is a dimensionless quantity. It can vary from 10^{-3} to 10^{-13} according to material properties and operating conditions. We notice that $\frac{dw_*}{dl} = 10^{-9}$ means that a layer of 1 μm thickness is worn during sliding for a distance 1 km. Based on the wear intensity characteristics, the system of wear resistance classes has been created in Russia to evaluate the wear resistance of friction components. Wear resistance is defined as $I = \left(\frac{dw_*}{dl}\right)^{-1}$. The lowest class of wear resistance is the third one $\left(I = 10^3 \div 10^4\right)$, the highest class is the twelfth one $\left(I = 10^{12} \div 10^{13}\right)$. The typical classes of wear resistance for some parts and friction units are presented in Table 6.1.

The following relation exists between $\frac{dw_*}{dt}$ and $\frac{dw_*}{dl}$, if $V = \text{const}$:

$$\frac{dw_*}{dt} = V \frac{dw_*}{dl}. \tag{6.3}$$

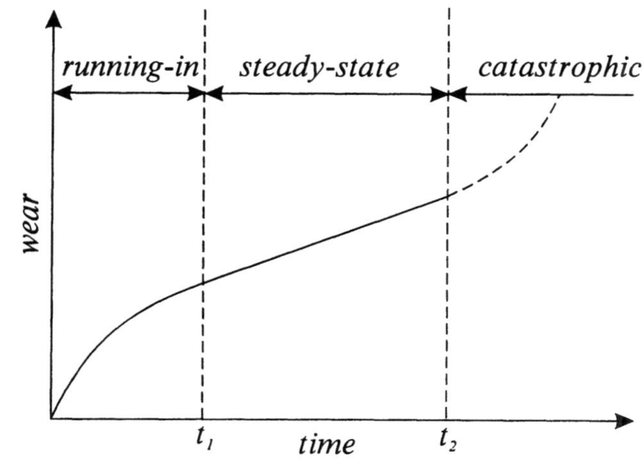

Figure 6.1: A typical dependence of wear on testing time.

Thus the wear characteristics of each friction component can be estimated from Eqs. (6.1) and (6.2). Are they characteristics of the material, or something else? We can answer this question by using a systematic approach to the investigation of tribological joints developed by Molgaard and Czichos (1977) and Czichos (1980). Generally the linear wear w_* is a function of structural parameters $\{S\}$ and input parameters $\{X\}$, i.e.

$$w_* = F(S, X), \tag{6.4}$$

where $\{S\}$ includes the following: structure elements (bodies in contact, interfacial and environment medium), properties of structure elements (aggregate state, geometric characteristics, volume, surface and bulk properties) and interaction of the elements; $\{X\}$ includes load, velocity, time, temperature, etc. Consequently, the wear characteristics of the material depend on individual properties of the material as well as on the properties of the system as a whole.

6.1.2 Experimental and theoretical study of the wear characteristics

A relation of the type (6.4) is the wear equation in its integral form. In the profound investigation by Meng and Ludema (1995) devoted to the history of the wear problems, it is noted that there are roughly 200 relations which can be classified as wear equations. It is well known that the wear characteristics depend on more than one hundred parameters.

There are two different methods for establishing these relations: empirical and mathematical simulation.

Empirical wear equations are established by extension of testing results. We will list some peculiarities of tribological tests on wear study.

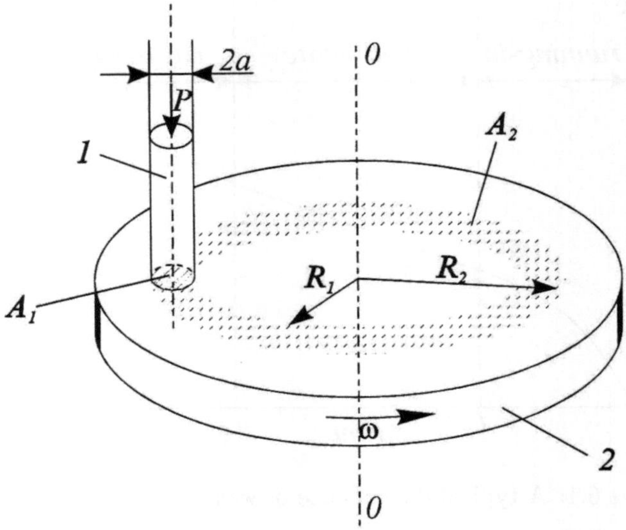

Figure 6.2: Scheme of pin-on-disk wear testing apparatus. A_2 is an annular wear scar.

Fig. 6.1 illustrates a typical dependence of linear wear on testing time.

The first period of the wear process $0 < t < t_1$ is called *running-in*. This is a very important stage of the wear process. During the running-in period, as a rule, the equilibrium (stable) surface roughness is provided, the chemical content of the surface (oxidation, diffusion) is established, and the temperature field of the friction pair is stabilized, i.e. the self-organizing processes of the system hold (Bershadsky, 1981, Polzer, Ebeling and Firkovsky, 1988 and Bushe, 1994). During the running-in period the wear intensity changes with time; this stage can last a long time.

The running-in period gives way to the *steady-state stage* of wear. For this time period ($t_1 < t < t_2$) the wear is directly proportional to the test time or the sliding distance, i.e. the wear rate (intensity) does not change. It is at this stage of wear that the wear characteristics which appear in the wear equation are registered.

In some cases, especially for inhomogeneous materials and for modified surfaces, there is a stage of *catastrophic wear* ($t > t_2$), when the wear rate increases radically.

When estimating wear rate, we must take into account that the rubbing surfaces of two interacting bodies may have different wear conditions. Let us consider the pin-on-disk friction testing apparatus (a common device for tribological tests) to illustrate this conclusion (see Fig. 6.2). When the pin (1) slides on the disk (2), the pin rubbing area is A_1, which coincides with the nominal area of contact $A_1 = \pi a^2$; for the disk, the rubbing surface is a ring with area $A_2 = \pi \left(R_2^2 - R_1^2\right)$. The time of the wear for the pin Δt_1 and for the disk Δt_2 will be also different.

During the test time interval Δt, any point of the pin surface is in friction

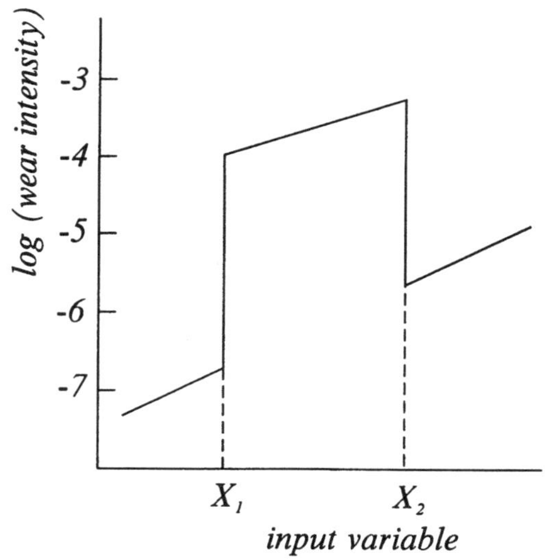

Figure 6.3: A typical dependence of wear intensity on an input parameter (for example, load, velocity, temperature, etc.); X_1 and X_2 are critical values of the parameter.

interaction for the whole of this time, i.e. for the pin the wear time Δt_1 is equal to Δt. For the disk, during the time interval Δt any point is in friction interaction with the pin during the time interval $\Delta t_2 = \dfrac{2a\Delta t}{\pi(R_1 + R_2)}$. Thus the wear rates of materials 1 and 2 in the testing apparatus (Fig. 6.2) can be estimated by the formulae:

$$\left(\frac{dw_*}{dt}\right)_1 = \frac{\Delta\,(v_w)_1}{\pi a^2 \Delta t_1}, \quad \left(\frac{dw_*}{dt}\right)_2 = \frac{\Delta\,(v_w)_2}{4a^2 \Delta t_1}, \tag{6.5}$$

where $\Delta\,(v_w)_1$ and $\Delta\,(v_w)_2$ are experimentally measured volumes of worn material of the pin and the disk, respectively.

When studying the dependence of wear on the input parameters of tribological system (loads, velocities, temperature, etc.), we can observe the phenomenon of abrupt variation of wear rate under smooth variation of input parameters. Fig. 6.3 illustrates this phenomenon. Points X_1 and X_2 are called critical or transition points of the tribosystem. At these values there is a change of wear mechanism, and the wear rate changes.

Based on the test results, we describe the relationship between wear characteristics and input variables. The wear equations may be also constructed by *a mathematical simulation* of the processes which occur at rubbing surfaces (see Chapter 5).

The simplest approach developed by Holm (1946) and Archard (1953) was based on the idea that the wear rate is proportional to the real area of contact of rough bodies. The coefficient of proportionality was estimated in wear tests.

Author	Wear equation	Material; friction part; conditions	Cause of wear
Lewis (1968)	$\frac{dv_w}{dt} = KPV$	filled PTFE; piston rings; unlubricated contact	adhesion
Khrushchov and Babichev (1970)	$\frac{dv_w}{dt} = K\frac{PV}{H}$	metals; unlubricated contact	micro-cutting
Rhee (1970)	$v_w = KP^\alpha V^\beta t^\gamma$	polymer-bonded friction materials (asbestos-reinforced polymers); breaks	adhesion with thermal process
Lancaster (1973)	$\frac{dv_w}{dt} = Kk_1k_2k_3k_4PV$, k_1, k_2, k_3, k_4 are wear-rate correction factors dependent on the operating conditions	filled thermoplastics; filled PTFE; dry rubbing bearings	
Larsen-Basse (1973)	$v'_w = K\frac{PV}{f}$, f is a frequency of impact, v'_w is a worn volume for one impact	carbide materials; drill bits	thermal fatigue, polishing of carbide grains (low drilling rates), transgranular fracturing (high drilling rates)
Moor, Walker and Appl (1978)	$\frac{dv_w}{dt} = KV^\beta P(v_c)$, v_c is a volume of rock removed per unit sliding distance, $\beta = 1.8$	diamond inserts; rotary drag bits	burning by superficial grafitization, breakage by impact, matrix errosion

K is specific coefficient for each wear equation, H is a hardness, P is a normal load.

Table 6.2: Empirical wear equations

Author	Wear equation	Mechanism of wear
Holm (1946)	$\frac{dw_*}{dt} = K\frac{pV}{H}$	adhesive
Archard (1953)	$\frac{dw_*}{dt} = K\frac{pV}{H}$	adhesive
Kragelsky (1965)	$\frac{dw_*}{dt} = Kp^\alpha V \ (\alpha > 1)$	fatigue
Rabinowicz (1965)	$\frac{dw_*}{dt} = K\frac{pV}{H}$	abrasive
Rabinowicz (1971)	$\frac{dw_*}{dt} = K\frac{pV}{H}$	fretting
Harricks (1976)	$\frac{dw_*}{dt} = KpV$	fretting

K is specific coefficient for each wear equation, H is a hardness.

Table 6.3: Theoretical models

A rich variety of wear equations based on fracture mechanics has been suggested in the last 20 years. These equations include the quantities relating to fatigue strength (Kragelsky, 1965), critical magnitude of energy absorbed by material (Fleischer, 1973), shear failure determined by a slip line analysis (Challen and Oxley, 1979), brittle fracture characteristics (Evans and Marshall, 1981). These theories considerably extend the number of parameters that have an influence on the wear, including the parameters which characterize the properties of materials.

As will be shown in § 6.2, for investigation of the kinetics of contact characteristics of junctions in wear process, we need to know the dependence of a wear rate on the contact pressure p and the relative sliding velocity V. Analysis of a number of wear equations obtained theoretically and experimentally shows that in many cases this dependence can be presented in the form

$$\frac{dw_*}{dt} = K_w p^\alpha V^\beta, \tag{6.6}$$

where K_w is the wear coefficient, and α and β are parameters which depend on material properties, friction conditions, temperature, etc.

We present some wear equations obtained in wear tests with different materials (Table 6.2) and in theoretical models (Table 6.3). Based on these results, we can evaluate the parameters α and β and the wear coefficient K_w in Eq. (6.6) for different mechanisms of wear.

6.2 Formulation of wear contact problems

The irreversible shape changes of bodies in contact arising from the wear of their surfaces, are taken into consideration for mathematical formulation of wear contact problems. The value of the linear wear w_* (change of the linear dimension of the body in the direction perpendicular to the rubbing surface) is often used to describe the wear quantitatively. Generally, the surfaces are worn nonuniformly, hence the linear wear $w_*(x, y)$ should be considered at each point (x, y) of the rubbing surface.

6.2.1 The relation between elastic displacement and contact pressure

We assume that the irreversible surface displacement $w_*(x, y)$ is small, and comparable to the elastic displacement $w(x, y)$. Hence for the determination of the stress state of the contacting bodies, the boundary conditions are posed on undeformed surfaces, neglecting both the elastic displacement $w(x, y)$ and the surface wear $w_*(x, y)$.

Under this assumption the pressure $p(x, y, t)$ within the contact region and the elastic displacement $w(x, y, t)$ for an arbitrary instant of time t are related by operator A which is analogous to the operator relating the pressure and elastic displacement in the corresponding contact problem when the wear does not occur, i.e.

$$w(x, y, t) = A\left[p(x, y, t)\right]. \tag{6.7}$$

For example, Eq.(6.7) has the following form for frictionless contact of a cylindrical punch and an elastic half-space

$$w(x, t) = \frac{2\left(1-\nu^2\right)}{\pi E} \int\limits_{-a(t)}^{a(t)} \ln|x - x'|\, p(x', t)\, dx' + \text{const}, \qquad -a(t) < x < a(t). \tag{6.8}$$

If the size of the contact zone does not change during the wear process, the operator A is time-independent; this occurs, for example, in the contact problem for the punch with a flat face and an elastic foundation. Otherwise the unknown contact area should be obtained at each instant of time from the condition

$$p(x, y, t)\Big|_{x, y \in \Gamma} = 0,$$

which holds on the boundary Γ of the contact region $\Omega(t)$. This condition is needed to ensure the continuity of the surface displacement gradient at the boundary of the contact zone, for a punch whose shape is described by a smooth function.

It must be noted that the requirement of a small value of $w_*(x, y, t)$ follows from the functional restrictions for components operating, for example, at precision

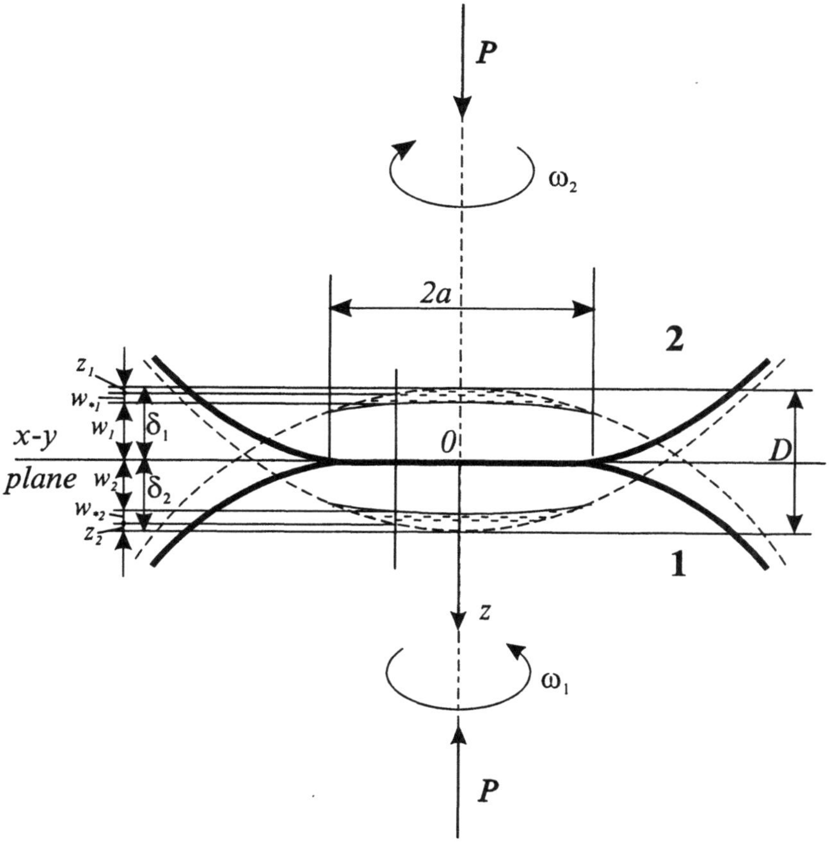

Figure 6.4: Elastic and wear displacements in the contact of two bodies.

junctions. For some wear contact problems, the value of the linear wear $w_*(x, y, t)$ is comparable to the size of the body in contact; now the relation (6.7) between the elastic displacement and pressure becomes more complex and time-dependent. In particular, it can depend on the geometry of the worn body. We describe such a problem in § 6.7 where we investigate a contact of a punch and an elastic half-space coated by a thin elastic layer.

6.2.2 Contact condition

We consider a contact of two elastic bodies (Fig. 6.4). We take the rectangular coordinate axis $Oxyz$ connected with the body 1. The origin O is the point where the surfaces touch at $t = 0$ if they are brought into contact by a negligibly small force. The Oz axis is chosen to coincide with the common normal to the two surfaces at O. The undeformed shapes of two surfaces are specified by the functions

$$z_1 = f_1(x,y), \quad z_2 = -f_2(x,y).$$

During the compression, the surface of each body is displaced parallel to Oz by an elastic displacement $w_i(x,y,t)$, $(i=1,2)$ (measured positive into each body) due to the contact pressure. So the following relation takes place within the contact zone at the initial instant of time $t=0$:

$$w_1(x,y,0) + w_2(x,y,0) = D(0) - [f_1(x,y) + f_2(x,y)], \tag{6.9}$$

where $D(t)$ is the approach of the bodies.

Let us consider any allowable changes in the relative position of the contacting bodies in friction process. We assume for the time being that there is no change of the body shapes due to their wear. If the contact condition (6.9) for any point of body 1 is valid after any relative displacement of body 2, we can use this equation to describe the contact condition at an arbitrary instant of time. Taking into account the shape changes of the bodies during wear process, we obtain

$$\begin{aligned} w_1(x,y,t) + w_{*1}(x,y,t) + w_2(x,y,t) + w_{*2}(x,y,t) \\ = D(t) - [f_1(x,y) + f_2(x,y)]\,. \end{aligned} \tag{6.10}$$

Wear contact problems with the contact condition in the form (6.10) are denoted as class $\mathcal{A}$ problems. Many practical problems fall into this category: the wear of axisymmetric bodies rotating about their common axis of symmetry; the wear in contact of a long cylinder, sliding back and forth along its generatrix on an elastic half-space. The last problem can be considered in a two-dimensional (plane) formulation.

We will classify the problem as type $\mathcal{B}$ if the form of the contact condition (6.9) changes because of relative displacement of the body 2 allowed by the considered friction process. For the problems of the type $\mathcal{B}$, the contact condition at an arbitrary instant of time depends on relative displacement of body 2. For example, if the punch with the shape function $z = f_2(x,y)$ moves in the direction of the y-axis with the constant velocity V over the elastic half-space (the body 1, which is worn due to friction), the contact condition for the fixed point (x,y) of the elastic half-space has the form

$$\begin{aligned} w_1(x,y,t)+w_{*1}(x,y,t) = D(t) - f(x,y-Vt), \quad a(x,t)<y<b(x,t), \\ t_* = \frac{b(x,t)-a(x,t)}{V}. \end{aligned} \tag{6.11}$$

Here $a(x,t)$ and $b(x,t)$ indicate the ends of the contact region, t_* is the contact time of the given point (x,y) in a single pass. We will consider wear contact problems of type $\mathcal{B}$ for different kinds of junctions in § 6.6 and § 8.2.

It is worth noting that the wear contact problem for one junction can be referred to class $\mathcal{A}$ or $\mathcal{B}$, depending on which component and its wear is under investigation. For example, for the junctions presented in Fig. 6.2, the contact problem is one of type $\mathcal{B}$ if we study the wear of the disk, while it is the problem of the type $\mathcal{A}$ if we analyze the wear of the pin and neglect the shape change of the disk surface in the wear process.

To complete the mathematical model of the wear contact problem, we must know the dependence of the linear wear $w_*(x,y,t)$ on the contact pressure $p(x,y,t)$ and on the sliding velocity $V(x,y,t)$. The dependence generally can be described by an operator involving these functions. Since the linear wear at the given point (x,y) at instant t is the total displacement, which is the accumulation of the elementary displacements which have taken place for instants $t' \leq t$, this operator is of hereditary type, and can be written as an integral operator

$$w_*(x,y,t) = \int_0^t K(t,t') f[p(x,y,t'),\, V(x,y,t')]\, dt'. \tag{6.12}$$

In wear contact problem formulation, we often use the simpliest forms of Eq. (6.12). It was pointed out in § 6.1 that, for different mechanisms of wear, the dependence of a wear rate on the contact pressure and the sliding velocity (the wear equation) has the following form

$$\frac{\partial w_*}{\partial t} = K_w p^\alpha(x,y,t) V^\beta(x,y,t), \tag{6.13}$$

It follows from Eq. (6.13) that the linear wear is determined by the formula

$$w_*(x,y,t) = K_w \int_0^t p^\alpha(x,y,t') V^\beta(x,y,t')\, dt'. \tag{6.14}$$

Eqs. (6.7), (6.10) (or Eq. (6.11)), (6.13) provide the complete system of equations for determining the contact pressure $p(x,y,t)$, the shape of the worn surface $w_*(x,y,t)$, and the elastic displacement $w(x,y,t)$. It must be noted that if the approach function $D(t)$ is not given, but we know the total normal load $P(t)$ applied to the contacting bodies, we can use the equilibrium equation to complete the system of equations

$$\iint_\Omega p(x,y,t)\, dxdy = P(t). \tag{6.15}$$

6.3 Wear contact problems of type $\mathcal{A}$

6.3.1 Steady-state wear for the problems of type $\mathcal{A}$

Let us examine the system of equations (6.7), (6.10) and (6.13) to investigate changes in contact characteristics for problems of type $\mathcal{A}$. At first we consider the case

$$V(x,y,t) = V_\infty(x,y), \quad \Omega(t) = \Omega_\infty, \quad \frac{dD(t)}{dt} = D_\infty, \tag{6.16}$$

i.e. these functions are time-independent in the wear process. Then the system of equations can be rewritten in the form

$$\begin{aligned} w(x,y,t) &= A[p(x,y,t)], && (6.17)\\ \frac{\partial w_*}{\partial t} &= K_w p^\alpha(x,y,t)\, V_\infty^\beta(x,y), && (6.18)\\ w(x,y,t)+w_*(x,y,t) &= D(0)+D_\infty t - f(x,y), && (6.19)\end{aligned}$$

where

$$w(x,y,t) = w_1(x,y,t) + w_2(x,y,t),$$
$$f(x,y) = f_1(x,y) + f_2(x,y),$$
$$w_*(x,y,t) = w_{*1}(x,y,t) + w_{*2}(x,y,t).$$

The system of equations (6.17), (6.18), (6.19) has a steady-state solution which determines the contact pressure $p_\infty(x,y) = \lim_{t\to+\infty} p(x,y,t)$ in the steady-state wear process

$$p_\infty(x,y) = \left[\frac{D_\infty}{K_w V_\infty^\beta(x,y)}\right]^{1/\alpha}. \tag{6.20}$$

From Eq. (6.19) we obtain the following condition for the steady-state wear process

$$\frac{\partial w_*(x,y,t)}{\partial t} = \frac{dD(t)}{dt} = D_\infty,$$

i.e. the steady-state wear is characterized by a uniform wear rate within the contact region. The equation of the shape of the worn surface $f_\infty(x,y)$ of the body 1 follows from Eqs. (6.17), (6.19) and (6.20)

$$\begin{aligned} f_\infty(x,y) = \lim_{t\to+\infty}\left[f_1(x,y) + w_*(x,y,t) - f_1(x^0,y^0) - w_*(x^0,y^0,t)\right] = \\ -f_2(x,y) + f_2(x^0,y^0) - A\left[p_\infty\right](x,y) + A\left[p_\infty\right](x^0,y^0), \end{aligned} \tag{6.21}$$

where $(x^0,y^0)\in\Omega_\infty$, $A\left[p_\infty\right](x,y)$ is the value of an operator A, calculated at the point (x,y), for the function p_∞ determined by Eq. (6.20).

Substituting Eq. (6.20) into the equilibrium equation (6.15) we obtain the formula for determining the steady-state normal load P_∞

$$P_\infty = \left(\frac{D_\infty}{K_w}\right)^{1/\alpha} \iint\limits_{\Omega_\infty} \frac{dxdy}{V_\infty^{\beta/\alpha}(x,y)}. \tag{6.22}$$

6.3.2 Asymptotic stability of the steady-state solution

Let us represent the general solution of Eqs. (6.17), (6.18) and (6.19) in the form

$$p(x,y,t) = p_\infty(x,y) + \varphi(x,y,t), \tag{6.23}$$

where $p_\infty(x, y)$ is the steady-state solution determined by Eq. (6.20). It is an asymptotically stable steady-state solution if the function $\varphi(x, y, t)$ satisfies the condition

$$\lim_{t \to +\infty} \varphi(x, y, t) = 0. \tag{6.24}$$

In the linear problem formulation ($\alpha = 1$) we can note the sufficient conditions for the representation of the solution in the forms of Eqs. (6.23) and (6.24). Under the assumption that the linear operator A is time-independent, i.e. validity of the relation

$$\frac{\partial}{\partial t} A[p] = A\left[\frac{\partial p}{\partial t}\right], \tag{6.25}$$

it follows from relations (6.17), (6.18), (6.19), (6.20) and (6.23) that the function $\varphi(x, y, t)$ satisfies the equation

$$A\left[\frac{\partial \varphi(x, y, t)}{\partial t}\right] = -K_w \varphi(x, y, t) V_\infty^\beta(x, y). \tag{6.26}$$

We shall seek the solution of this equation in the form

$$\varphi(x, y, t) = \psi(x, y) T(t). \tag{6.27}$$

Then we obtain

$$\dot{T}(t) A[\psi(x, y)] = -K_w V_\infty^\beta(x, y) \psi(x, y) T(t)$$

or

$$\dot{T}(t) + \lambda T(t) = 0, \tag{6.28}$$

$$q(x, y) - \lambda A_1[q(x, y)] = 0, \tag{6.29}$$

where

$$q(x, y) = K_w V_\infty^\beta(x, y) \psi(x, y),$$
$$A_1[q(x, y)] = A\left[\frac{q(x, y)}{K_w V_\infty^\beta(x, y)}\right]. \tag{6.30}$$

We denote the system of eigenvalues of Eq. (6.29) by $\Lambda \equiv \{\lambda_n\}_{n=1}^\infty$. It follows from Eq. (6.28) that

$$T(t) = T(0) \exp(-\lambda_n t). \tag{6.31}$$

To find the function $\psi(x, y, t)$ we should study the spectrum Λ of the operator A_1. The particular solutions $\varphi(x, y, t)$ of Eq. (6.26) satisfy the condition (6.24) if all $\lambda_n > 0$. This occurs if the operator A_1 is self-adjoint and positive semi-definite (Tricomi, 1957).

In § 6.4 and § 6.5 we will investigate some wear contact problems, in which the operator A_1 satisfies the sufficient conditions listed here for existence of asymptotically stable steady-state wear.

A necessary condition for the asymptotic stability of the steady-state solution in a non-linear wear contact problem ($\alpha \neq 1$ in Eq. (6.18)) is discussed in § 8.4.

6.3.3 General form of the solution

We assume that A_1 is a totally continuous, self-adjoint, positive semi-definite linear operator. As a consequence, the system of its eigenfunctions $U_n(x,y)$ is complete and orthonormalized in the space of continuous functions. The eigenvalues λ_n of this operator are positive.

According to Eqs. (6.23), (6.27) and (6.31), we can write the contact pressure at an arbitrary instant of time in the form

$$p(x,y,t) = p_\infty(x,y) + \frac{1}{K_w V_\infty^\beta(x,y)} \sum_{n=1}^{\infty} A_n U_n(x,y) \exp(-\lambda_n t). \qquad (6.32)$$

The coefficients A_n are found by the expansion of the contact pressure at the initial instant of time $t = 0$ in the series of eigenfunctions $U_n(x,y)$

$$p(x,y,0) = p_\infty(x,y) + \frac{1}{K_w V_\infty^\beta(x,y)} \sum_{n=1}^{\infty} A_n U_n(x,y). \qquad (6.33)$$

The shape of the worn surface at an arbitrary instant of time is determined by the equation obtained from Eq. (6.18) for $\alpha = 1$ and Eq. (6.32)

$$w_*(x,y,t) = D_\infty t + \sum_{n=1}^{\infty} \frac{A_n}{\lambda_n} U_n(x,y)\left(1 - \exp(-\lambda_n t)\right). \qquad (6.34)$$

If the functions $V(x,y,t)$, $\Omega(t)$, $D(t)$ are time-dependent and satisfy the conditions $\lim\limits_{t\to+\infty} V(x,y,t) = V_\infty(x,y)$, $\lim\limits_{t\to+\infty} \Omega(t) = \Omega_\infty$ and $\lim\limits_{t\to+\infty} \frac{dD(t)}{dt} = D_\infty$ (or $\lim\limits_{t\to+\infty} P(t) = P_\infty$), then the solution of the system of Eqs. (6.7), (6.10) and (6.13) approaches to that determined by Eqs.(6.32) and (6.34) as $t \to +\infty$. So the necessary conditions for the existence of a steady-state regime of wear process for the contact problems of the type $\mathcal{A}$ is the stabilization of the external characteristics (approach of the contacting bodies $D(t)$, normal load P(t) etc.) in time. If $P_\infty = 0$ or $D_\infty = 0$, then the contact pressure $p_\infty(x,y) = 0$.

6.4 Contact of a circular beam and a cylinder

Let us examine the problem of type $\mathcal{A}$, in which A (see Eq. (6.7)) is a time-independent differential operator, and use the method described in § 6.3 for determining the changes of contact characteristics in wear process.

6.4.1 Problem formulation

We will investigate a contact of an initially bent circular beam 1 and the inside surface of a rigid cylinder 2 (Fig. 6.5). The beam takes the form of an open circular ring; the size of the gap at the cut is negligibly small. In the course of displacement

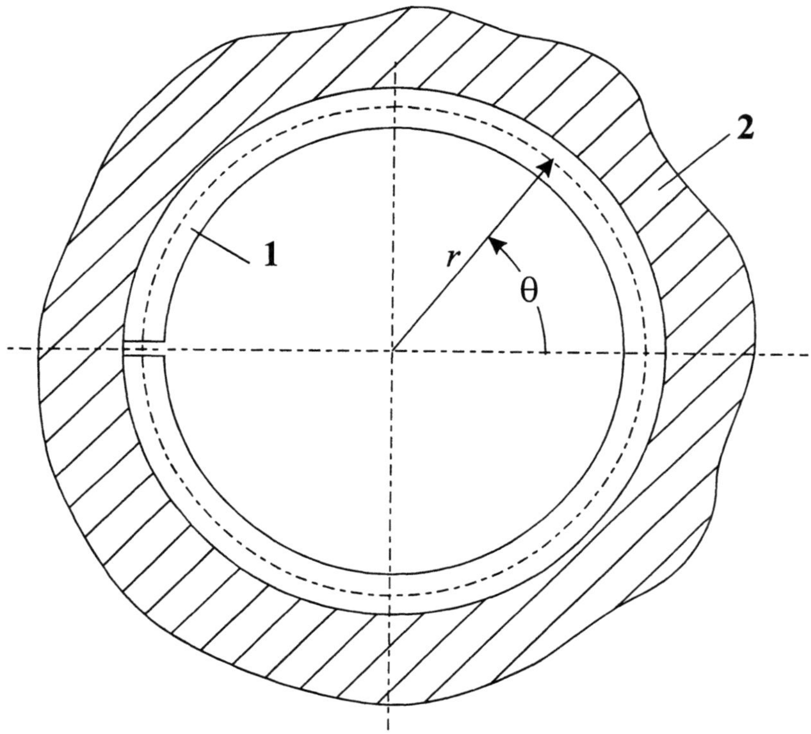

Figure 6.5: Scheme of contact of an open ring inserted into a cylinder liner.

of the ring along the cylinder generatrix, wear occurs both at the surface of the cylinder and at the surface of the ring.

We assume that the rate of wear $\frac{\partial w_*(\theta,t)}{\partial t}$ of the ring and the cylinder surfaces at any point is proportional to the pressure $p(\theta,t)$ between the ring and the cylinder

$$\frac{\partial w_*(\theta,t)}{\partial t} = K_w p(\theta,t).$$

Here θ is the polar angle (see Fig. 6.5); the wear coefficient K_w can depend on sliding velocity, temperature, etc.

As the result of wear, the thickness of the ring will decrease. In determining the radial elastic deflection $u_r(\theta,t)$ of the ring we neglect these variations and assume that the moment of inertia J of the ring remains roughly constant while it is in operation.

Under this assumption, the radial deflection $u_r(\theta,t)$ can be obtained from the following equation which is valid for bending of circular beams of small curvature

(Timoshenko, 1943):

$$\frac{1}{r^2} \cdot \frac{\partial^2 u_r(\theta,t)}{\partial\theta^2} + \frac{u_r(\theta,t)}{r^2} = -\frac{M(\theta,t)}{EJ}. \tag{6.35}$$

Here $M(\theta,t)$ is the bending moment in the beam, which is taken to be positive if it sets up compressive stresses in the exterior fibers of the beam; r is the radius of curvature of the ring, E is the Young's modulus.

We assume that the ends of the beam at the gap site ($\theta = \pm\pi$) are free of forces, i.e., at these points the bending moment and tensile forces are equal to zero. Then the bending moment at an arbitrary cross section of the ring is set up by the pressure $p(\theta,t)$ between the cylinder and the ring, i.e.

$$M(\theta,t) = -r^2 \int_{-\pi}^{\theta} p(\theta',t)\sin(\theta-\theta')\,d\theta', \quad -\pi \le \theta \le \pi. \tag{6.36}$$

Thus, the pressure $p(\theta,t)$, the radial deflection $u_r(\theta,t)$ and the total wear of the ring and the cylinder $w_*(\theta,t)$ are determined from the following system of equations

$$\frac{\partial^2 u_r(\theta,t)}{\partial\theta^2} + u_r(\theta,t) = \frac{r^4}{EJ}\int_{-\pi}^{\theta} p(\theta',t)\sin(\theta-\theta')d\theta', \tag{6.37}$$

$$\frac{\partial w_*(\theta,t)}{\partial t} = K_w p(\theta,t), \quad u_r(\theta,t) = u_r(\theta,0) - w_*(\theta,t). \tag{6.38}$$

The last equation is the condition for the contact of the ring and the cylinder.

Let us introduce the following dimensionless variable and functions

$$t_1 = \frac{K_w EJt}{r^4}, \quad p_1(\theta,t_1) = \frac{p(\theta,t)r^3}{EJ},$$
$$u_{r1}(\theta,t_1) = \frac{u_r(\theta,t)}{r}, \quad w_{*1}(\theta,t) = \frac{w_*(\theta,t)}{r}.$$

Then the system of equations (6.37) and (6.38) can be rewritten as

$$\frac{\partial^2 u_{r1}(\theta,t_1)}{\partial\theta^2} + u_{r1}(\theta,t_1) = \int_{-\pi}^{\theta} p_1(\theta',t_1)\sin(\theta-\theta')\,d\theta', \tag{6.39}$$

$$\frac{\partial u_{r1}(\theta,t_1)}{\partial t_1} = -p_1(\theta,t_1). \tag{6.40}$$

6.4.2 Solution

We apply the differential operator $\dfrac{\partial^2(\cdot)}{\partial\theta^2} + (\cdot)$ to Eq. (6.39), and obtain

$$\frac{\partial^4 u_{r1}(\theta,t_1)}{\partial\theta^4} + 2\frac{\partial^2 u_{r1}(\theta,t_1)}{\partial\theta^2} + u_{r1}(\theta,t_1) = p_1(\theta,t_1). \tag{6.41}$$

The equation for determining the ring deflection $u_{r1}(\theta, t_1)$ follows from Eqs. (6.40) and (6.41), namely

$$\frac{\partial^4 u_{r1}(\theta,t_1)}{\partial\theta^4} + 2\frac{\partial^2 u_{r1}(\theta,t_1)}{\partial\theta^2} + \frac{\partial u_{r1}(\theta,t_1)}{\partial t_1} + u_{r1}(\theta,t_1) = 0. \tag{6.42}$$

We will solve Eq. (6.42) by the method of a separation of variables. The unknown function $u_{r1}(\theta, t_1)$ can be written in the form

$$u_{r1}(\theta,t_1) = U(\theta)T(t_1), \quad T(0) = 1. \tag{6.43}$$

The functions $T(t_1)$ and $U(\theta)$ are determined from the equations

$$\frac{dT}{dt_1} + \lambda^2 T = 0, \tag{6.44}$$

$$\frac{d^4U}{d\theta^4} + 2\frac{d^2U}{d\theta^2} + U(1-\lambda^2) = 0, \tag{6.45}$$

where λ is the unknown parameter. The solutions of Eqs. (6.44) and (6.45) are

$$\begin{aligned} T(t_1) &= \exp(-\lambda^2 t_1), \\ U(\theta) &= A\sinh\Lambda_1\theta + B\cosh\Lambda_1\theta + C\sin\Lambda_2\theta + D\cos\Lambda_2\theta, \\ \Lambda_1 &= \sqrt{\lambda-1}, \quad \Lambda_2 = \sqrt{\lambda+1}, \quad \lambda > 1. \end{aligned} \tag{6.46}$$

The function $u_r(\theta, t)$ satisfies the condition $u_r(-\theta, t) = u_r(\theta, t)$, so that

$$A = C = 0. \tag{6.47}$$

The coefficients B and D can be found by satisfying the equilibrium equations for the ring. The equilibrium equation for the forces applied to the ring is

$$\int_{-\pi}^{\pi} p_1(\theta,t_1)\cos\theta\, d\theta = 0. \tag{6.48}$$

Using Eqs. (6.40), (6.43), (6.46), (6.47) and (6.48), we obtain

$$B\int_{-\pi}^{\pi}\cosh\Lambda_1\theta\cos\theta\, d\theta + D\int_{-\pi}^{\pi}\cos\Lambda_2\theta\cos\theta\, d\theta = 0.$$

Upon integrating, we can rewrite this equation as

$$B\Lambda_1\sinh\Lambda_1\pi + D\Lambda_2\sin\Lambda_2\pi = 0. \tag{6.49}$$

The second equation for determining the coefficients B and D can be obtained from the following condition

$$M(-\pi,t) = M(\pi,t) = 0. \tag{6.50}$$

We consider the equation (6.50) at the instant $t = 0$ and take into account Eq. (6.35), and obtain

$$\left.\frac{d^2U}{d\theta^2}\right|_{\theta=\pi} + U(\pi) = 0,$$

or, substituting Eq. (6.46) and taking into account Eqs. (6.47), we have

$$B \cosh \Lambda_1 \pi - D \cos \Lambda_2 \pi = 0. \tag{6.51}$$

The system of equations (6.49) and (6.51) is used to find the coefficients B and D. The system has a solution different from zero, if eigenvalues λ_n satisfy the characteristic equation

$$\Lambda_2 \sin \Lambda_2 \pi \cosh \Lambda_1 \pi + \Lambda_1 \cos \Lambda_2 \pi \sinh \Lambda_1 \pi = 0. \tag{6.52}$$

Substituting the coefficients B and D determined by Eqs. (6.49) and (6.51) for $\lambda_n > 1$ in Eq. (6.46), we obtain the particular solutions in the form

$$U_n(\theta) = \frac{\cos \Lambda_{2n}\pi}{\cosh \Lambda_{1n}\pi} \cosh \Lambda_{1n}\theta + \cos \Lambda_{2n}\theta, \;\; \Lambda_{1n} = \sqrt{\lambda_n - 1}, \;\; \Lambda_{2n} = \sqrt{\lambda_n + 1}. \tag{6.53}$$

It is easy to check that $\lambda = 1$ does not satisfy Eq. (6.45). For $\lambda < 1$ the solution of Eq. (6.45) can be written in the form

$$U(\theta) = A \cos L_2 \theta + B \cos L_1 \theta, \;\; \left(L_1 = \sqrt{1-\lambda}, \;\; L_2 = \sqrt{1+\lambda}\right). \tag{6.54}$$

The system of equations for determining the coefficients A and B follows from the relationships (6.48) and (6.50). The characteristic equation of the system is

$$- L_2 \sin L_2 \pi \cos L_1 \pi + L_1 \sin L_1 \pi \cos L_2 \pi = 0. \tag{6.55}$$

It is evident that $\lambda_0 = 0$ is the solution of Eq. (6.55). The second root of Eq. (6.55) is $\lambda_1 = 0.80$ calculated to the second decimal place. The particular solutions corresponding to the eigenvalues λ_0 and λ_1 have the form

$$\begin{aligned} U_0(\theta) &= \cos \theta, \\ U_1(\theta) &= \cos L_{21}\theta + \frac{\cos L_{21}\pi}{\cos L_{11}\pi} \cos L_{11}\theta, \\ L_{11} &= \sqrt{1-\lambda_1}, \quad L_{21} = \sqrt{1+\lambda_1}. \end{aligned} \tag{6.56}$$

The functions $U_n(\theta)$ determined by Eqs. (6.53) and (6.56) are mutually orthogonal. To prove this, we consider two particular solutions (6.53) for $\lambda_n \neq \lambda_m$

$$\int_{-\pi}^{\pi} U_m(\theta) U_n(\theta) d\theta = \frac{4 \cos \Lambda_{2n}\pi \cos \Lambda_{1m}\pi}{\lambda_m^2 - \lambda_n^2} [F(\lambda_m) - F(\lambda_n)],$$

$$F(\lambda) = \lambda[\Lambda_1 \tanh \Lambda_1 \pi + \Lambda_2 \tan \Lambda_2 \pi].$$

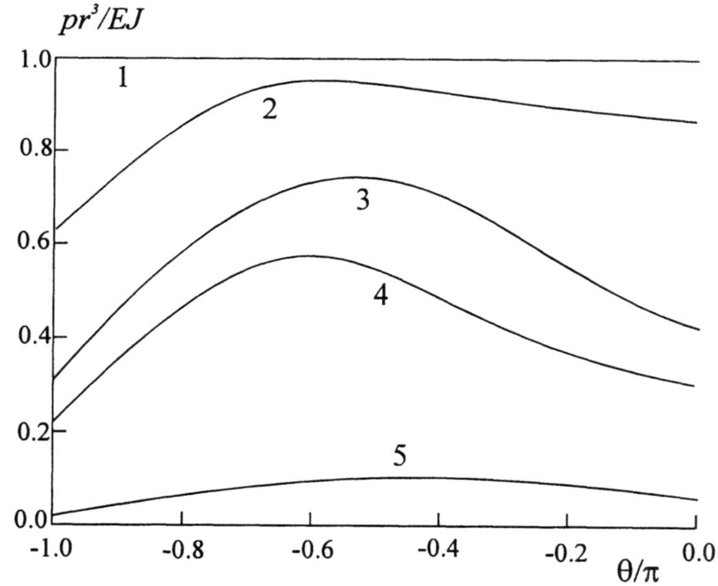

Figure 6.6: The time variation caused by wear of pressure distribution in contact of an open ring with a cylinder liner; $t_1 = 0$ (curve 1), $t_1 = 0.1$ (curve 2), $t_1 = 0.5$ (curve 3), $t_1 = 1.0$ (curve 4), $t_1 = 5.0$ (curve 5); $t_1 = K_w EJt/r^4$.

The right side of this relationship is equal to zero in view of Eq. (6.52). The orthogonality of the other particular solutions can be proved in a similar manner.

Expanding the known function $u_{r1}(\theta, 0)$ (which is determined by the shape of the ring in the free state) into a series in the complete orthonormal system of functions $U_n(\theta)$, we find the coefficients A_n:

$$u_{r1}(\theta, 0) = A_0 \cos\theta + \sum_{n=1}^{\infty} A_n U_n(\theta).$$

Then taking into account Eqs. (6.43) and (6.46) we obtain the relationship for determining the ring deflection $u_{r1}(\theta, t_1)$ at succeeding instants in time

$$u_{r1}(\theta, t_1) = A_0 \cos\theta + \sum_{n=1}^{\infty} A_n U_n(\theta) \exp\left(-\lambda_n^2 t_1\right). \tag{6.57}$$

The equation for determining the pressure $p_1(\theta, t_1)$ follows from Eqs. (6.40) and (6.57), namely

$$p_1(\theta, t_1) = \sum_{n=1}^{\infty} A_n \lambda_n^2 U_n(\theta) \exp\left(-\lambda_n^2 t_1\right), \tag{6.58}$$

The expressions for functions $U_n(\theta)$ are given by formulae (6.53) and (6.56).

The analysis of the characteristic equation (6.52) shows that the eigenvalues λ_n form a rapidly increasing sequence: $\lambda_1 = 0.80$, $\lambda_2 = 2.32$, $\lambda_3 = 6.69$, $\lambda_4 = 13.16$, $\lambda_5 = 21.63$, $\lambda_6 = 32.12$, etc., (with an accuracy of 0.005). This makes possible to sum only the first few terms of series (6.58) to determine the pressure distribution for instants of time not close to zero.

Fig. 6.6 illustrates the pressure between the ring and the cylinder for different times. The initially uniform pressure distribution becomes nonuniform in the wear process. Wear can produce a gap between the ring and the cylinder.

The solution obtained here can be applied to study sealing properties of a piston ring, and to evaluate its useful life.

6.5 Contact problem for an elastic half-space

In this part we develop a general method for solving 2-D and 3-D wear contact problems of type $\mathcal{A}$, for the case of a constant contact region in a wear process (elasticity operator A (6.7) is time-independent). The linear relation between a wear rate and a contact pressure is used; this allows us to reduce the problems to linear integral equations.

6.5.1 Problem formulation

Consider a punch rotating or sliding back and forth over an elastic half-space. The shape function of the punch is described by the equation $z = f(x, y)$. In a system of coordinates attached to the punch, the relation between the elastic displacement $w(x, y, t)$ in the z-axis direction and the contact pressure $p(x, y, t)$ (see Eq. (6.7)) has the form of the integral equation

$$w(x, y, t) = \iint_{\Omega} K(x, y, x', y')p(x', y', t)\, dx'dy'. \tag{6.59}$$

The kernel $K(x, y, x', y')$ does not depend on time, so that Eq. (6.59) holds at each instant of time. As was mentioned in § 6.2, this assumption is valid if the wear displacement and the elastic displacement of the half-space surface are small and of comparable size. In this case, we can consider both relative to the undeformed surface of the elastic half-space. Eq. (6.59) also holds at an arbitrary instant of time if only the punch experiences wear. There is no restriction on the magnitude of the punch linear wear $w_*(x, y, t)$ in this case.

The kernel $K(x, y, x', y')$ is generally symmetric and positive. The kernel symmetry is explained by the fact that it is a function of the distance between the point with coordinates (x, y) where the displacement is measured and the point with coordinates (x', y') where the normal load $p(x', y', t)\, dx'dy'$ is applied. To prove the kernel positiveness, let us consider the functional $J[q]$

$$J[q] = \iint_{\Omega} q(x, y) \left[\iint_{\Omega} K(x, y, x', y')q(x', y')\, dx'dy' \right] dxdy,$$

where $q(x,y)$ is any continuous function not identically zero within the region Ω. The functional $J(q)$ can be rewritten in the form which follows from Eq. (6.59)

$$J(q) = \iint\limits_{\Omega} q(x,y) w_q(x,y)\, dxdy.$$

Thus the functional $J(q)$ represents the total work done by an arbitrary pressure $q(x,y)$ on the corresponding displacements $w_q(x,y)$ of the points of the contact region, $(x,y) \in \Omega$. If the pressure is not zero, the work is always nonnegative. So $J(q) \geq 0$ for any function $q(x,y)$, not identically zero. This establishes that the kernel is positive semi-definite.

The contact condition (6.10) of the punch ($w_1(x,y) = 0$) and the elastic half-space at an arbitrary instant of time can be written as

$$w(x,y,t) + w_*(x,y,t) = D(t) - f(x,y). \tag{6.60}$$

Here $w_*(x,y,t)$ is the irreversible displacement due to wear of the punch or elastic foundation in the direction of the z-axis. We assume that the function $w_*(x,y,t)$ satisfies the wear equation (6.18).

The equations (6.18), (6.59) and (6.60) are used for determining the contact pressure $p(x,y,t)$, the elastic displacement $w(x,y,t)$ and the wear displacement $w_*(x,y,t)$ if the approach $D(t)$ is a known function.

If the normal load $P(t)$ applied to the punch is given, then to determine the unknown function $D(t)$ we must add the equilibrium equation to the system of Eqs. (6.10), (6.59) and (6.60)

$$P(t) = \iint\limits_{\Omega} p(x,y,t)\, dxdy. \tag{6.61}$$

Based on the analysis presented in §6.3, we can write the necessary conditions of the existence of the steady-state wear regime described by Eqs. (6.20) and (6.21). There is steady-state wear if the rate $\frac{dD}{dt}$ of the approach of the contacting bodies and the normal load $P(t)$ have the asymptotic values

$$\lim_{t\to+\infty} \frac{dD(t)}{dt} = \lim_{t\to+\infty} \frac{\partial w_*(x,y,t)}{\partial t} = D_\infty, \tag{6.62}$$

and

$$\lim_{t\to+\infty} \frac{dP}{dt} = P_\infty,$$

where P_∞ is determined by Eq. (6.22).

$$P_\infty = \left(\frac{D_\infty}{K_w}\right)^{1/\alpha} \iint\limits_{\Omega} \frac{dxdy}{V_\infty^{\beta/\alpha}(x,y)}.$$

If $P_\infty = 0$ or $D_\infty = 0$, then the contact pressure $p_\infty(x,y) = 0$.

The equation of the shape of the worn punch surface $f_\infty(x,y)$ follows from Eqs. (6.21) and (6.59)

$$f_\infty(x,y) = -\iint\limits_{\Omega}[K(x,y,x',y') - K(x^0,y^0,x',y')]\left[\frac{D_\infty}{K_w V_\infty^\beta(x',y')}\right]^{1/\alpha} dx'dy', \tag{6.63}$$

where $(x^0, y^0) \in \Omega$.

From this equation and the analysis presented in § 6.3 it follows that the kinetics of the wear process depends essentially on the type of punch motion.

6.5.2 Axisymmetric contact problem

Consider an axisymmetric contact problem for a punch of annular form in plan, rotating about its axis with a constant angular velocity ω, and pressed into an elastic half-space (Fig. 6.7). The shape of the punch is described by the equation $z = f(r)$. The force $P(t)$ and moment $M(t)$ applied to the punch are generally time-dependent functions. A solution of the problem can be used to calculate the wear of such junctions as thrust sliding bearings, end face seals, clutches, disk brakes and others.

The contact occurs within the annular region $a \le r \le b$. We assume that the inner a and external b radii do not change during the wear process. This is precisely so for a punch with a flat base, and approximately true if the variations of the contact region due to wear are small compared to its width.

As the punch rotates, tangential stresses $\tau_{z\theta}$ appear within the contact region. They coincide in direction with the direction of rotation, i.e. they are perpendicular to the radius of the contact region and

$$\tau_{z\theta}(r,t) = \mu p(r,t),$$

where $p(r,t)$ is the normal pressure within the contact region, and μ is the coefficient of friction. Because of the wear process, all components of stress and displacement are functions of time t.

The stress state of the elastic half-space at an arbitrary instant in time satisfies the following boundary conditions:

- within the contact region $r \in [a,b]$

$$u_z = w(r,t),\ \tau_{z\theta} = \mu p(r,t), \tau_{zr} = 0, \tag{6.64}$$

- outside the contact region $r \notin [a,b]$

$$\sigma_z = \tau_{z\theta} = \tau_{zr} = 0.$$

Here $w(r,t)$ is the elastic displacement of the half-space in the z-axis direction at any instant in time.

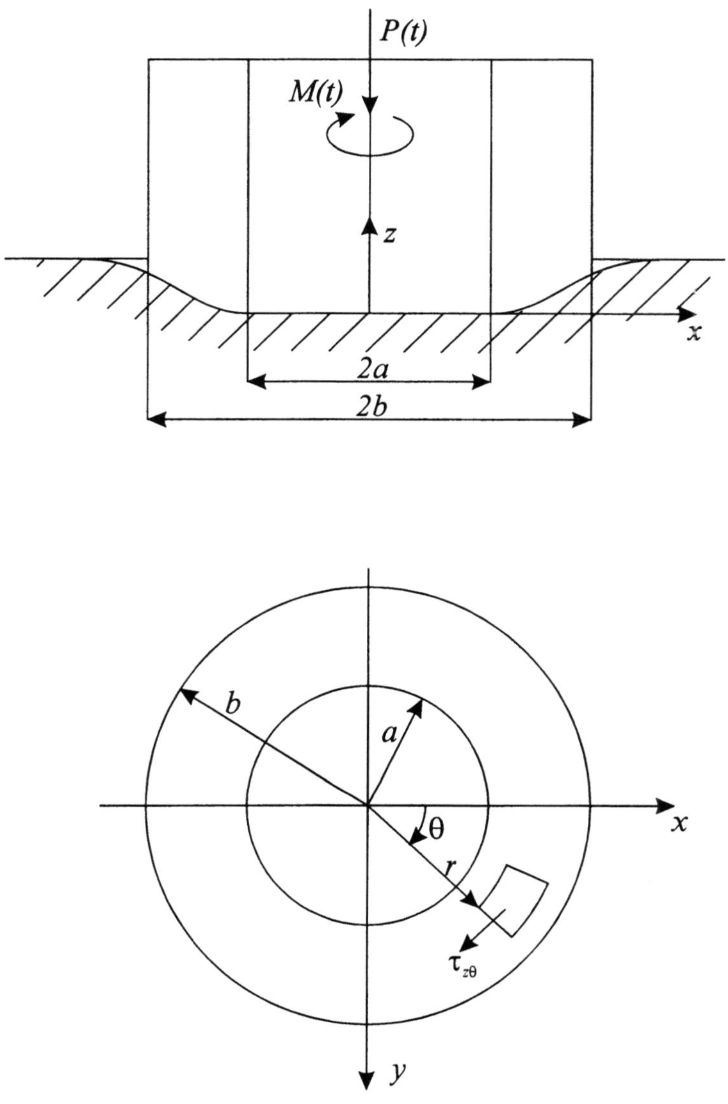

Figure 6.7: Scheme of contact of an annular cylindrical punch rotating on an elastic half-space surface.

Galin (1953) showed that the stress state corresponding to this boundary conditions can be broken down into two independent states: $\sigma_z = \sigma_z^{(1)} + \sigma_z^{(2)}$, $\tau_{z\theta} = \tau_{z\theta}^{(1)} + \tau_{z\theta}^{(2)}$, etc., satisfying the boundary conditions (problem 1)

$$\begin{aligned} u_z^{(1)} = w(r,t),\ \tau_{z\theta}^{(1)} &= \tau_{zr}^{(1)} = 0, \quad r \in [a,b], \\ \sigma_z^{(1)} = \tau_{z\theta}^{(1)} &= \tau_{zr}^{(1)} = 0, \quad r \notin [a,b]. \end{aligned} \tag{6.65}$$

and (problem 2)

$$\begin{aligned} u_z^{(2)} = 0,\ \tau_{z\theta}^{(2)} &= -\mu\sigma_z^{(1)},\ \tau_{zr}^{(2)} = 0, \quad r \in [a,b], \\ \sigma_z^{(2)} = \tau_{z\theta}^{(2)} &= \tau_{zr}^{(2)} = 0, \quad r \notin [a,b]. \end{aligned} \tag{6.66}$$

Eq. (6.65) shows that $\sigma_z^{(1)}$, $\tau_{z\theta}^{(1)}$, etc. are determined by the solution of the frictionless contact problem for the punch and the elastic half-space. The solution of the contact problem with the boundary conditions (6.66) shows that $u_z^{(2)} = 0$ and $\sigma_z^{(2)} = 0$ at the elastic half-space surface. So the relationship between the normal displacement $w(r,t) = u_z = u_z^1$ and the contact pressure $p(r,t) = -\sigma_z = -\sigma_z^{(1)}$ follows from the solution of the problem 1, and has the form

$$w(r,t) = \frac{1}{\pi E^*} \int_a^b \int_0^{2\pi} \frac{p(r',t) r' dr' d\varphi}{\sqrt{r^2 + r'^2 - 2rr' \cos\varphi}}, \tag{6.67}$$

where

$$E^* = \frac{E}{1-\nu^2}.$$

The tangential stress at the half-space surface is determined by the equation following from the solution of problem 2:

$$\tau_{z\theta} = \tau_{z\theta}^{(2)} = \mu p(r,t). \tag{6.68}$$

The shape of the elastic half-space surface changes during the wear process. We use the wear equation in the form (6.18) to determine the wear displacement $w_*(r,t)$ in the z-axis direction. This equation for $\beta = 1$ is written as

$$\frac{\partial w_*}{\partial t} = K_w \omega r p(r,t), \tag{6.69}$$

We assume that, from $t = 0$ to time t, the punch shifts by a distance $D(t)$ along its axis, and that there is no change in the position of the punch axis. Then at an arbitrary instant in time, the contact condition for the punch and the half-space has the form

$$w(r,t) + w_*(r,t) = D(t) - f(r). \tag{6.70}$$

Substituting Eq. (6.67) into Eq. (6.70) and taking Eq. (6.69) into account, we obtain an integral equation for determining the contact pressure in the wear process

$$\frac{1}{\pi E^*}\int_a^b\int_0^{2\pi}\frac{p(r',t)r'dr'd\varphi}{\sqrt{r^2+r'^2-2rr'\cos\varphi}} = D(t)-f(r)-K_w\omega r\int_0^t p(r,\tau)d\tau. \tag{6.71}$$

Based on the general method described in §6.3, we introduce the new function $q(r,t) = rp(r,t)$, which we seek in the following form:

$$q(r,t) = q_\infty + \sum_{n=1}^{\infty} q_n(r)\exp\left(-\lambda_n\omega t\right). \tag{6.72}$$

Substituting Eq. (6.72) into Eq. (6.71) we obtain the following equation

$$\sum_{n=1}^{\infty}\left[1-\exp\left(-\lambda_n\omega t\right)\right]\left[\frac{K_w}{\lambda_n}q_n(r)- \frac{1}{\pi E^*}\int_a^b\int_0^{2\pi}\frac{q_n(r')\,dr'd\varphi}{\sqrt{r^2+r'^2-2rr'\cos\varphi}}\right] = D(t)-D(0)-K_w\omega q_\infty t. \tag{6.73}$$

Let us look at various possible cases of this problem. If the punch does not move along its axis, i.e. $D(t) = D(0)$, Eq. (6.73) shows that the contact pressure approaches zero ($q_\infty = 0$). To find the unknown functions $\bar{q}_n(\rho) = \frac{q_n(\rho b)}{bE^*}$ and the values λ_n we have a homogeneous Fredholm integral equation of the second kind $\left(\rho = \frac{r}{b}\right)$

$$\bar{q}_n(\rho) - \lambda_n\int_{a/b}^{1} H(\rho,\rho')\bar{q}_n(\rho')\,d\rho' = 0 \tag{6.74}$$

with symmetric positive semi-definite kernel

$$H(\rho,\rho') = \frac{1}{\pi E^* K_w}\int_0^{2\pi}\frac{d\varphi}{\sqrt{\rho^2+\rho'^2-2\rho\rho'\cos\varphi}} = \frac{4}{\pi E^* K_w(\rho+\rho')}K\left(\frac{2\sqrt{\rho\rho'}}{\rho+\rho'}\right), \tag{6.75}$$

where $K(x)$ is the complete elliptic integral of the first kind. In the asymptotic case $\frac{b-a}{a} \ll 1$, i.e. if the ring width is far less of its radius, the kernel $H(\rho,\rho')$ takes the simple form

$$H(\rho,\rho') = \frac{2}{\pi E^* K_w}\ln\frac{8a}{|\rho-\rho'|b}.$$

The eigenvalues λ_n determined by Eq. (6.74) are all real and positive since the kernel (6.75) is real, symmetric and positive semi-definite. The eigenfunctions of Eq. (6.74) are orthogonal by virtue of the symmetry of the kernel.

The contact pressure $p(r,0)$ at the initial instant in time can be found by solving the frictionless contact problem for the axisymmetric annular punch and an elastic half-space. This problem has been investigated by Gubenko and Mossakovsky (1960), Collins (1963), Aleksandrov (1967), Gladwell and Gupta (1979); see also the monographs by Galin (1976) and Gladwell (1980). For instance, if the punch has a flat base ($f(r) = f = \text{const}$) and the annular width is much more then its inner radius $\frac{b-a}{a} \gg 1$ the relation given by Gubenko and Mossakovsky (1960) can be used ($a < r < b$)

$$p(r,0) = \frac{2fE^*}{\pi^2}\left[\frac{1}{\sqrt{b^2-r^2}}\arctan\frac{b\sqrt{r^2-a^2}}{a\sqrt{b^2-r^2}} + \frac{a}{b\sqrt{r^2-a^2}}\right]. \tag{6.76}$$

Expanding the known function $\bar{q}(\rho,0) = \rho p(\rho b, 0)/E^*$ into a series in the complete orthonormal system of eigenfunctions $U_n(\rho)$ of Eq. (6.74), we find the coefficients A_n:

$$\bar{q}(\rho,0) = \sum_{n=1}^{\infty} A_n U_n(\rho).$$

Then the contact pressure $\bar{p}(\rho,t) = p(\rho,t)/E^*$ at succeeding instants in time is calculated from the formula

$$\bar{p}(\rho,t) = \frac{1}{\rho}\sum_{n=1}^{\infty} A_n U_n(\rho)\exp(-\lambda_n \omega t). \tag{6.77}$$

The linear-wear case, i.e. $D(t) = D(0) + D_\infty t$, also necessitates solution of the integral equation (6.74). The solution of the problem takes the following form

$$\bar{p}(\rho,t) = \frac{1}{\rho}\left[\bar{q}_\infty + \sum_{n=1}^{\infty} A_n U_n(\rho)\exp(-\lambda_n \omega t)\right], \tag{6.78}$$

where $\bar{q}_\infty = \dfrac{D_\infty}{K_w \omega b E}$. Using the equilibrium equation (6.61) we obtain the normal load function $P(t)$ in this case

$$P(t) = P_\infty + \sum_{n=1}^{\infty} A_n P_n \exp(-\lambda_n \omega t),$$

where

$$P_\infty = \frac{2\pi D_\infty (b-a)}{K_w \omega}, \tag{6.79}$$

$$P_n = 2\pi b^2 E^* \int_{a/b}^{1} U_n(\rho)\, d\rho.$$

If the known functions $D(t)$ or $P(t)$ have another form, and satisfy the conditions

$$D(t) = D(0) + D_\infty t + D_*(t), \quad D_*(t) \le A \exp(-\lambda_1 \omega t),$$
$$P(t) = P_\infty + P_*(t), \quad P_*(t) \le B \exp(-\lambda_1 \omega t),$$

where A and B are some constants, the problem can be solved using a similar technique (Goryacheva, 1988) and is reduced to the investigation of the inhomogeneous Fredholm integral equations.

The method can also be used to solve the wear contact problem for a punch which has a circular contact region of radius b. However, in this case Eq. (6.69) shows that the displacement due to wear will be zero at the center of the contact region. This should lead to increasing contact pressure at this point; this in turn will cause irreversible plastic deformation at the center of the contact region. Thus, although irreversible changes of surface shape occur over the whole contact region, the solution based on the theory of elasticity given below will be valid for the whole contact zone except for a small region of radius a near its center. The eigenfunctions $U_n(\rho)$ in Eq. (6.78) can be found from the analysis of Eq. (6.74) with the symmetric and positive semi-definite kernel (6.75) for $\frac{a}{b} \ll 1$.

The initial contact pressure $p(r,0)$ can be determined by the formula (Galin, 1953):

$$p(r,0) = \frac{E^*}{4\pi} \int_0^b \Delta f(r') L(r,r') dr',$$

where

$$L(r,r') = \int_0^{2\pi} \frac{2r'}{\pi\sqrt{r^2 + r'^2 - 2rr'\cos\varphi}} \arctan \frac{\sqrt{b^2 - r^2}\sqrt{b^2 - r'^2}}{b\sqrt{r^2 + r'^2 - 2rr'\cos\varphi}}\, d\varphi,$$

$$\Delta = \frac{1}{r'} \cdot \frac{\partial}{\partial r'} + \frac{\partial^2}{\partial r'^2}.$$

Kellog's method (see, for example, Mikhlin and Smolitsky, 1967) was used to determine eigenfunctions $U_k(r)$ and eigenvalues λ_k of the Fredholm equation (6.74) with the real symmetric and positive semi-definite kernel (6.75). Successive approximations at the k-th step were calculated from the formula

$$\tilde{U}_k^{(n+1)}(\rho) = \int_{a/b}^{1} H(\rho,\rho') \left[U_k^{(n)}(\rho') - U_k^{(n)}(\rho)\right] d\rho' + U_k^{(n)}(\rho) \int_{a/b}^{1} H(\rho,\rho')\, d\rho'.$$

n	$k_n\ \left(\frac{a}{b}=0.0005\right)$	$k_n\ \left(\frac{a}{b}=0.5\right)$	n	$k_n\ \left(\frac{a}{b}=0.0005\right)$	$k_n\ \left(\frac{a}{b}=0.5\right)$
1	0.03643	0.18286	12	1.91324	6.62237
2	0.06730	0.90155	13	2.17578	7.48047
3	0.15221	1.57259	14	2.42712	8.23339
4	0.29421	2.22609	15	2.76887	8.46120
5	0.46936	2.79990	16	2.91857	9.62898
6	0.63901	3.41897	17	3.32462	11.06820
7	0.77450	3.83968	18	3.75936	12.86998
8	0.99863	4.37473	19	4.22465	15.48172
9	1.25300	5.06344	20	4.80057	19.90402
10	1.39264	5.67423	21	8.06343	30.73539
11	1.69633	6.18522			

Table 6.4: The eigenvalues of integral equation (6.74)

Here the first integral has no singularity at $\rho' = \rho$ and can be calculated numerically, the second integral is calculated analytically. The function $\tilde{U}_k^{(0)}(\rho)$ for each k-th step was taken in the orthogonal complement to the linear hull of the eigenfunctions $U_1(\rho)$, $U_2(\rho)$,...,$U_{k-1}(\rho)$, corresponding to the eigenvalues $0 < \lambda_1 < \lambda_2 < \ldots < \lambda_{k-1}$, which were found at the previous steps. Then the eigenvalue λ_k is determined as

$$\lambda_k = \lim_{n\to\infty} \lambda_k^{(n)} = \lim_{n\to\infty} \frac{\left\| U_k^{(n)} \right\|_{L^2}}{\left\| U_k^{(n+1)} \right\|_{L^2}}.$$

Table 6.4 shows the numerical results of eigenvalue calculations for the cases $\frac{a}{b} = 5\cdot 10^{-4}$ and $\frac{a}{b} = 5\cdot 10^{-1}$. The values $k_n = \lambda_n/(\pi E^* K_w)$ increase rapidly with n. This makes it possible to consider just the first few terms of the series (6.72) in determining the contact pressure for large time.

Fig. 6.8 illustrates the contact pressure distribution under the ring punch with flat face at the initial instant of time (curve 1) and in the steady-state wear, i.e. $t \to +\infty$ (curve 2). Note that the singularity of the pressure distribution at the ends of the contact zone, which is present when $t = 0$ disappears for $t > 0$.

The proposed method can be used to analyze the wear both of the elastic foundation and of the punch. The shape of the worn punch surface in the steady-

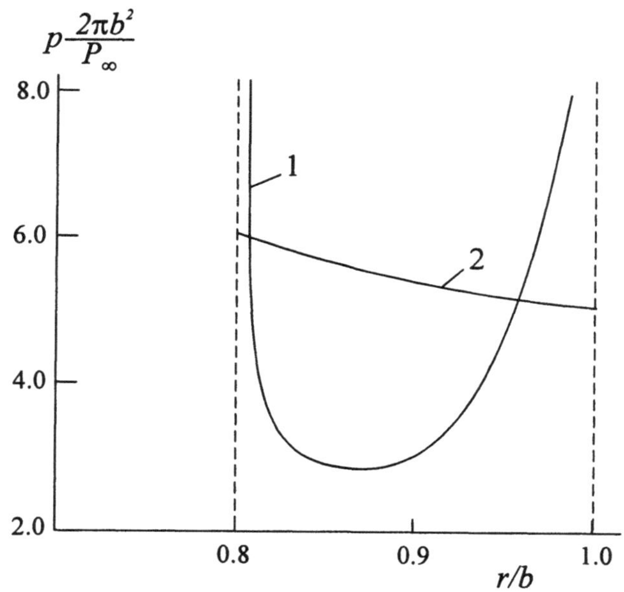

Figure 6.8: The initial (curve 1) and steady-state (curve 2) pressure distributions in contact of a flat-ended annular cylindrical punch with an elastic half-space.

state wear is calculated from the formula which follows from Eq.(6.63), namely

$$f_\infty(r) - f_\infty(a) = \frac{2P_\infty}{\pi^2 E^*(b-a)} \left[\int_{a/r}^{1} K(x)dx - \int_{a/b}^{r/b} \frac{K(x)}{x} dx \right] \tag{6.80}$$

where $K(x)$ is a complete elliptic integral of the first kind.

6.5.3 The case $V(x, y) = V_\infty$

Now we investigate a wear contact problem for a strip punch (3) sliding back and forth on a surface of an elastic layer (1). This problem is considered in a two-dimensional formulation (Fig. 6.9). The solution of the problem can be used for the prediction of the durability of different types of slideways.

The integral equation for the problem is written as

$$\frac{2}{\pi E^*} \int_{a}^{a} H_1 \left(\frac{x - x'}{h_0} \right) p(x', t)\, dx' = D(t) - f(x) - K_w V_\infty^\beta \int_0^t p(x, t')\, dt'.$$

Here $f(x)$ is a shape of the punch contacting surface, $D(t)$ is the punch penetration into the layer due to wear, h_0 is the layer thickness, V_∞ is the punch sliding velocity,

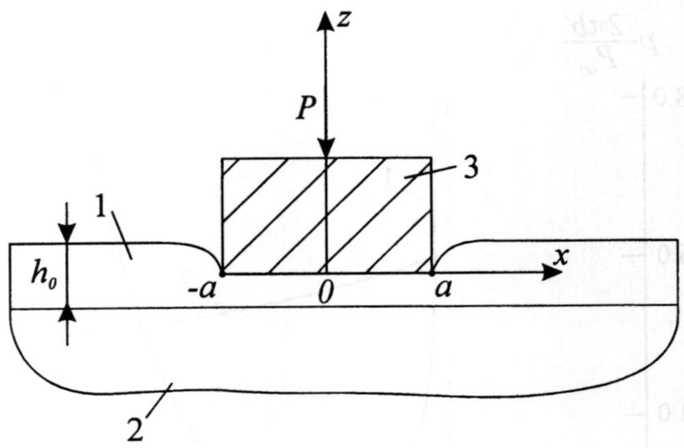

Figure 6.9: Scheme of contact of a cylindrical punch and a layered elastic half-space.

$p(x,t)$ is the pressure distribution, K_w, β are the parameters in the wear equation which is described by the relation

$$\frac{\partial w_*(x,t)}{\partial t} = K_w p(x,t) V_\infty^\beta .$$

The kernel $H_1(y)$ can be represented as Galin (1976))

$$H_1(y) = \int_0^{+\infty} \frac{L(u)}{u} \cos(uy)\, du,$$

where

$$L(u) = \frac{\cosh 2u - 1}{\sinh 2u + 2u}$$

in the problem for the elastic layer placed on the rigid substrate (2) in the absence of tangential stresses between the layer and substrate;

$$L(u) = \frac{2\kappa \sinh 2u - 4u}{2\kappa \cosh 2u + 4u^2 + 1 + \kappa^2}, \qquad \kappa = 3 - 4\nu$$

in the problem for the elastic layer bonded to a rigid substrate.

This wear contact problem can also be reduced to a Fredholm integral equation by the method described in § 6.5.2 (Goryacheva, 1988). Here we give only the formula for the worn punch shape f_∞ in the steady state wear which follows from Eq.(6.63)

$$f_\infty(x) - f_\infty(0) = \frac{P_\infty}{\pi E^* a} \int_{-a}^{a} \left[H_1\left(\frac{x'-x}{h_0}\right) - H_1\left(\frac{x'}{h_0}\right)\right] dx' . \qquad (6.81)$$

Using the expansion of the kernel $H_1(y)$ into series, given by Aleksandrov (1968), we can reduce this relationship to the following form which is valid for $\left(\frac{a}{h_0}\right)^4 \ll 1$ $\left(\xi = \frac{x}{a}\right)$

$$f_{\infty 1}(\xi) = \frac{[f_\infty(a\xi) - f_\infty(0)]\pi E^*}{P_\infty} = -(1-\xi)\ln(1-\xi) + (1+\xi)\ln(1+\xi) + 1.432\left(\frac{a\xi}{h_0}\right)^2.$$

Note that if the punch has the shape function given by Eq. (6.81) at the initial instant of time, and the contact region is restricted by the shape of the punch and is independent of time, this initial punch shape does not change in the wear process. Wear is uniform within the contact region. This conclusion follows from the analysis presented in § 6.23. So the specimen with the shape described by Eq. (6.81) provides the uniform wear condition, and can be used for the study of wear equation in tests.

The same conclusion could be drawn from the analysis of Eq. (6.80) which holds for a spin motion of an annular punch.

6.6 Contact problems of type $\mathcal{B}$

The mathematical formulation of the wear contact problem of type $\mathcal{B}$ includes Eqs. (6.7), (6.11) and (6.12). It follows from this system of equations that the contact conditions at any given point of the body 1 change in time, so the linear wear is determined by integrating the contact pressure function within the region of the contact, depending on the type of motion. Note that for some problems of this type, wear takes place only during a limited time.

We will illustrate the method of solution of the problems of type $\mathcal{B}$ by considering two particular problems.

6.6.1 The wear of an elastic half-space by a punch moving translationally

We consider the punch moving in the direction of the x-axis with a constant velocity V (Fig. 6.10). The contact of the punch and the elastic half-space takes place within the region $\Omega \equiv \{x, y : x \in (-a, a),\ \ y \in (-b, b)\}$ in the system of coordinates (x, y) attached to the moving punch. The punch has a rectangular cross-section at the plane $x = \text{const}$. The normal load P is applied to the punch.

The wear of the elastic half-space occurs when the punch moves. The shape changes of the half-space surface can be determined from the wear equation (6.13). In the system of coordinates x_1, y_1, z_1, related to the half-space surface, it can be written as

$$\frac{\partial w_*(x_1, y_1, t)}{\partial t} = K_w p^\alpha(x_1, y_1, t) V^\beta. \tag{6.82}$$

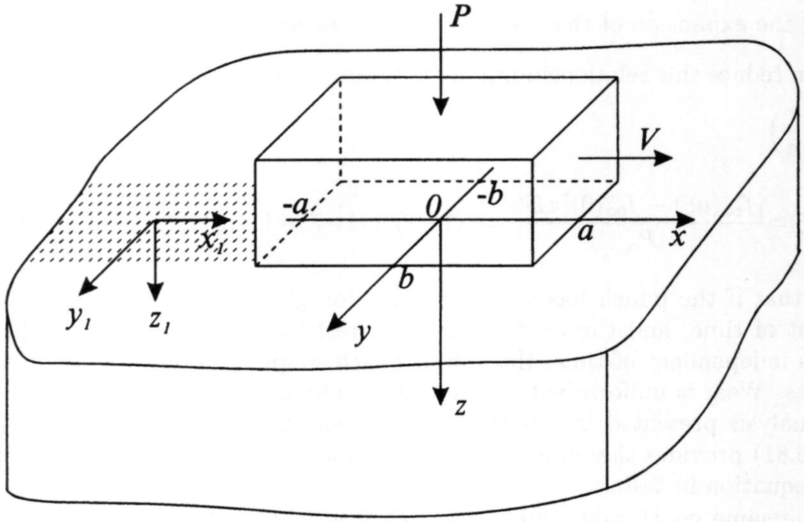

Figure 6.10: Wear of an elastic half-space by a punch of the rectangular form in plane.

It was proved by Galin (1953), that the pressure $p(x,y)$ within a contact area Ω which is a long rectangle ($a \gg b$) can be presented by the relation

$$p(x,y) = \frac{p_1(x)}{\pi\sqrt{b^2 - y^2}} \tag{6.83}$$

where the function $p_1(x)$ depends linearly on the elastic displacement $w(x,0) = w_1(x)$ in the z-axis direction. Thus

$$p_1(x) = kw_1(x), \tag{6.84}$$

where

$$k = \frac{\pi E}{2\left(1 - \nu^2\right)\log\left(a/b\right)}.$$

This relation is similar to that found for a Winkler foundation model.

By virtue of Eq. (6.84) and the fact that the punch moves along the x-axis, the wear contact problem can be considered in a two-dimensional formulation on the coordinate plane $y = 0$.

We will investigate the steady-state wear process and fix some point $(x_1, 0)$ of the boundary of the half-space, take $t = 0$ to be the instant that point $(x_1, 0)$ arrives at the contact ($x_1 = a$), and denote by $t(x)$ the instant at which the point $(x_1, 0)$ will have coordinates $(x, 0)$ in the (x, y) system. Then we obtain

$$t(x) = \frac{a - x}{V},$$

We introduce the functions $W_*(x)$ and $\tilde{p}(x)$ which are time independent in the system of coordinates x, z, by

$$w_*(a, 0, t(x)) = W_*(x), \quad W_*(x) = 0 \quad \text{if } x > a,$$

$$p(a, 0, t) = \tilde{p}(a - Vt).$$

Using these relations and Eq. (6.82), we obtain the following relationship between $W_*(x)$ and $\tilde{p}(x)$

$$W_*(x) = K_w V^\beta \int_0^{t(x)} \tilde{p}^\alpha(a - Vt')\, dt' = K_w V^{\beta-1} \int_x^a \tilde{p}^\alpha(s)\, ds. \tag{6.85}$$

The contact condition of the punch and the worn half-space at the section $y = 0$ has the following form in the moving system of coordinates x, z

$$W_*(x) + w_1(x) = D - f(x), \tag{6.86}$$

where $f(x)$ is the shape function of the punch at the plane $y = 0$, D is its penetration.

On differentiating Eqs. (6.84), (6.85) and (6.86) and substituting Eqs. (6.84) and (6.85) into Eq. (6.86), we obtain

$$K_* \tilde{p}^\alpha(x) - \frac{\pi b}{k} \tilde{p}'(x) = f'(x), \tag{6.87}$$

where

$$K_* = K_w V^{\beta-1}. \tag{6.88}$$

Eq. (6.87) and the equilibrium equation

$$\int_{-a}^{a} \tilde{p}(x)\, dx = P_0$$

provide the complete system of equations to determine the function $p(x)$. The value P_0 can be determined from Eq. (6.83) if the normal load P applied to the punch is known.

The solution of the problem for the case $\alpha = 1$ has the following dimensionless form

$$\tilde{p}_1(\xi) = \left[C - \int_{-1}^{\xi} f_1'(\xi') \exp(-\kappa\xi')\, d\xi' \right] \exp(\kappa\xi), \tag{6.89}$$

$$C = \frac{P_{01} - \int_{-1}^{1} [1 - \exp(\kappa(1 - \xi'))]\, f_1'(\xi')\, d\xi'}{2 \sinh \kappa},$$

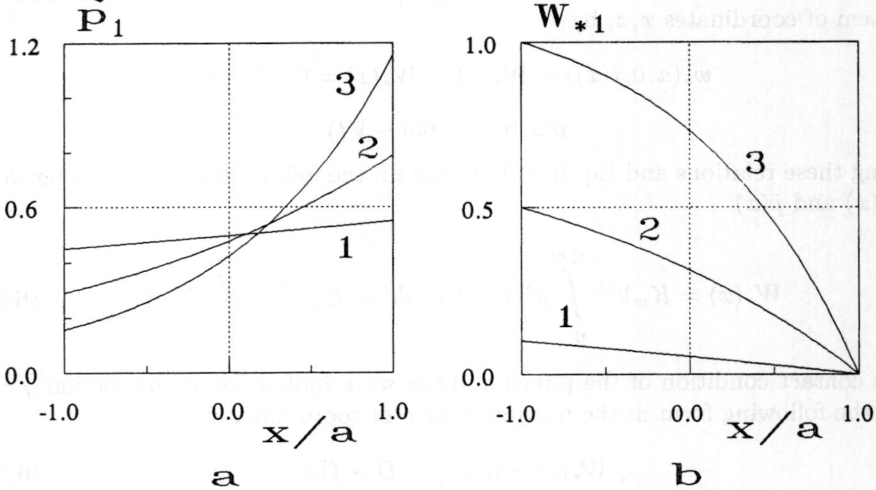

Figure 6.11: The steady-state pressure $\tilde{p}_1$ (a) and the shape of the worn surface W_{*1} (b) of an elastic half-space within the contact region for different values of parameter $\kappa = \dfrac{K_w V^{\beta-1} E a}{2(1-\nu^2)\, b \log a/b}$: $\kappa = 0.1$ (curves 1), $\kappa = 0.5$ (curves 2), $\kappa = 1$ (curves 3) and $\alpha = 1$.

where

$$\tilde{p}_1(\xi) = \frac{\pi b \tilde{p}(a\xi)}{ka}, \quad \xi = \frac{x}{a}, \quad \kappa = \frac{K_* a k}{\pi b}, \quad P_{01} = \frac{K_* P_0}{a}, \quad f_1(\xi) = \frac{f(\xi a)}{a}.$$

Fig. 6.11 illustrates the contact pressure $\tilde{p}_1(\xi)$ and the shape of the worn surface $W_{*1}(\xi) = \dfrac{W_*}{a}$ of the half-space within the contact region, for the contact problem of a punch with flat face ($f'(x) = 0$) and the elastic half-space. Based on Eqs. (6.85) and (6.89), the function $W_*(x)$ is calculated from the formula

$$W_{*1}(\xi) = \frac{P_{01}}{2\sinh\kappa}\left(\exp(\kappa) - \exp(\kappa\xi)\right).$$

It is interesting to note that Eq. (6.87) can also be used to find the shape of the moving punch which has uniform wear in the steady-state wear of the elastic half-space. As mentioned above, the investigation of punch wear relates to problems of type $\mathcal{A}$. The steady-state wear of the punch moving translationally with a constant velocity V occurs only if the contact pressure is distributed uniformly within the contact region, and does not change in the wear process, i.e. $p(x,y) = p_0$, where $p_0 = \dfrac{P}{4ab}$ (see § 6.3). Then the equation for the punch shape $f_0(x)$ which will not

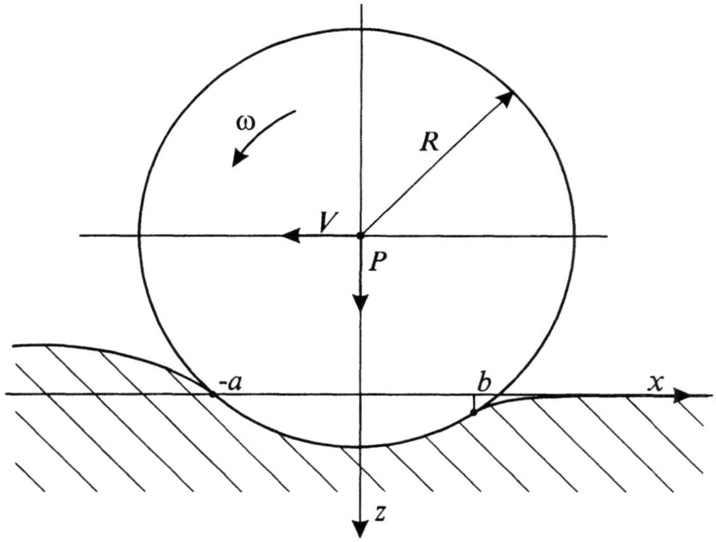

Figure 6.12: Wear of a half-plane by a disk executing translational and rotational motion.

change in the wear process follows from Eq. (6.87)

$$f_0(x) = K_* p_0^\alpha x + \text{const},$$

where K_* is determined by Eq. (6.88).

6.6.2 Wear of a half-plane by a disk executing translational and rotational motion

A more complicated contact problem of type $\mathcal{B}$ is considered by Soldatenkov (1989).

A rigid disk of radius R is pressed into an elastic half-plane, and moves translationally to the left along it (see Fig. 6.12) with a constant velocity $V > 0$, while at the same time rotating with a constant angular velocity ω. The positive direction of rotation is shown in the figure. The normal force P is applied to the disk.

We will take into account the wear of the half-plane by the disk. We assume that the linear wear $w_*(x_1, t)$ is determined from Eq. (6.13), which can be written as

$$\frac{\partial w_*(x_1, t)}{\partial t} = K_w |\omega R - V| p(x_1, t). \tag{6.90}$$

Here (x_1, z_1) is the coordinate frame fixed in the half-plane.

We will investigate the steady-state wear following the procedure described above. Based on Eq. (6.90), we establish the relationship between the linear wear $W_*(x)$ and contact pressure $p(x)$ determined in the moving system of coordinates

x, z. Assuming that the point x_1 arrives at the contact zone, i.e. $x_1 = -a$, at the instant $t = 0$, we denote the instant of time $t(x)$ at which the point x_1 will have coordinate x in the moving system. Then we have

$$t(x) = \frac{a+x}{V},$$
$$w_*(-a, t(x)) = W_*(x), \quad W_*(x) = 0 \quad \text{if} \quad x \le -a,$$
$$p(-a, t) = \tilde{p}(-a + Vt),$$

$$W_*(x) = K_w|\omega R - V| \int_0^{t(x)} p(-a + Vt')\, dt' = K_w \frac{|\omega R - V|}{V} \int_{-a}^{x} p(s)\, ds. \tag{6.91}$$

The contact condition between the disk and the half-plane is

$$W_*'(x) + w'(x) = -f'(x), \quad x \in [-a, b], \tag{6.92}$$

where $w(x)$ is the elastic displacement of the half-plane along the z-axis (in the x, z system); $f(x)$ is the shape of the contacting surface of the disk which can be represented by the equation $f(x) = \dfrac{x^2}{2R}$ valid for $a + b \ll R$; the prime denotes differentiation with respect to x.

Under the assumption that the wear $W_*(x)$ and the velocity V are small, the derivative $w'(x)$ can be expressed by the relation corresponding to the static problem of deformation of a half-plane (see Galin (1953) or Johnson (1987))

$$w'(x) = K \int_{-a}^{b} p(s) \frac{ds}{s - x} + \pi \vartheta K \tau_{xz}(x), \quad (z = 0), \tag{6.93}$$

$$K = \frac{2\,(1 - \nu^2)}{\pi E}, \quad \vartheta = \frac{1 - 2\nu}{2(1 - \nu)},$$

where $\tau_{xz}(x)$ is the tangential contact stress, which can be expressed in terms of the contact pressure in accordance with Coulomb's law (see § 3.1):

$$\tau_{xz}(x) = (\mu p(x) + \tau_0) \operatorname{sgn}(\omega R - V).$$

Substituting Eqs. (6.91) and (6.93) into Eq. (6.92) leads to the following equation for $p(x)$:

$$m p(x) + K \int_{-a}^{b} p(s) \frac{ds}{s - x} = F(x), \tag{6.94}$$

$$F(x) = -f'(x) - \pi \vartheta K \tau_0 \operatorname{sgn}(\omega R - V), \tag{6.95}$$

$$m = \pi \vartheta K \mu \operatorname{sgn}(\omega R - V) + \frac{K_w}{V} |\omega R - V|. \tag{6.96}$$

In addition, we have the equilibrium equation

$$P = \int_{-a}^{b} p(x)\,dx. \tag{6.97}$$

The solution $p(x)$ of Eq. (6.94) can readily be obtained by using the technique described in § 3.2 and in Muskhelishvili's (1946) and Johnson's (1987) books. Assuming that function $p(x)$ belongs to the Hölder class within $[-a, b]$ and is bounded at the ends of the contact region, we have

$$p(x) = \frac{m}{m^2 + (\pi K)^2} F(x) - \frac{K}{m^2 + (\pi K)^2} (a + x)^{1/2-\eta} (b - x)^{1/2+\eta} \times$$
$$\times \int_{-a}^{b} \frac{F(s)\,ds}{(a + s)^{1/2-\eta} (b - s)^{1/2+\eta} (s - x)}, \quad x \in [-a, b] \tag{6.98}$$

under the condition

$$\int_{-a}^{b} \frac{F(x)\,dx}{(a + x)^{1/2-\eta} (b - x)^{1/2+\eta}} = 0, \tag{6.99}$$

$$\eta = \frac{1}{\pi} \arctan \frac{m}{\pi K}, \quad |\eta| < \frac{1}{2}. \tag{6.100}$$

Taking into account Eq. (6.95), we reduce Eqs. (6.98) and (6.99) to

$$p(x) = \frac{(a + x)^{1/2-\eta} (b - x)^{1/2+\eta}}{Rm\sqrt{1 + \left(\pi \dfrac{K}{m}\right)^2}}, \tag{6.101}$$

$$\frac{a - b}{a + b} = 2\eta + \frac{2\pi\vartheta K}{a + b} R\tau_0 \operatorname{sgn}(\omega R - V). \tag{6.102}$$

From Eqs. (6.97)–(6.101) we obtain

$$(a + b)^2 = \frac{2PRK}{\left(\dfrac{1}{4} - \eta^2\right)}. \tag{6.103}$$

Eqs. (6.101)–(6.103) completely specify the distribution of the contact pressure $p(x)$. By integrating Eq. (6.101) in accordance with Eq. (6.91), we can determine the wear distribution $W_*(x)$. In the general case, $W_*(x)$ is expressed in terms of a hypergeometric function.

Let us analyze Eqs. (6.101)–(6.103), and consider some cases.

First we note that in the absence of wear, $K_w = 0$ and $\omega = 0$, Eqs. (6.101)–(6.103) coincide with the solution obtained in § 3.2 for the contact problem for a parabolic punch with limiting friction.

Wear affects both the contact pressure distribution $p(x)$ and the position of the contact region. In particular, Eqs. (6.96), (6.100), (6.102) and (6.103) show that wear causes the center of the contact region to shift in the direction of translation of the disk (opposite to the x-axis), thus decreasing the displacement of the center of the contact region arising from friction forces for $\omega R - V < 0$, and increasing this displacement for $\omega R - V > 0$.

Of interest is the case $\dfrac{\omega R}{V} = 1 - \dfrac{\pi\vartheta K\mu}{K_w}$, for $\dfrac{\omega R}{V} < 1$, when, by virtue of Eq. (6.96), $m = 0$, and hence $\eta = 0$. In this case the contact pressure has the same distribution as in the case of parabolic punch in the frictionless contact with the elastic half-plane (Galin, 1953). The only exception is that the center of the contact region is shifted by an amount $\pi\vartheta KR\tau_0$ opposite to the direction of translation of the disk (in the x direction). The corresponding wear distribution $W_*(x)$, in accordance with Eq. (6.91), has the form

$$W_*(x) = \frac{K_w|\omega R - V|}{\pi RVK}\left\{\frac{1}{2}\left(x - \frac{b-a}{2}\right)\sqrt{(a+x)(b-x)}+\right.$$

$$\left.\frac{(a+b)^2}{8}\left[\arcsin\left(2\frac{x+a}{a+b} - 1\right) + \frac{\pi}{2}\right]\right\}.$$

Note also that, in the absence of rotation of the disk ($\omega = 0$), solution (6.101)–(6.103) is independent of the translation velocity V of the disk.

The solution obtained here can be used to analyze the process of wearing of a material by an abrasive tool.

6.7 Wear of a thin elastic layer

The method described in § 6.4 can be used to investigate the wear kinetics of a thick layer bonded to an elastic foundation. If the irreversible displacement of the layer surface due to wear is commensurate with the elastic displacement, and much less than the thickness of the layer, we can use the same relationship between elastic displacement and contact pressure as in the contact problem without wear. In this case the operator A in Eq. (6.7) does not depend on time under the supplementary assumption that the contact area remains constant during the wear process.

However it is not possible to use this method to investigate the wear process of thin coatings. For thin coatings, the wear displacement can be commensurate with the thickness of the coating. For instance, it is important in practice to know the lifetime of the coating, which is estimated by the time when the wear displacement at any point is equal to the thickness of the coating.

It is difficult to obtain the exact solution of this problem, because we do not know the operator A (see Eq. (6.7)) for the contact problem with a complex shaped boundary. Below, we examine an approximate solution which makes it possible to analyze the kinetics of changes of all the contact characteristics and the coating thickness during the wear process.

6.7.1 Problem formulation

We investigate a contact of a cylindrical punch and a layered elastic foundation. The coordinate system $(Oxyz)$ is connected with the punch (3) (see Fig. 6.9), which has a shape function $z = f(x)$, $f(-x) = f(x)$. This problem can be considered as two-dimensional. We assume that the elastic modulus of the coating (1) less than that of the foundation (2). The coating is modeled as an elastic strip lying on the elastic half-plane without friction (problem 1), or bonded to the elastic half-plane (problem 2).

The strip is worn by the punch sliding along the y-axis. We assume that $h(x,0) = h_0$, and the wear rate $\dfrac{\partial h(x,t)}{\partial t}$ is proportional to the contact pressure $(p(x,t))^\alpha$:

$$\frac{\partial h(x,t)}{\partial t} = -K_w \left[\frac{p(x,t)}{p^*}\right]^\alpha, \tag{6.104}$$

where K_w is the wear coefficient and p^* is a standard pressure.

The punch is loaded by a constant normal force P. The tangential stress within the contact region is directed along the y-axis, so that $\tau_{xz} = 0$. The component τ_{yz} of the tangential stress does not influence the contact pressure distribution, which can be found as the solution of a plane contact problem. The component τ_{yz} influences the wear rate and can be taken into account by the wear coefficient K_w.

The contact condition of the points of the punch and the worn strip surface for $x \in (-a, a)$ at any instant in time has a form

$$h(x,t) - h(0,t) + (w(x,t) - w(0,t)) = f(x), \quad f(0) = 0, \tag{6.105}$$

where $w(x,t)$ is the elastic displacement of the strip surface in the direction of the z-axis, a is the half-contact width, which is assumed to be fixed in the wear process.

The displacement gradient $w'_x(x,0) = \dfrac{\partial w(x,0)}{\partial x}$ of the elastic strip loaded by a normal pressure $p(x,0)$ can be obtained from the following equation given by Aleksandrova (1973)

$$\int_{-a}^{a} p(x',0) K\left(\frac{x'-x}{h_0}\right) dx' = \pi h_0 E_1^* w'_x(x,0), \quad |x| \le a,$$

$$E_i^* = \frac{E_i}{2\left(1-\nu_i^2\right)}, \tag{6.106}$$

where E_i, ν_i are Young's moduli and Poisson's ratios of the strip ($i = 1$) and the half-plane ($i = 2$), respectively.

The kernel of the integral equation (6.106) has the form

$$K(t) = \int_0^{+\infty} N(u) \sin tu \, du, \tag{6.107}$$

where for problem 1

$$N(u) = \frac{\cosh 2u - 1 + n\,(2u + \sinh 2u)}{2u + \sinh 2u + n\,(\cosh 2u - 1 - 2u^2)},$$

and for problem 2

$$N(u) = \frac{M + 4u \exp(-2u) - L \exp(-4u)}{M - (1 + 4u^2 + LM)\exp(-2u) + L\exp(-4u)},$$

$$M = \frac{\mu_2 \kappa_1 + n\mu_1}{n\mu_1 - \mu_2}, \qquad L = \frac{n\mu_1 \kappa_2 - \mu_2 \kappa_1}{n\mu_1 \kappa_2 + \mu_2},$$

where

$$\mu_i = 1 - \nu_i, \quad \kappa_i = 3 - 4\nu_i, \quad n = \frac{E_1^*}{E_2^*}.$$

We can use the representation of the kernel $K(t)$ given by Aleksandrova (1973), which is valid for large t and small n

$$K(t) \sim \frac{n}{t} - A\pi\delta'(t), \tag{6.108}$$

where

$$A = \frac{1}{2} \text{ – for problem 1,}$$

$$A = \frac{2(1-n)(\kappa - 1)(2 - n + n\kappa)}{(\kappa+1)^2} \text{ – for problem 2} \tag{6.109}$$

if $\nu_1 = \nu_2 = \nu, \quad \kappa_1 = \kappa_2 = \kappa$,

$\delta(t)$ – Dirac's function.

Substituting Eq. (6.108) in Eq. (6.106), we can obtain the integral equation

$$\frac{Ah}{E_1^*} p'(x,0) + \frac{1}{\pi E_2^*} \int_{-a}^{a} \frac{p(x',0)\, dx'}{x' - x} = -w'(x,0). \tag{6.110}$$

Aleksandrova (1973) showed that this equation holds for a thin strip $\left(\frac{h}{a} \ll 1\right)$, and $n \le 2$.

Integrating both sides of Eq. (6.110) with respect to x, we obtain

$$-\frac{Ah_0}{E_1^*}\left[p(x,0)-p(0,0)\right]+\frac{1}{\pi E_2^*}\int_{-a}^{a} p(x',0)\ln\left|\frac{x'-x}{x'}\right|dx' = w(x,0)-w(0,0). \tag{6.111}$$

The first term in the left side of this equation can be considered as the displacement of the strip surface which behaves like a Winkler elastic foundation with proportionality coefficient $k=\dfrac{Ah_0}{E_1^*}$. This interpretation of the first term in Eq. (6.111) makes sense if it is examined together with the second term, which is the substrate displacement $w_2(x,0)$.

It was proved by Soldatenkov (1994) that for slight relative change of the strip thickness ($h'(x)\ll 1$), Eq. (6.111) still holds, except that the first term takes the form

$$w_1(x,t)=-\frac{A}{E_1^*}\left[h(x,t)p(x,t)-h(0,t)p(0,t)\right]. \tag{6.112}$$

Eq. (6.111) with the first term $w_1(x,t)$ in the form of Eq. (6.112) is the generalization of the foregoing interpretation of Eq. (6.111) to the case of variable $h(x,t)$. It can be written as

$$-\frac{A}{E_1^*}\left[h(x,t)p(x,t)-h(0,t)p(0,t)\right]+\frac{1}{\pi E_2^*}\int_{-a}^{a} p(x',t)\ln\left|\frac{x'-x}{x'}\right|dx' = w(x,t)-w(0,t). \tag{6.113}$$

Substituting Eq. (6.113) into Eq. (6.105), we obtain the equation for determining the contact pressure at an arbitrary instant of time

$$h(x,t)-h(0,t)-\frac{A}{E_1^*}\left[h(x,t)p(x,t)-h(0,t)p(0,t)\right]+\frac{1}{\pi E_2^*}\int_{-a}^{a} p(x',t)\ln\left|\frac{x'-x}{x'}\right|dx' = f(x). \tag{6.114}$$

The strip thickness at any instant of time is determined from Eq. (6.104)

$$h(x,t)=h_0-K_w\int_0^t\left[\frac{p(x,t')}{p^*}\right]^{\alpha}dt'. \tag{6.115}$$

In addition we have the equilibrium equation

$$\int_{-a}^{a} p(x,t)\,dx = P. \tag{6.116}$$

6.7.2 The dimensionless analysis

We note that if the function $f(x)$ is symmetrical, we obtain symmetrical solutions $p(x,t)$ and $h(x,t)$ of Eqs. (6.114)–(6.116). We assume that the shape of the punch allows the contact region to be constant during the wear process (Fig. 6.9).

We introduce the dimensionless quantities

$$\xi = \frac{x}{a},\ \tau = t\frac{K_w}{a},\ \bar{P} = \frac{P}{aE_2^*},$$

$$\bar{h}_0 = \frac{h_0}{a},\ \bar{h}(\xi,\tau) = \frac{1}{a}h\left(\xi a, \tau\frac{a}{K_w}\right),\ R = \frac{A}{n}, \tag{6.117}$$

$$\bar{p}(\xi,\tau) = \frac{1}{E_2^*}p\left(\xi a, \tau\frac{a}{K_w}\right),\ \bar{f}(\xi) = \frac{1}{a}f(\xi a).$$

The system of equations (6.115)–(6.116) can be rewritten in the dimensionless form

$$\bar{h}(\xi,\tau) = \bar{h}_0 - \int_0^{\tau}\left[\frac{\bar{p}(\xi,\tau')}{\bar{p}^*}\right]^{\alpha} d\tau', \tag{6.118}$$

$$\bar{h}(\xi,\tau) - \bar{h}(0,\tau) - R\left[\bar{h}(\xi,\tau)\bar{p}(\xi,\tau) - \bar{h}(0,\tau)\bar{p}(0,\tau)\right] +$$
$$\frac{1}{\pi}\int_{-1}^{1}\bar{p}(\xi',\tau)\ln\left|\frac{\xi'-\xi}{\xi'}\right| d\xi' = \bar{f}(\xi), \tag{6.119}$$

$$\int_{-1}^{1}\bar{p}(\xi,\tau)\,d\xi = \bar{P}. \tag{6.120}$$

The solution of the wear contact problem is found from this system of equations.

6.7.3 Calculation techniques and numerical results.

To solve equations (6.118)–(6.120) we convert from continuous time to discrete time by breaking the time down into small intervals (τ_k, τ_{k+1}): $\tau_{k+1} = \tau_k + \Delta\tau$, $\tau_0 = 0$, $k = 0, 1, \ldots$ Then the system (6.118)–(6.120) can be approximated by the following relations

$$\bar{h}_{k+1}(\xi) = \bar{h}_k(\xi) - \left[\frac{\bar{p}_k(\xi)}{\bar{p}^*}\right]^{\alpha}\Delta\tau,\quad \bar{h}_0(\xi) = \bar{h}_0. \tag{6.121}$$

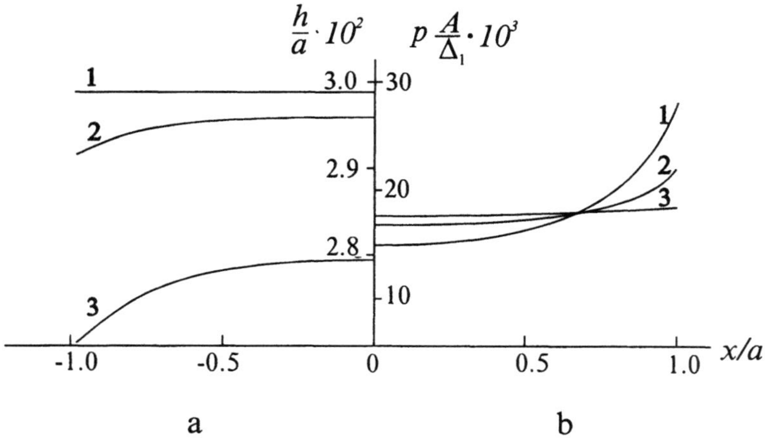

Figure 6.13: Profile of the worn surface of the layer (a) and pressure distribution (b) within the contact region during the wear process: $\tau = 0$ (curve 1), $\tau = 0.15$ (curve 2), $\tau = 0.64$ (curve 3).

$$\bar{h}_k(\xi) - \bar{h}_k(0) - R\left[\bar{h}_k(\xi)\bar{p}_k(\xi) - \bar{h}_k(0)\bar{p}_k(0)\right] +$$
$$\frac{1}{\pi}\int_{-1}^{1} \bar{p}_k(\xi')\ln\left|\frac{\xi' - \xi}{\xi'}\right| d\xi' = \bar{f}(\xi), \tag{6.122}$$

$$\int_{-1}^{1} \bar{p}_k(\xi)\, d\xi = \bar{P}, \tag{6.123}$$

which become the relations (6.118)–(6.120) as $\Delta\tau \to 0$. The function $\bar{p}_k(\xi)$ found from Eqs. (6.122) and (6.123) determines in accordance with Eq. (6.121), the function $\bar{h}_{k+1}(x)$ at the following moment. As a result we obtain the pressure distribution at various instants of discrete time in the strip wear process.

For the solution of the system of equations (6.122) and (6.123), we use the method of transformation of integral equations to finite-dimensional systems of linear equations (Kantorovich and Krylov, 1952).

For the numerical calculations, we assume that the strip is bonded to the substrate (problem 2) and that the rigidity of the strip is less than the substrate one. This case can be applied to investigate the wear of solid lubricant coatings.

For the calculation, we took the shape function $\bar{f}(\xi) = 10^{-3}\xi^2$ and the following values of the dimensionless parameters: $\alpha = 1.4$, $R = 3.8$, $\bar{p}^* = 0.26$, $\bar{h}_0 = 3 \cdot 10^{-2}$, $\bar{P} = 9 \cdot 10^{-3}$.

Fig. 6.13 illustrates the contact pressure distribution and the worn surface profile at various times. In the wear process the contact pressure equalizes, i.e.

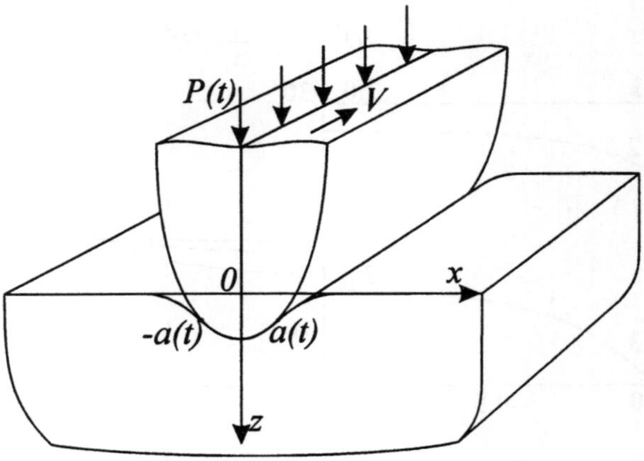

Figure 6.14: Scheme of the contact with time-dependent contact region in the wear process.

running-in of the rubbing surface occurs. The results show that equalization of the contact pressure is followed by the stabilization of the worn surface profile.

The analysis of the dependence of the running-in time on the initial layer thichness is presented in Goryacheva and Soldatenkov (1983). They also investigated the wear of the coating during the running-in time and the lifetime of coating for various values of the parameters of the problem.

6.8 Problems with a time-dependent contact region

Contact problems with a fixed contact region have been considered in § 6.3–§ 6.7. The contact area was determined by the punch shape, and did not change during the wear process. Generally the assumption of a constant contact region is roughly satisfied for small changes of the shape of contacting bodies in wear process.

If the contact region changes during the wear process, i.e. $a = a(t)$, the operator A in Eq. (6.7) depends on time. The contact problem becomes nonlinear even if the wear process (6.12) is linear. To analyze the main features of the wear process for this case, we use the simplest model of an elastic body which is a Winkler elastic foundation.

6.8.1 Problem formulation

We consider a cylindrical punch ($z = f(x)$, where $f(x)$ is a differentiable function) which moves over the elastic foundation in the direction of y-axis with the velocity V (see Fig. 6.14). We assume that the elastic displacement $w(x, t)$ in the direction of the z-axis is determined by the formula

$$w(x,t) = kp(x,t), \tag{6.124}$$

where $k = \dfrac{h}{K}$, K is the elastic modulus of the foundation, h is its depth, $p(x,t)$ is a contact pressure.

We assume that the wear of the punch surface is considerably less than the wear of the foundation, so we take into account only the wear of the foundation. The wear equation is considered in the linear form

$$\frac{\partial w_*(x,t)}{\partial t} = K_w p(x,t), \tag{6.125}$$

where the wear coefficient K_w can depend on the velocity V, temperature, coefficient of friction, etc.

The condition that the points of the punch and foundation coincide within the contact zone $(-a(t), a(t))$ at an arbitrary instant of time is written as

$$w(x,t) + w_*(x,t) = D(t) - f(x), \quad -a(t) \le x \le a(t). \tag{6.126}$$

Here $D(t)$ is the displacement of the punch along the z-axis.

The force $P(t)$ is applied to the punch (see Fig. 6.14), so the following equilibrium equation must be satisfied at an arbitrary instant of time

$$\int_{-a(t)}^{a(t)} p(x,t)\,dx = P(t). \tag{6.127}$$

The contact pressure is equal to zero at the ends of the contact region because of smoothness of the punch shape, so

$$p(\pm a(t), t) = 0. \tag{6.128}$$

The equations (6.124)–(6.128) are used to find the unknown functions $p(x,t)$, $w(x,t)$, $w_*(x,t)$, $a(t)$ and $D(t)$.

6.8.2 The cases of increasing, decreasing and constant contact region

Let us consider the Eq. (6.126) at the end of the contact region $a(t)$. Taking into account Eqs. (6.124) and (6.128), we obtain

$$w_*(a(t),t) = D(t) - f(a(t)). \tag{6.129}$$

Subtracting Eq. (6.129) from Eq. (6.126) gives

$$w(x,t) + w_*(x,t) - w_*(a(t),t) = f(a(t)) - f(x). \tag{6.130}$$

After differentiation Eq. (6.130) with respect to time and use of Eqs. (6.124) and (6.125) we obtain

$$k\frac{\partial p(x,t)}{\partial t} + K_w p(x,t) - K_w p(a,t) - \frac{\partial w_*(x,t)}{\partial a}\cdot\frac{da}{dt} = \frac{df}{da}\cdot\frac{da}{dt}. \tag{6.131}$$

Eq. (6.131) is valid within the contact region $(-a(t), a(t))$. Upon integrating Eq. (6.131) over this region, taking into account Eq. (6.128) and the relationship

$$\frac{d}{dt}\int_{-a}^{a} p(x,t)\,dx = \int_{-a}^{a} \frac{\partial p(x,t)}{\partial t}\,dx,$$

we have

$$kP'(t) + K_w P(t) = \left[f'(a) + \frac{\partial w_*(x,t)}{\partial a}\right]\frac{da}{dt}2a. \tag{6.132}$$

The conditions corresponding to the cases of increasing, decreasing and constant contact region can be obtained based on Eq. (6.132).

Let us consider the case of increasing contact region, i.e. $\frac{da}{dt} > 0$, and find the restriction imposed on the function $P(t)$. Eqs. (6.125) and (6.128) show that the relation $w_*(a(t),t) = 0$ is valid for an arbitrary moment of time. Differentiating this identity with respect to time, we obtain

$$\frac{\partial w_*}{\partial t} + \frac{\partial w_*}{\partial a}\cdot\frac{da}{dt} = 0.$$

It follows from Eqs. (6.125) and (6.128) that $\frac{\partial w_*}{\partial t} = 0$. So

$$\frac{\partial w_*}{\partial a}\cdot\frac{da}{dt} = 0.$$

Then it follows from Eq. (6.132) that the rate of the contact width increase is calculated by

$$\frac{da}{dt} = \frac{kP'(t) + K_w P(t)}{2af'(a)}. \tag{6.133}$$

If the assumption is made that $f'(a) > 0$, the following conditions should be satisfied for the increase of the contact area

$$\frac{P'(t)}{P(t)} > -\frac{K_w}{k}. \tag{6.134}$$

As an example, let us consider the contact of a smooth punch with shape $\frac{x^2}{2R}$ which is loaded by the constant force $P(t) = P_0$. It is evident that the condition (6.134) is fulfilled. To find the contact width at an arbitrary instant of time we can use the equation which follows from Eq. (6.133) in this particular case

$$\frac{da}{dt} = \frac{K_w P_0 R}{2a^2}.$$

Upon integrating this equation we obtain

$$a^3 - a_0^3 = \frac{3}{2}P_0RK_wt, \quad a_0^3 = \frac{3}{2}P_0Rk$$

or

$$a = a_0\sqrt[3]{1 + \frac{K_w}{k}t}.$$

Let us find now the condition on the function $P(t)$ which provides constant contact width, i.e. $a(t) = a_0$. Eq. (6.132) for $\frac{da}{dt} = 0$, gives the following relation

$$kP'(t) + K_wP(t) = 0,$$

So the contact width is constant if the load changes exponentially with time

$$P(t) = P_0 \exp\left(-\frac{K_w}{k}t\right).$$

Differentiating Eq. (6.129) with respect to time, we obtain

$$\frac{\partial w_*(a(t),t)}{\partial t} + \left(\frac{\partial w_*(a(t),t)}{\partial a} + f'(a)\right)\frac{da}{dt} = D'(t),$$

If $\frac{da}{dt} = 0$, from this equation, taking into account Eqs. (6.125) and (6.128), it follows that $D'(t) = 0$. So, for the smooth punch, the constant contact width occurs if the approach between the punch and the foundation does not change during the wear process. The contact pressure $p(x,t)$ is determined by the equation which follows from Eq. (6.131)

$$k\frac{\partial p(x,t)}{\partial t} + K_wp(x,t) = 0.$$

So the contact pressure tends to zero if $t \to \infty$ and, as follows from Eq. (6.126), the shape of the worn surface is the same as the initial shape of the punch $f(x)$.

It is easy to show in a similar manner, that the contact width decreases, i.e. $\frac{da}{dt} < 0$, if the load $P(t)$ satisfies the equation

$$\frac{P'(t)}{P(t)} < -\frac{K_w}{k}.$$

It should be noted that this analysis holds for the simple model described above. Similar analyses can be applied to investigate more complicated contact problems with time-dependent contact region.

The example of the solution of the wear contact problem with increasing contact width in wear process is given in § 8.1.

Chapter 7

Wear of Inhomogeneous Bodies

7.1 Variable wear coefficient

Different technical methods used for hardening of surfaces change their properties and essentially influence the character of the surface wear during the friction process. Local surface hardening (laser processing, ion implantation, etc) produces a structural inhomogeneity and, as a consequence, nonuniform wear. This leads to a waviness which often improves the performance of friction pairs. For example, it is well known that in the imperfect lubrication regime artificial hollows are created on the friction surface, thereby increasing the oil capacity of the surface, which in turn reduces the wear and the danger of seizure (see Garkunov, 1985). The presence of such pockets makes it possible to limit the presence of wear products in the friction zone, thereby improving the wear resistance of the junction and the stability of its tribotechnical characteristics.

We will present a mathematical model for the study of wear kinetics and shape changes for surfaces with variable wear coefficient. This study makes it possible for us to analyze the worn surface shape dependence on geometric and tribotechnical hardening parameters and to discuss the problem of the choice of these parameters in order to make the worn surface have certain geometric properties.

7.1.1 Problem formulation

We consider the problem of the wear of an elastic half-space with variable wear resistance by a rigid body (a punch). We assume that the contact region Ω does not change during the punch movement. Note that the problem of wear of a punch with a variable wear resistance in contact with an elastic half-space may be considered in a similar manner. We expect that the wear rate $w(x, y, t)$ is related to the contact pressure $p(x, y, t)$ and the sliding velocity $V(x, y)$ at the half-space surface $z = 0$ by the formula

$$\frac{\partial w(x,y,t)}{\partial t} = K_w(x,y)\left[\frac{p(x,y,t)}{p^*}\right]^\alpha \left[\frac{V(x,y)}{V^*}\right]^\beta, \tag{7.1}$$

where p^* and V^* are characteristic values of pressure and velocity, respectively, α, β are parameters which depend on material properties, friction conditions, etc., and $K_w(x,y)$ is the wear coefficient ($K_w(x,y) > 0$) which is assumed to be dependent on the coordinates (x,y).

In any specific problem the displacement $u_z(x,y,t)$ of the half-space surface is related to the contact pressure by means of an operator A:

$$u_z(x,y,t) = A[p(x,y,t)]. \tag{7.2}$$

We assume that operator A is independent of time.

Eqs. (7.1) and (7.2) with the contact condition

$$u_z(x,y,t) + w(x,y,t) = D(t) - f_0(x,y) \tag{7.3}$$

provide the complete system for the analysis of wear kinetics of a half-space surface for a given initial shape of the punch $f_0(x,y)$ and approach function $D(t)$.

For a known total load $P(t)$ applied to the body and an unknown function $D(t)$ we must add to Eqs. (7.1)–(7.3) the equilibrium equation

$$\iint\limits_\Omega p(x,y,t)\,dxdy = P(t). \tag{7.4}$$

If either the function dD/dt which is the rate of surface approach, or the load function $P(t)$, possesses an asymptote, that is

$$\lim_{t\to+\infty}\frac{dD}{dt} = \lim_{t\to+\infty}\frac{dw(x,y,t)}{dt} = D_\infty,$$

or

$$\lim_{t\to+\infty} P(t) = P_\infty,$$

then the system of equations (7.1)–(7.3) (or (7.1)–(7.4)) permits the stationary solution

$$p_\infty(x,y) = p^*\left[\frac{D_\infty (V^*)^\beta}{K_w(x,y)V^\beta(x,y)}\right]^{1/\alpha}. \tag{7.5}$$

At the given asymptotic value P_∞ of the normal load, the constant D_∞ is determined from the equilibrium condition (7.4)

$$D_\infty = \frac{P_\infty^\alpha}{(p^*)^\alpha (V^*)^\beta}\left[\iint\limits_\Omega \frac{dxdy}{K_w^{1/\alpha}(x,y)V^{\beta/\alpha}(x,y)}\right]^{-\alpha}. \tag{7.6}$$

As is known, a solution with arbitrary initial conditions converges to the stationary solution (7.5) if and only if the latter is asymptotically stable. It is shown in § 6.3 that the operator A has to satisfy definite conditions to ensure asymptotic stability of the solution (7.5) and so the existence of the steady-state stage of the wear process. Sufficient conditions for the asymptotic stability of the solution (7.5) at $\alpha = 1$ and a constant wear rate coefficient K_w were established in § 6.3.

We will consider the operator A of the following types

- for the 2-D periodic contact problem

$$A[p(x,t)] = -\frac{2(1-\nu^2)}{\pi E}\int_0^l p(x',t)\ln 2\left|\sin\frac{\pi(x'-x)}{l}\right| dx', \tag{7.7}$$

where l is the period,

- for the 3-D contact problem

$$A[p(x,y,t)] = \frac{(1-\nu^2)}{\pi E}\iint_\Omega p(x',y',t)\phi(x',y',x,y)dx'dy', \tag{7.8}$$

where

$$\phi(x',y',x,y) = \frac{1}{[(x-x')^2+(y-y')^2]^{1/2}}. \tag{7.9}$$

These operators are positive semi-definite, and so ensure the asymptotic stability of the stationary solution (7.5), in the linear ($\alpha = 1$) and non-linear ($\alpha \neq 1$) cases.

The shape of the worn surface corresponding to the stationary solution (7.5), can be represented as a sum of a function $f(x,y)$ which is independent of time (stationary shape) and the time-dependent function $D(t)$. From Eqs. (7.2), (7.3) and (7.5) we obtain

$$\begin{gathered} f(x,y) = f_0(x,y) + A[p_\infty(x,y)] = \\ f_0(x,y) + A\left[\left(\frac{(p^*)^\alpha (V^*)^\beta D_\infty}{K_w(x,y)V^\beta(x,y)}\right)^{1/\alpha}\right]. \end{gathered} \tag{7.10}$$

The stationary shape $f(x,y)$ depends on the wear coefficient $K_w(x,y)$ and the type of the punch motion, i.e. the function $V(x,y)$.

Thus, if the restrictions needed for the existence of an asymptotically stable steady-state stage of the wear process are satisfied, the expression for the pressure $p(x,y)$ at an arbitrary instant of time can be written in the form

$$p(x,y,t) = p_\infty(x,y) + p_1(x,y,t),$$

where $p_\infty(x,y)$ is determined by Eq. (7.5) and

$$\lim_{t\to+\infty} p_1(x,y,t) = 0.$$

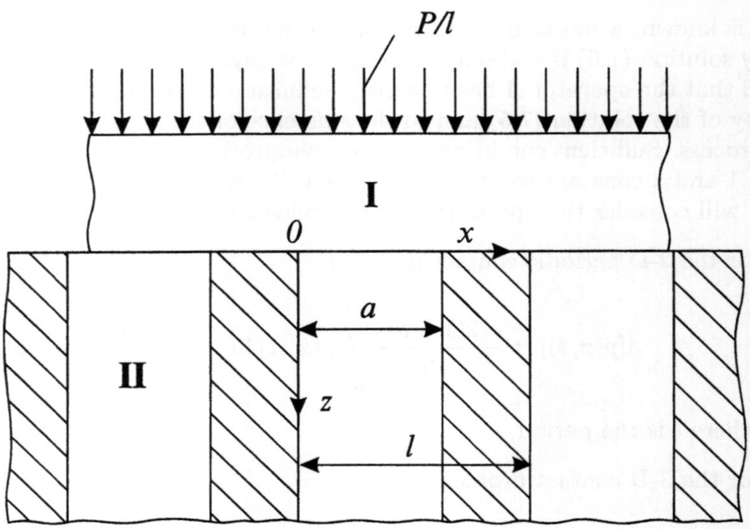

Figure 7.1: Scheme of contact of the flat punch and an elastic body hardened inside strips.

Thus the wear process is divided into two stages: running-in and steady-state. The steady-state stage is described by Eqs. (7.5), (7.6) and (7.10).

We will now determine the shape of the worn surface and contact characteristics for the steady-state stage of the wear process for surfaces hardened inside strips, circles, etc.

7.1.2 Steady-state wear stage for the surface hardened inside strips

We consider the 2-D periodic contact problem for an elastic half-space II and a punch I with a flat base (see Fig. 7.1). The punch moves back and forth along the y-axis in the plane $z = 0$. Within the strips $(nl+a \leq x < (n+1)l, -\infty < y < +\infty)$ the elastic body is subjected to local hardening, which in turn is determined by structural effects. For this reason the wear coefficient $K_w(x, y)$ is variable along the x-axis. The elastic characteristics E and ν of the half-space, which are, as a rule, structure-insensitive, will be considered as constant.

For definiteness we assume that only the surface of the elastic half-space wears in the friction process. For the case under consideration, Eq. (7.1) takes the form

$$\frac{\partial w(x,t)}{\partial t} = K_w(x) \left[\frac{p(x,t)}{p^*} \right]^\alpha . \tag{7.11}$$

We assume that the wear rate coefficient $K_w(x)$ is a step function:

$$K_w(x) = \begin{cases} K_{w1}, & x \in [nl, a+nl], \\ K_{w2}, & x \notin [nl, a+nl], \end{cases} \tag{7.12}$$

where K_{w1} and K_{w2} are the wear rate coefficients outside and inside the hardened zones $[nl+a, (n+1)l]$, respectively ($K_{w1} > K_{w2}$).

The problem is periodic with period l. Since there is a complete contact of the two bodies in the plane $z = 0$, the initial pressure is distributed uniformly, i.e. $p(x,0) = P(0)/l$ ($-\infty < x < +\infty$). During wear there is change of the initially plane surface of the half-space and redistribution of the pressure $p(x,t)$.

Since motion occurs in the direction perpendicular to the $x0z$ plane, we can neglect the influence of the friction force on the contact pressure distribution and use the operator A in the form (7.7). The wear of the surface $w(x,t)$ and the pressure $p(x,t)$ at an arbitrary instant of time are periodic functions. They can be determined from Eqs. (7.2), (7.4), (7.11) and Eq. (7.3) which takes the form

$$u_z(x,t) + w(x,t) = D(t).$$

From Eqs. (7.5), (7.6) and (7.10) we obtain the expressions for the pressure p_∞, the wear rate D_∞ and the shape $f(x)$ of the worn surface for the steady-state stage of the wear process:

$$p_\infty(x) = \begin{cases} p^* \left[\dfrac{D_\infty}{K_{w1}}\right]^{1/\alpha}, & x \in [nl, a+nl], \\ p^* \left[\dfrac{D_\infty}{K_{w2}}\right]^{1/\alpha}, & x \notin [nl, a+nl], \end{cases} \tag{7.13}$$

$$D_\infty = \left\{ \frac{P_\infty}{p^* \left[a K_{w1}^{-1/\alpha} + (l-a) K_{w2}^{-1/\alpha} \right]} \right\}^\alpha, \tag{7.14}$$

$$f(x) = -\frac{2(1-\nu^2)p^*}{\pi E} \left[\left(\frac{D_\infty}{K_{w1}}\right)^{1/\alpha} \int_0^a \ln 2 \left| \sin \frac{\pi(x'-x)}{l} \right| dx' + \left(\frac{D_\infty}{K_{w2}}\right)^{1/\alpha} \int_a^l \ln 2 \left| \sin \frac{\pi(x'-x)}{l} \right| dx' \right], \tag{7.15}$$

where P_∞ is the load applied to one period in the steady-state stage.

We introduce the dimensionless parameters

$$\bar{a} = \frac{a}{l}, \quad m = \frac{K_{w2}}{K_{w1}}, \quad m < 1 \tag{7.16}$$

and use the Lobachevsky function $L(y)$

$$L(y) = -\int_0^y \ln\cos t dt, \quad |y| \le \frac{\pi}{2}$$

to reduce Eq. (7.15) to the form

$$f(x) = \frac{2(1-\nu^2)D_\infty^{1/\alpha} p^* l}{\pi E K_{w2}^{1/\alpha}} \left(m^{1/\alpha} - 1\right) \left\{ (1-\bar{a})\ln 2 - \frac{1}{\pi}\left[L\left(\frac{\pi}{2} - \frac{\pi x}{l}\right) + L\left(\frac{\pi}{2} - \pi\bar{a} + \frac{\pi x}{l}\right)\right]\right\}. \tag{7.17}$$

For further calculation it is convenient to represent the function $L(y)$ in series form (see Gradshteyn and Ryzhik, 1971):

$$L(y) = y\ln 2 - \frac{1}{2}\sum_{n=1}^{\infty}(-1)^{n-1}\frac{\sin 2ny}{n^2}. \tag{7.18}$$

Using Eqs. (7.14), (7.17) and (7.18), we obtain finally

$$f(x) = -\frac{2(1-\nu^2)P_\infty m_1}{\pi^2 E(1-\bar{a}m_1)} \sum_{n=1}^{\infty} \frac{\sin \pi n\bar{a} \cos \pi n\left(\bar{a} - \frac{2x}{l}\right)}{n^2}, \tag{7.19}$$

where

$$m_1 = 1 - m^{1/\alpha}.$$

It follows from Eq. (7.19) that the function $f(x) \equiv 0$ for $\bar{a} = 0$ and $\bar{a} = 1$. This means that the surface of the elastic body remains plane during the wear process if there is no local hardening. For the remaining values of $\bar{a}$ the function $f(x)$ is periodic with period l.

The values of the function $f(x)$ at the points $x = 0$, $x = a$ and $x = l$ are determined by

$$f(0) = f(a) = f(l) = -\frac{(1-\nu^2)P_\infty m_1}{\pi^2 E(1-\bar{a}m_1)} \sum_{n=1}^{\infty} \frac{\sin 2n\pi\bar{a}}{n^2}. \tag{7.20}$$

Fig. 7.2 illustrates the function $f(x)$ (see Eq. (7.19)) for different values of the parameters m_1 and $\bar{a}$. This function describes the shape of the worn surface which becomes wavy due to the wear process.

Using the derivative $f'(x)$

$$f'(x) = -\frac{4(1-\nu^2)P_\infty m_1}{\pi E l(1-\bar{a}m_1)} \sum_{n=1}^{\infty} \frac{\sin \pi n\bar{a} \sin \pi n\left(\bar{a} - \frac{2x}{l}\right)}{n},$$

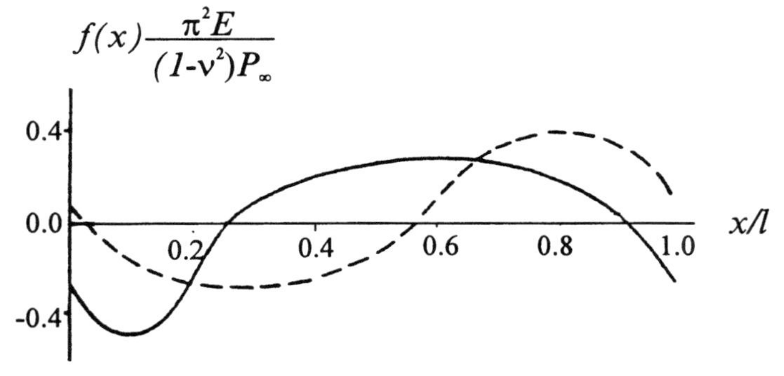

Figure 7.2: The steady-state shape of the worn surface for $m_1 = 0.5$, $\bar{a} = 0.2$ (solid line) and $m_1 = 0.3$, $\bar{a} = 0.6$ (dashed line).

we obtain the extremal values of the function $f(x)$ which are at the points $x = \frac{a}{2} + kl$ and $x = \frac{a+l}{2} + kl$ $(k = 0, \pm 1, \pm 2, \ldots)$, where $f'(x) = 0$:

$$f\left(\frac{a}{2}\right) = -\frac{2(1-\nu^2)P_\infty m_1}{\pi^2 E\,(1-\bar{a}m_1)} \sum_{n=1}^{\infty} \frac{\sin n\pi\bar{a}}{n^2},$$
$$f\left(\frac{a+l}{2}\right) = -\frac{2(1-\nu^2)P_\infty m_1}{\pi^2 E\,(1-\bar{a}m_1)} \sum_{n=1}^{\infty} (-1)^n \frac{\sin n\pi\bar{a}}{n^2}. \tag{7.21}$$

So the maximum difference in values of the function $f(x)$ is determined by

$$\Phi = \left| f\left(\frac{a}{2}\right) - f\left(\frac{a+l}{2}\right) \right| = \frac{4(1-\nu^2)P_\infty\, m_1}{\pi^2 E\,(1-\bar{a}m_1)} \sum_{n=1}^{\infty} \frac{\sin(2n-1)\pi\bar{a}}{(2n-1)^2}. \tag{7.22}$$

The plots of the function $\Phi(\bar{a})\frac{\pi^2 E}{4(1-\nu^2)P_\infty}$ (dashed lines) for various values of the parameter m_1 are presented in Fig. 7.3.

The volume of the valleys on the surface characterizes its oil capacity in contact interaction. We find the area S, enclosed between the curve $z = f(x)$ and the straight line $z = f\left(\frac{a+l}{2}\right)$, over the one period l:

$$S = lf\left(\frac{a+l}{2}\right) - \int_0^l f(x)dx. \tag{7.23}$$

It follows from Eq. (7.19), that

$$\int_0^l f(x)dx = 0.$$

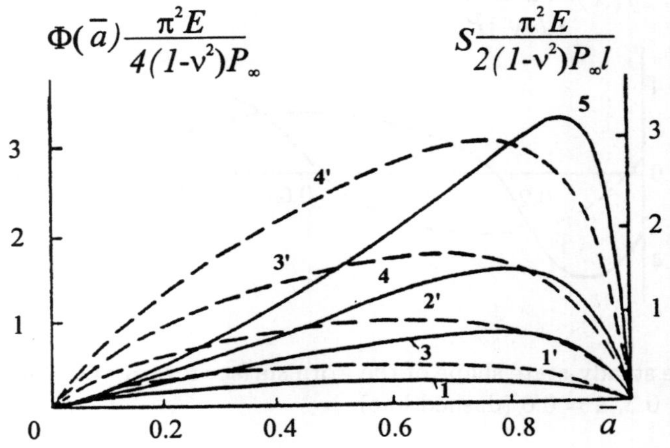

Figure 7.3: Functions $\Phi(\bar{a})$, Eq. (7.22), (dashed lines) and $S(\bar{a})$, Eq. (7.24), (solid lines) at $m_1 = 0.1$ (curve 1′), $m_1 = 0.2$ (curve 1), $m_1 = 0.3$ (curve 2′), $m_1 = 0.5$ (curves 3, 3′), $m_1 = 0.7$ (curves 4, 4′), $m_1 = 0.9$ (curve 5).

So we obtain from Eqs. (7.21) and (7.23) the value of S in a single period

$$S = \frac{2(1-\nu^2)P_\infty m_1 l}{\pi^2 E\,(1-\bar{a}m_1)} \sum_{n=1}^{\infty} (-1)^{n-1} \frac{\sin n\pi\bar{a}}{n^2}. \tag{7.24}$$

The volume of the valleys in the worn surface can be characterized by the value of S. Fig. 7.3 illustrates the dependence of the dimensionless area $S\dfrac{\pi^2 E}{2(1-\nu^2)P_\infty l}$ on the parameter $\bar{a}$ for various values of m_1 (solid lines). The results show that with variation of the parameter $\bar{a}$ from 0 to 1, i.e. with reduction of the width of the strip subjected to local hardening from l (total treatment of the surface) to 0 (untreated surface), there is initially an increase of the volume of the valleys in the worn surface and then a reduction of this volume to zero. For $\bar{a} = \bar{a}^*$ the volume of the valleys is maximal. The magnitude $\bar{a}^*$ depends on the ratio m of the wear coefficients of the hardened and unhardened zones and lies in the range $0.6 < \bar{a}^* < 1$ with variation of m from 0 to 1. The value of $\bar{a}^*$ increases as m decreases. Thus, to achieve a particular volume of the valleys in the worn surface we can select the ratio of the wear coefficients m and, for a chosen value of m, select the required width of the strip subjected to local hardening.

The area (volume) of the valleys in the worn surface, determined by Eq. (7.24) is numerically equal to the minimal amount of material worn during the running-in time T

$$S \le \int_0^l w(x,T)\,dx.$$

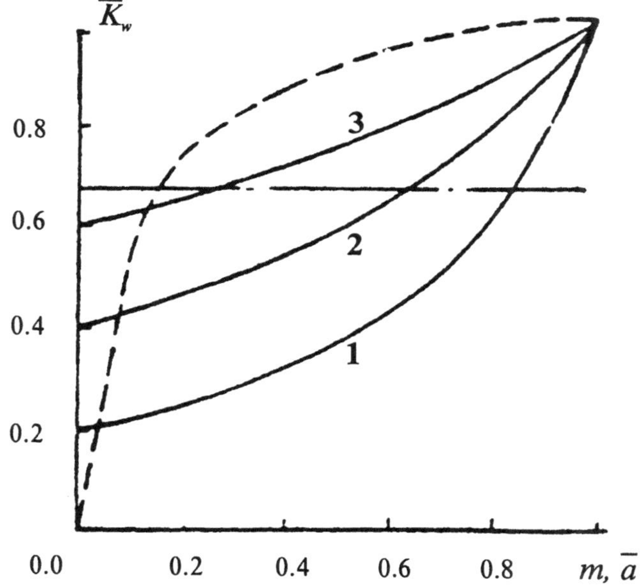

Figure 7.4: Dependence of the effective wear coefficient on parameter m at $\bar{a} = 0.9$ (dashed line) and on parameter $\bar{a}$ (solid lines) at $m = 0.2$ (curve 1), $m = 0.4$ (curve 2), $m = 0.6$ (curve 3); $\alpha = 1.5$.

The wear rate in the steady-state stage is characterized by the effective wear coefficient D_∞ (Eq. (7.14)), which can be represented in the form:

$$D_\infty = mK_{w1} \left\{ \frac{P_\infty}{p^* l \left[1 - \bar{a}\left(1 - m^{1/\alpha}\right)\right]} \right\}^\alpha .$$

This function shows that a given wear rate value can be achieved by appropriate choice of parameters $\bar{a}$ and m.

Fig. 7.4 illustrates the dependence of the dimensionless effective wear coefficient

$$\bar{K}_w = \frac{D_\infty}{K_{w1}} \left(\frac{p^* l}{P_\infty}\right)^\alpha = \frac{m}{\left[1 - \bar{a}\left(1 - m^{1/\alpha}\right)\right]^\alpha}$$

on the parameter $\bar{a}$ for three given values of m (solid lines) and on the parameter m for $\bar{a} = 0.9$ (dashed line). Intersection of these curves with dot-dashed line, $\bar{K}_w = 0.66$, gives some values of parameters $\bar{a}$ and m providing this fixed value of the wear rate in the steady-state stage.

Based on this analysis we can conclude that, during wear of a surface hardened inside strips, there arises an operational waviness, the parameters of which depend on the ratio of the wear coefficients of the hardened and unhardened zones and

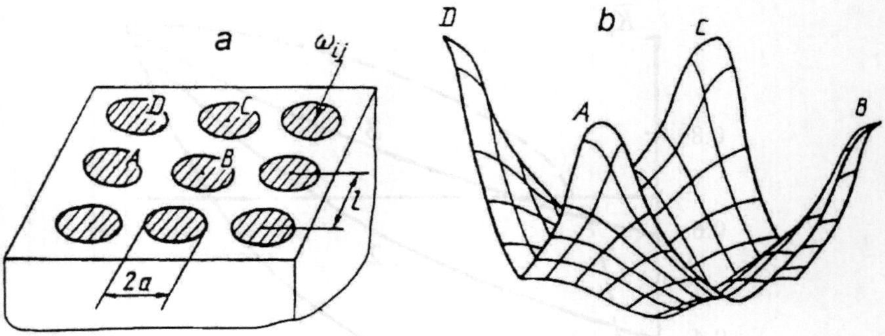

Figure 7.5: Scheme of the hardened domain arrangement at a surface of a half-space (a) and a shape of the worn surface at one period (b).

their characteristic dimensions. The volume of the valleys in the worn surface is larger, the larger the difference between the wear coefficients of the different zones. Their maximal values depend on the relative characteristic dimension of the hardened zone. The valleys reach maximal volume for $0.6 < \bar{a} < 1$ in the entire range of values of m.

Achievement of a specified value of the effective wear rate coefficient $\bar{K}_w$ can be realized either by varying the degree of hardening of the material (for fixed $\bar{a}$) or by varying the ratio of the dimensions of the hardened and unhardened zones (with fixed m) or, finally, by a combination of these methods.

7.1.3 Steady-state wear stage for a surface hardened inside circles

We consider a contact between an elastic half-space and a punch with a flat base ($f_0(x,y) = 0$) moving translationally on the half-space surface in various directions at a constant speed. We assume that the contact region Ω coincides with the plane $z = 0$. The half-space surface is hardened within circular domains ω_{ij} of radius a, arranged around the nodal points of a square lattice (Fig. 7.5(a)). The set of domains is denoted by $\omega = \sum_{i,j=1}^{\infty} \omega_{ij}$.

We assume that, due to hardening, the wear coefficient is a step function

$$K_w(x,y) = \begin{cases} K_{w1}, & (x,y) \in \Omega \setminus \omega, \\ K_{w2}, & (x,y) \in \omega. \end{cases} \tag{7.25}$$

We introduce the dimensionless parameter $m = \dfrac{K_{w2}}{K_{w1}}$ which characterizes the extent of hardening and the parameter $\bar{a} = \dfrac{a}{l}$ which is a geometric characteristic of hardening. The parameters vary in the ranges $0 < \bar{a} \leq 1/2$ and $m_0 \leq m < 1$, where m_0 is a limit value of m due to processing technology of the half-space surface.

We will establish the dependence of wear rate and geometric characteristics of the worn surface on the hardening parameters m and $\bar{a}$. From Eqs. (7.5), (7.6), (7.25) we obtain the pressure distribution $p_\infty(x,y)$ and the effective wear coefficient $\bar{K}_w$ in the steady-state stage of the wear process

$$\frac{p_\infty(x,y)}{p^*} = \begin{cases} \bar{p}_1, & (x,y) \in \Omega \setminus \omega, \\ \bar{p}_2, & (x,y) \in \omega, \end{cases} \tag{7.26}$$

$$\bar{K}_w = \frac{D_\infty}{K_{w1}} \left(\frac{p^* l^2}{P_\infty}\right)^\alpha \left(\frac{V^*}{V}\right)^\beta = \frac{1}{1 + \pi \bar{a}^2 m_2}, \tag{7.27}$$

where

$$\bar{p}_1 = \bar{K}_w^{1/\alpha} \left(\frac{P_\infty}{p^* l^2}\right), \quad \bar{p}_2 = \left(\frac{\bar{K}_w}{m}\right)^{1/\alpha} \left(\frac{P_\infty}{p^* l^2}\right), \quad m_2 = m^{-1/\alpha} - 1, \tag{7.28}$$

P_∞ is a load per period.

The shape of the worn surface in the steady-state stage of the wear process can be deduced from Eqs. (7.8), (7.10) and (7.26):

$$\bar{f}(x,y) = \frac{f(x,y)}{l} = \frac{1}{l} \iint\limits_\omega \Delta\bar{p}\phi(x',y',x,y)dx'dy' + C_1, \tag{7.29}$$

where

$$C_1 = \frac{1-\nu^2}{\pi E} \cdot \frac{\bar{p}_1 p^*}{l} \iint\limits_\Omega \phi(x',y',x,y)dx'dy',$$
$$\Delta\bar{p} = \frac{1-\nu^2}{\pi E} p^*(\bar{p}_2 - \bar{p}_1) = \frac{P_\infty m_2}{l^2(1+\pi\bar{a}^2 m_2)} \cdot \frac{1-\nu^2}{\pi E}, \tag{7.30}$$

where the function $\phi(x',y',x,y)$ is determined by Eq. (7.9).

To determine the shape of the worn surface $\bar{f}_1(x,y) = \bar{f}(x,y) - C_1$, we use an approximating formula

$$\bar{f}_1(x,y) = \sum_{k,l} \frac{\Delta\bar{p}}{l} \iint\limits_{\omega_{kl}} \phi(x',y',x,y)dx'dy' +$$
$$\sum_{i,j=1}^\infty \frac{\bar{P}l}{[(x-x_{ij})^2 + (y-y_{ij})^2]^{1/2}}, \tag{7.31}$$

where ω_{kl} is the hardened domain near the point (x,y). The pressure distribution $\Delta\bar{p}$ within ω_{kl} is taken into account. For the remote domains $\omega_{ij} \neq \omega_{kl}$ we replace the pressure $\Delta\bar{p}$ by concentrated forces $\bar{P} = \Delta\bar{p}\pi\bar{a}^2$ applied at the centres (x_{ij}, y_{ij})

of domains ω_{ij}. This allows the replacement of integration by summation for the domains ω_{ij}. This replacement is based on the analysis presented in Chapter 2. The calculations show that the error due to the replacement is of order $O(a^2/(x^2+y^2))$. For example, at $\bar{a} = 0.5$ the error is 0.9% at a distance $x^2 + y^2 = 2l^2$.

Eq. (7.31) for the points $x = y$ may be reduced to

$$\begin{aligned}
&\Phi_1(\xi) = \bar{f}_1(x(\xi), y(\xi)) - \bar{f}_1(x(0), y(0)) = \gamma(\xi)\Delta\bar{p}\pi\bar{a}^2 - \\
&4\Delta\bar{p}\left[4F(\bar{a}) + \sqrt{2}F\left(\frac{\bar{a}}{\sqrt{2}}\right) - \sqrt{2}(1-\xi)F\left(\frac{\bar{a}}{\sqrt{2}(1-\xi)}\right) - \right. \\
&2\left((1-\xi)^2+\xi^2\right)^{1/2} F\left(\frac{\bar{a}}{((1-\xi)^2+\xi^2)^{1/2}}\right) - \\
&\left. 2\left((1+\xi)^2+\xi^2\right)^{1/2} F\left(\frac{\bar{a}}{((1+\xi)^2+\xi^2)^{1/2}}\right)\right] + 2\Delta\bar{p}\pi\bar{a} - \Psi(\xi),
\end{aligned} \tag{7.32}$$

where

$$\Psi(\xi) = \begin{cases} 4\bar{a}\Delta pE\left(\dfrac{\sqrt{2}\xi}{\bar{a}}\right), & \sqrt{2}\xi \le \bar{a}, \\[2ex] 4\sqrt{2}\Delta\bar{p}\xi F\left(\dfrac{\bar{a}}{\sqrt{2}\xi}\right), & \sqrt{2}\xi > \bar{a}, \end{cases}$$

$$\xi = \frac{1}{2} - \frac{x}{l} = \frac{1}{2} - \frac{y}{l}, \quad \left(0 < \xi \le \frac{1}{2}\right),$$

$$\begin{aligned}
\gamma(\xi) = \sum_{k=1}^{\infty}\sum_{n=1}^{\infty}&\left(\frac{1}{[(k-1+\xi)^2+(n-1+\xi)^2]^{1/2}} + \right. \\
&\frac{1}{[(k-1+\xi)^2+(n-\xi)^2]^{1/2}} + \frac{1}{[(k-\xi)^2+(n-1+\xi)^2]^{1/2}} + \\
&\left.\frac{1}{[(k-\xi)^2+(n-\xi)^2]^{1/2}} - \frac{4}{[(k-1)^2+n^2]^{1/2}}\right) - \\
&\left(\frac{1}{\sqrt{2}(1-\xi)} + \frac{2}{[(1-\xi)^2+\xi^2]^{1/2}} + \frac{2}{[(1+\xi)^2+\xi^2]^{1/2}} - \frac{1}{\sqrt{2}} - 4\right)
\end{aligned} \tag{7.33}$$

where $K(t)$ and $E(t)$ are the complete elliptic integrals of the first and the second kinds, respectively, $F(t) = E(t) - (1-t^2)K(t)$.

The series (7.33) converges, since its each term is of the order $1/\left(k^2+n^2\right)^{3/2}$.

The expressions for an arbitrary point (x, y), which are not provided here due to their cumbersome form, are similar to Eq. (7.32). The plot of the function $\bar{f}_1(x, y)$ for one period of the lattice is presented in Fig. 7.5 (b).

The amplitude $L = \Phi_1\left(\frac{1}{2}\right)$, and the area of diagonal section

$$S = \int_0^{1/2} \Phi_1(\xi)d\xi \tag{7.34}$$

are the important geometric characteristics of the worn surface. From Eqs. (7.30), (7.32) and (7.33), on expanding the elliptic integrals for small parameters, we can obtain the following expression for L:

$$L = \frac{\lambda \bar{a} m_2}{1 + \pi \bar{a}^2 m_2}\left(1 - 1.1332\bar{a} - 0.2285\bar{a}^3 + O(\bar{a}^5)\right),$$

where

$$\lambda = \frac{2(1-\nu^2)P_\infty}{El^2}.$$

The simplest analysis of the dependence of L on the parameters m_2 and $\bar{a}$ shows that

1) at fixed $\bar{a}$ the value L reaches a maximum at $m = m_0$;

2) for any m, there exists a value $\bar{a}^*$ $(0 < \bar{a}^* < 1/2)$ at which the amplitude L is maximal.

The plots of the function $L(\bar{a})$ for different values of m_2 are depicted by the solid lines in Fig. 7.6.

The dashed lines in the figure show the dependence of the hollow section area S on the parameter $\bar{a}$ at various values of m_2 obtained numerically from Eqs. (7.32) and (7.34).

The analysis of the function (7.27) shows that the effective wear coefficient $\bar{K}_w$ in the steady-state stage of the wear process is equal to 1 for $\bar{a} = 0$ (nonhardened surface) and decreases as the radius $\bar{a}$ of the hardened domain, or the parameter m_2 increases (m decreases).

Thus, the variation of the parameters m and $\bar{a}$ within the limits admissible by technology makes it possible to control the tribological and geometric characteristics of the wavy surfaces generated due to wear.

7.1.4 The shape of the worn surface of an annular punch for various arrangements of hardened domains

Wear contact problems for surfaces with bounded contact region possess additional peculiarities due to the edge effects. We find the steady-state shape of the worn surface of an annular-end punch. We established earlier that the shape of the worn surface depends on the arrangement of hardened domains, as well as on the character of relative motion of the friction surfaces.

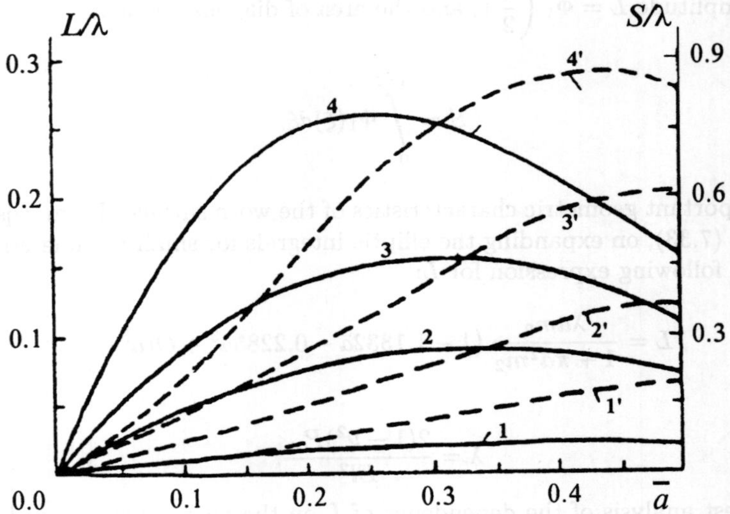

Figure 7.6: Dependencies of the amplitude L (solid lines) and the section area S (dashed lines) of the hollow at the worn surface on hardening parameters: $m_2 = 0.1$ (curves 1, 1′), $m_2 = 0.5$ (curves 2, 2′), $m_2 = 1$ (curves 3, 3′), $m_2 = 2$ (curves 4, 4′).

We consider first a punch with a flat base hardened inside the set of domains $\omega = \sum_{i=1}^{N} \omega_i$ (see Fig. 7.7 (a)). The i-th domain ω_i is an annulus with the inner radius r_i and thickness ρ, r_1 and $r_N + \rho$ are the inner and outer radii of the punch, respectively, N is the number of hardened domains. The punch is pressed into the elastic half-space by the load P_∞ and moves translationally in the various directions so that $|V| = \text{const}$.

The wear coefficient $K_w(x, y)$ is determined by Eq. (7.25).

In the steady-state stage of the wear process the pressure is distributed according to Eqs. (7.26) and (7.28). From Eq. (7.28) we obtain

$$\frac{\bar{p}_1}{\bar{p}_2} = m^{1/\alpha}$$

$$\bar{p}_2 = \bar{p}_1 + \Delta\bar{p},$$

where

$$\Delta\bar{p} = \bar{p}_1 \left(\frac{1}{m^{1/\alpha}} - 1 \right) = \bar{p}_1 m_2.$$

The pressure $\bar{p}_1$ is found from the equilibrium condition

$$\bar{p}_1 \pi \left[(r_N + \rho)^2 - r_1^2 \right] + \bar{p}_1 m_2 \pi \rho \left(2 \sum_{i=1}^{N} r_i + N\rho \right) = \frac{P_\infty}{p^*}.$$

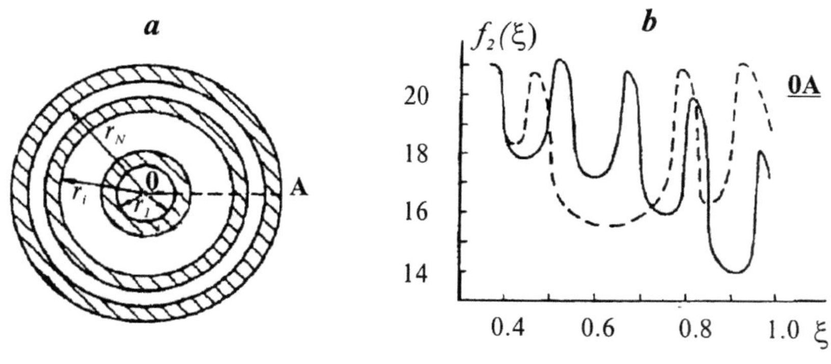

Figure 7.7: Scheme of hardened domain arrangement (a) and the shape of the worn surface at the radial cross-section (b) for uniform distribution of the hardened annular domains (solid line) and for the case $\xi_1 = 0.35$, $\xi_2 = 0.47$, $\xi_3 = 0.78$, $\xi_4 = 0.925$, $\xi_5 = 0.975$ (dashed line), $\rho/r_1 = 0.2$.

The shape of the worn surface can be obtained from Eqs. (7.8), (7.10) and (7.26):

$$f(r) = \frac{4(1-\nu^2)p^*}{\pi E}\bar{p}_1 \int\limits_{r_1}^{r_N+\rho} K\left(\frac{2\sqrt{rr'}}{r+r'}\right)\frac{r'dr'}{r+r'}+$$
$$\frac{4(1-\nu^2)p^*}{\pi E}\bar{p}_1 m_2 \sum_{i=1}^{N}\int\limits_{r_i}^{r_i+\rho} K\left(\frac{2\sqrt{rr'}}{r+r'}\right)\frac{r'dr'}{r+r'}. \tag{7.35}$$

This expression was simplified and written in means of elliptic integrals in Goryacheva and Torskaya (1992).

Fig. 7.7(b) illustrates the shape of the worn surface at a radial cross-section for different arrangement of the hardened domain. The dependence $f_2(\xi) = \dfrac{f(\xi)\pi E}{2(1-\nu^2)p^*\bar{p}_1(r_N+\rho)}$ $\left(\xi = \dfrac{r}{r_N+\rho}\right)$ when hardened domains are arranged uniformly along the radius, is depicted by the solid line ($N = 5$, $\xi_1 = 0.35$). Due to the boundedness of the contact region, the hardened domains wear nonuniformly (the edge effect). The location of the hardened domains significantly affects the stationary shape of the surface. By varying the parameters r_i, we may obtain the shape of the worn surface satisfying, for instance, the condition $f(\xi_i+\rho/2) = \text{const}$ ($\xi_i = r_i/(r_N+\rho)$). It is represented by the dashed line in Fig. 7.7(b).

As an example of another character of hardening we consider the contact between a rigid annular punch rotating about its axis and an elastic half-space which has a surface which does not wear. The flat surface of the punch is hardened inside N sectors $\omega_k \equiv \{r_1 < r < r_2,\quad 2\pi k/N - \theta_1/2 \le \theta \le 2\pi k/N + \theta_1/2\}$ as shown in Fig. 7.8(a) (shaded domains). From Eqs. (7.5) and (7.25) we can deduce the

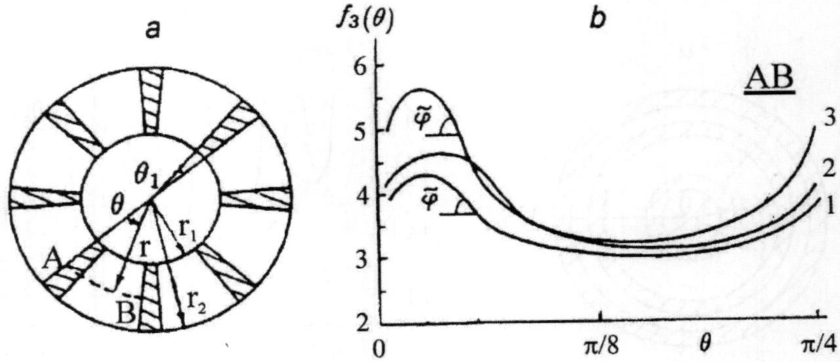

Figure 7.8: Scheme of hardened domain arrangement (a) and the shape of the worn surface at the cross-section $r^*/r_2 = 0.75$ (b) for $m_3 = 2$, $\theta_1 = 8°$ (curve 1), $m_3 = 2$, $\theta_1 = 12°$ (curve 2), $m_3 = 3$, $\theta_1 = 8°$ (curve 3).

following expression for the steady-state pressure distribution

$$\frac{p_\infty(r,\theta)}{p^*} = \begin{cases} \bar{p}_1(r), & (r,\theta) \in \Omega - \omega, \\ \bar{p}_2(r), & (r,\theta) \in \omega. \end{cases} \tag{7.36}$$

where

$$\bar{p}_1(r) = \left(\frac{C}{K_{w1} r^\beta}\right)^{1/\alpha}, \quad \bar{p}_2(r) = \left(\frac{C}{K_{w2} r^\beta}\right)^{1/\alpha},$$

Ω is the region ($\rho_1 \le \rho_2$, $0 < \theta \le 2\pi$), $\omega = \sum_{k=1}^{N} \omega_k$ and the constant C is determined from the equilibrium condition

$$P_\infty = \iint_\Omega p_\infty(r,\theta) r dr d\theta. \tag{7.37}$$

The shape of the worn surface was calculated from Eqs. (7.8), (7.10) and (7.36) in Goryacheva and Torskaya (1992). Fig. 7.8 (b) illustrates the worn surface shape

$$f_3(\theta) = f(\theta, r^*) \cdot \left(\frac{K_{w_1}}{C}\right)^{\frac{1}{\alpha}} \cdot \frac{\pi E}{(1-\nu^2)\, p^*}$$

for $\dfrac{r_1}{r_2} = 0.5$, $\dfrac{r^*}{r_2} = 0.75$, $\alpha = \beta$ and different parameters $m_3 = m^{1/\alpha}$ and θ_1. The results of calculations show that the treatment parameters (m and θ_1) influence the cavity profile characteristics (its amplitude, slope angle $\tilde{\varphi}$); this fact opens the possibility of using local hardening to produce specific surface formations in wear processes.

Thus the characteristics of the geometry of the worn surface depend strongly on the geometry of the contact pair, the relative motion of the parts and the local geometrical and tribotechnical hardening parameters.

7.2 Wear in discrete contact

The discrete character of the contact interaction plays a significant role in the wear process. Wear changes the surface macro- and microgeometry; at the same time the geometric and mechanical characteristics of the surface together with the contact conditions determine the surface wear.

In what follows we propose a mathematical model of a discrete contact wear. This model is based on the results presented in Chapter 2 and it can be used to study the process of running-in of bodies with surface microgeometry and also to investigate wear of inhomogeneous surfaces with rigid inclusions.

7.2.1 Mathematical model

We consider a system of N cylindrical punches, each with a flat base of radius a, which moves over the surface of an elastic half-space (see Fig.2.9). The system of punches is interconnected and is acted upon by a normal load $P(t)$. The punches are arranged arbitrarily inside a nominal region Ω.

We assume that in the process of friction the half-space wears so that its surface always remains flat and the wear of the punches leads to a gradual decrease of their heights. The wear rate at each contact spot is related to the load P_j acting on it and to a sliding velocity V_j, by a power–law relationship

$$\frac{dw_j}{dt} = K_w \left(\frac{P_j}{P^*}\right)^{\alpha} \left(\frac{V_j}{V^*}\right)^{\beta}, \quad j = 1, 2, \ldots, N, \tag{7.38}$$

where w_j is the linear wear of the j-th punch in the center of its own contact area ($w_j(0) = 0$), P^* and V^* are characteristic values of the load and the sliding velocity, respectively, and K_w is a coefficient that is equal to the linear wear rate at $P_j = P^*$, $V_j = V^*$.

From the contact condition of the j-th punch of the system with the elastic half-space it follows that

$$u_j(t) = h_j - w_j(t) - D(t), \tag{7.39}$$

where $u_j(t)$ is the indentation of the j-th punch at an arbitrary time t ($u_j(0) = D_j$), h_j is the initial height distribution of punches, and $D(t)$ is the approach of bodies under load ($D_0 = D(0) \neq 0$) (see Fig. 2.9).

In § 2.3, based on the discrete contact model, we deduced the relationship (2.35) between the indentation of each punch and the load distributed between the punches in the system. This relation at an arbitrary instant of time can be written as

$$u_j(t) = \frac{1-\nu^2}{2aE} P_j(t) + \frac{1-\nu^2}{\pi a E} \sum_{i=1,\, i\neq j}^{N} P_i(t) \arcsin \frac{a}{l_{ij}}. \tag{7.40}$$

Here E and ν are the Young's modulus and Poisson ratio of the elastic half-space, l_{ij} is the distance from the fixed j-th punch to the i-th punch. We should note that Eq. (7.40) holds exactly at the initial instant $t = 0$ when the contacting surface of each punch is flat. In the wear process, because of the nonuniform pressure distribution on the contact area, the shape of the contacting surface of each punch changes in relation to its location in Ω. In what follows we assume that the changes are negligibly small (they amount chiefly to rounding of the corners, where the greatest contact pressure occurs at the initial instant) and so we use Eq. (7.40) at any instant of time.

Eqs. (7.38)–(7.40) and the equilibrium condition

$$P(t) = \sum_{i=1}^{N} P_i(t) \tag{7.41}$$

provide a complete system of equations for studying the wear kinetics of the interconnected punches located at arbitrary distances l_{ij} from each other.

7.2.2 Model analysis

Differentiating Eqs. (7.39) and (7.40) with respect to time and taking account of Eq. (7.38), we can transform the resultant system of equations as follows:

$$\dot{q}_i(\bar{t}) + \sum_{j=1,\, j\neq i}^{N} b_{ij}\dot{q}_j(\bar{t}) = -\dot{D}_1(\bar{t}) - \Gamma_i q_i^{\alpha}(\bar{t}), \quad i = 1, 2, \ldots, N, \tag{7.42}$$

$$\sum_{i=1}^{N} q_i(\bar{t}) = \frac{1-\nu^2}{2aD_0E}P(\bar{t}), \tag{7.43}$$

where

$$q_i(\bar{t}) = \frac{1-\nu^2}{2aD_0E}P_i\left(\frac{\bar{t}D_0}{K_w}\right), \quad b_{ij} = \frac{2}{\pi}\arcsin\frac{a}{l_{ij}}, \quad (i \neq j),$$

$$\Gamma_i = \left(\frac{V_i}{V^*}\right)^{\beta}\left[\frac{2aD_0E}{(1-\nu^2)P^*}\right]^{\alpha}, \quad D_1(\bar{t}) = \frac{D\left(\frac{\bar{t}D_0}{K_w}\right)}{D_0}, \quad \bar{t} = t\frac{K_w}{D_0}, \tag{7.44}$$

and $\dot{(\cdot)}$ denotes the derivative $\dfrac{d}{d\bar{t}}$.

At a given initial height distribution h_i of punches, the initial values of $q_i(0)$ are known from Eqs. (7.39), (7.40) and (7.41) at $t = 0$.

We examine the solution of system (7.42) in accordance with different ways in which the problem can be formulated.

The case $\dot{D}_1(\bar{t}) = -\delta$. If in the process of wear the system of punches moves along the normal to the half-space surface with constant velocity, i.e., $D_1(\bar{t}) =$

$-\delta\bar{t}+1$, then Eqs. (7.42) become

$$\dot{q}_i(\bar{t}) + \sum_{j=1,\ j\neq i}^{N} b_{ij}\dot{q}_j(\bar{t}) = \delta - \Gamma_i q_i^{\alpha}(\bar{t}), \quad i = 1, 2, \ldots, N \tag{7.45}$$

and Eq. (7.43) serves to determine the behavior of the total load $P(t)$ acting on the system of punches.

We represent the solution of the system (7.45) in the form

$$q_i(\bar{t}) = \left(\frac{\delta}{\Gamma_i}\right)^{1/\alpha} + \phi_i(\bar{t}), \quad (i = 1, 2, \ldots, N), \tag{7.46}$$

where the functions $\phi_i(\bar{t})$ satisfy the following system of equations:

$$\dot{\phi}_i(\bar{t}) + \sum_{i=1, i\neq j}^{N} b_{ij}\dot{\phi}_j(\bar{t}) = -c_i\phi_i(\bar{t}) + O(\phi_i^2(\bar{t})), \tag{7.47}$$

$$c_i = \alpha\delta^{(\alpha-1)/\alpha}\Gamma_i^{1/\alpha} \quad (c_i > 0). \tag{7.48}$$

Since the last term in Eq. (7.47) has the estimate $O(||\phi||^2)$ where $||\phi||^2 = \phi_1^2(\bar{t}) + \phi_2^2(\bar{t}) + \ldots + \phi_N^2(\bar{t})$, the system (7.47) has a solution $\phi_i(t) \equiv 0$ that is asymptotically stable, if this property is displayed by the corresponding linear system with constant coefficients (see, for example, Cesari, 1959 and Petrovsky, 1973), having the following matrix form

$$B\dot{\phi} = -C\phi, \tag{7.49}$$

where B is a symmetric matrix with positive elements ($b_{ii} = 1$, and b_{ij} $(i \neq j)$ are defined by Eq. (7.44)), C is a diagonal matrix with elements $c_{ii} = c_i$ determined by Eq. (7.48) and $c_{ij} = 0$ if $i \neq j$.

We will show at first that matrix B is positive definite, i.e.,

$$(Bx, x) > 0 \tag{7.50}$$

for all x satisfying the condition $||x|| \neq 0$. We assume that x is a vector whose components, to within the multiplier $(1-\nu^2)/(2aD_0E)$, are the forces P_j acting on the punches ($j = 1, 2, \ldots, N$). Then the components of vector Bx constitute the elastic displacements of the corresponding punches. Therefore the scalar product (Bx, x) is the work of nonzero forces on the corresponding elastic displacements, which is always positive; inequality (7.50) is thus proved.

In view of Eq. (7.50), matrix B is nonsingular, and has an inverse matrix B^{-1}. Therefore Eq. (7.49) is equivalent to the equation

$$\dot{\phi} = -B^{-1}C\phi. \tag{7.51}$$

Then we consider the function $V = (Bx, x)$, which is positive definite by virtue of Eq. (7.50) and is continuously differentiable. In view of Eq. (7.51), and taking into account that

$$(Bx, -B^{-1}Cx) = (x, -BB^{-1}Cx) = -(x, Cx),$$

since $B = B^T$ by virtue of the symmetry of B, we can write the derivative of V in the form

$$\dot{V} = \frac{dV}{d\bar{t}} = (-Cx, x) + (Bx, -B^{-1}Cx) = -2(x, Cx).$$

Thus, we can specify a continuous function $W = 2(x, Cx) > 0$ for all x ($\|x\| \neq 0$), such that the derivative of the Lyapunov function V, by virtue of system (7.51), satisfies the condition $\dot{V} = -W$. In accordance with Lyapunov's lemma (see Petrovsky, 1973), the solution $\phi = 0$ of the system (7.51), or the equivalent system (7.49), is asymptotically stable.

So we can assert that there exists an asymptotically stable stationary solution of the system (7.45):

$$q_{i\infty} = (\delta/\Gamma_i)^{1/\alpha} = \frac{(1-\nu^2)P^*}{2aD_0E}\left(\frac{V^*}{V_i}\right)^{\beta/\alpha}\delta^{1/\alpha} \tag{7.52}$$

Note, that the particular case $\delta = 0$ corresponds to the solution of the problem for the system of punches arranged at the fixed distance from the half-space.

The case $P(t) = P_\infty$. In practice the total load $P(t)$ applied to the system of punches is given, rather than the punch displacement. We consider this case, and assume for the sake of being definite, that $P(t) = P_\infty$. As before, we will seek the solution of Eqs. (7.42) and (7.43) in the form (7.46), where the constant δ has the form which follows from Eq. (7.43):

$$\delta = \left[\frac{P_\infty(1-\nu^2)}{2aD_0E\sum\limits_{j=1}^{N}(1/\Gamma_j)^{1/\alpha}}\right]^\alpha, \tag{7.53}$$

while the functions $\phi_i(t)$ $(i = 1, 2, \ldots, N)$ satisfy the system

$$\dot{\phi}_i(\bar{t}) + \sum_{i=1,\ i\neq j}^{N} b_{ij}\dot{\phi}_j(\bar{t}) = -c_i\phi_i(\bar{t}) - \dot{D}_1(\bar{t}) + \delta + O(\phi_i^2(\bar{t})), \tag{7.54}$$

$$\sum_{i=1}^{N}\phi_i(\bar{t}) = 0, \tag{7.55}$$

where coefficients c_i are determined from Eq. (7.48). We divide the i-th equation of the system (7.54) by the constant $c_i > 0$ and then add up the N resultant equations. Taking into account Eq. (7.55), we obtain

$$\sum_{i=1}^{N}\left(\frac{1}{c_i} + \sum_{j=1}^{N}\frac{b_{ij}}{c_j}\right)\dot{\phi}_i(\bar{t}) = \left[-\dot{D}_1(\bar{t}) + \delta\right]\sum_{i=1}^{N}\frac{1}{c_i} + O(\|\phi\|^2),$$

which implies that

$$\delta - \dot{D}_1(\bar{t}) = \sum_{i=1}^{N} a_i \dot{\phi}_i(\bar{t}) + O(\|\phi\|^2), \tag{7.56}$$

$$a_i = \left(\frac{1}{c_i} + \sum_{j=1}^{N} \frac{b_{ij}}{c_j}\right) \Big/ \sum_{k=1}^{N} (1/c_k) .$$

Using Eq. (7.56), we reduce the system (7.54) to

$$\dot{\phi}_i(\bar{t}) + \sum_{j=1,\ i\neq j}^{N} b_{ij}\dot{\phi}_j(\bar{t}) - \sum_{j=1}^{N} a_j \dot{\phi}_j(\bar{t}) + c_i \phi_i(\bar{t}) = O(\|\phi\|^2). \tag{7.57}$$

The asymptotic stability of the zero solution of the system (7.57) has been proved in Goryacheva (1988) using Lyapunov's method.

From Eqs. (7.46) and (7.53) we write the stationary solution of the system (7.42) and (7.43) for $P(t) = P_\infty$ as

$$q_{i\infty} = \frac{(1-\nu^2)P_\infty}{2aD_0E \sum\limits_{j=1}^{N} (1/V_j)^{\beta/\alpha}} \cdot \frac{1}{(V_i)^{\beta/\alpha}}, \tag{7.58}$$

which is asymptotically stable. Note that, in the case $P(t) \neq$ const, the solution of the system (7.42) and (7.43) tends to the solution (7.58) as $t \to +\infty$ if the load $P(t)$ applied to the system of punches tends to the constant value, i.e. $P(t) \to P_\infty$ as $t \to +\infty$.

The stationary, or steady-state solution $q_{i\infty}$ given in Eq. (7.58), depends upon the total load applied to the system of punches, the sliding velocity and the position and the size of punches, and is independent of the initial values $q_i(0)$. The initial values have an influence upon the time when the system gets into the steady-state wear condition (running-in time).

7.2.3 Running-in stage of wear process

As an example, we investigate the process of running-in of a system of N punches of radius a arranged inside a circular region Ω at sites of a hexagonal lattice with a constant pitch l. All the punches are initially at the same level. This model has been described in § 2.3, where the indentation of a limited system of punches was investigated. We assume that the system is slipping back and forth along the surface of the elastic half-space, in such a way that the average slip velocities of the punches are the same, i.e. $V_1 = V_2 = \ldots = V_N$. The system is acted upon by a constant load P_∞.

The degree of redistribution of the loads applied to the punches is determined by the ratio $q_{\min}/q_{\max}$, where $q_{\min}$ is the minimum load per punch (for the present model it corresponds to the load applied to the central punch), and $q_{\max}$ is the maximum load per punch (in the present case it is the load applied to the punch

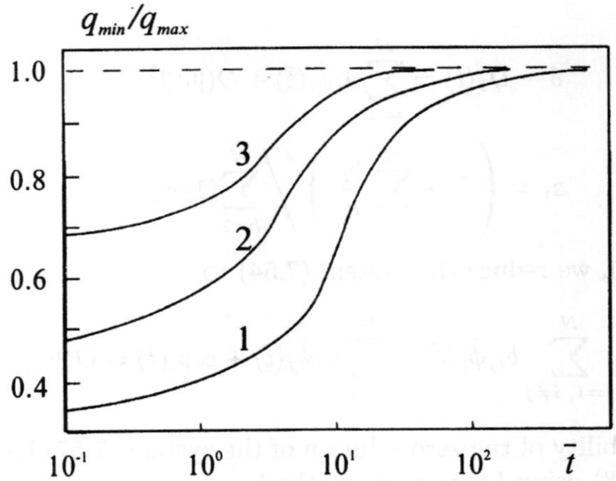

Figure 7.9: Effect of punch density on running-in time: $a/l = 0.355$ (curve 1), $a/l = 0.2$ (curve 2), $a/l = 0.1$ (curve 3) at $\alpha = 1$, $\Gamma = 0.16$; the dashed line corresponds to the steady-state condition.

located at the vicinity of the contour of Ω). In the steady-state regime for the reciprocating motion of the system of punches, $q_{\min}/q_{\max} = 1$.

The running-in time T is evaluated from the condition

$$\left|\frac{q_{\min}(T)}{q_{\max}(T)} - 1\right| < \varepsilon,$$

where ε is a small value given in advance ($\varepsilon \ll 1$).

Fig. 7.9 illustrates the influence of the punch density on the running-in time for the model consisting of 55 punches ($N = 55$). In calculations we considered $\alpha = 1$, $\Gamma = 0.16$, where $\Gamma = \Gamma_1 = \ldots = \Gamma_N$ is determined by Eq. (7.44). The results show that the lower the density, the shorter is the time needed to reach the equilibrium slate. This was to be expected, since at a low punch density the system is close to the equilibrium state even at the initial moment of time $t = 0$.

Fig. 7.10 depicts the dependence of the running-in stage on the parameter α. It follows from Eq. (7.38) that parameter α influences the wear rate of each punch. The higher is α, the more are the differences in the wear rate of the punches acted on by different loads. So for the cases under consideration (the initial conditions are the same for all cases) the running-in time is less at the higher α.

The analytical results presented here are in a good agreement with the experimental results on the models which are described in Goryacheva and Dobychin (1988).

In the general case of an arbitrary system of punches, the running-in time T

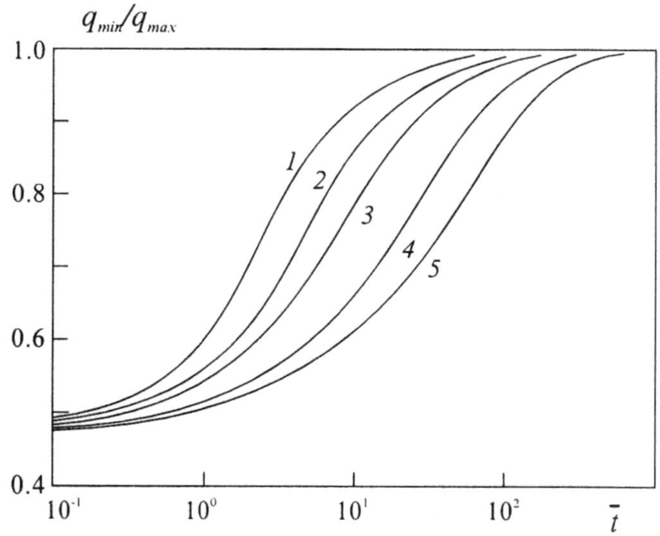

Figure 7.10: Running-in stage of the wear process for different values of the parameter α: $\alpha = 2$ (curve 1), $\alpha = 1.5$ (curve 2), $\alpha = 1$ (curve 3), $\alpha = 0.7$ (curve 4), $\alpha = 0.5$ (curve 5) at $a/l = 0.2$, $\Gamma = 0.16$.

can be evaluated from the condition

$$\sum_{i=1}^{N} \left[\frac{q_i(T) - q_{i\infty}}{q_{i\infty}} \right]^2 < \varepsilon.$$

We can investigate the wear kinetics of a system of punches engaged in rotational motion about some fixed point in a similar way.

The experimental and analytical results show that the running-in time is much less than the time needed to wear the punches at the given value. So, most of the time the steady-state wear occurs.

7.2.4 Steady-state stage of wear process

Along with the stationary load distribution given by Eq. (7.52) or Eq. (7.58), which occurs in the steady-state stage of the wear process, we must consider the shape of the worn surface which is characterized by the punch heights, $h_i - w_i(t)$. The stationary load distribution ensures an equal wear rate $dw_i/dt = \delta K_w$ for each punch of the system in accordance with Eq. (7.38). Hence we can describe the shape of the worn surface at the steady-state stage by a function $h_{i\infty} + \delta K_w t$. The function $h_{i\infty}$ describes the stationary shape of the worn surface. From Eqs. (7.39),

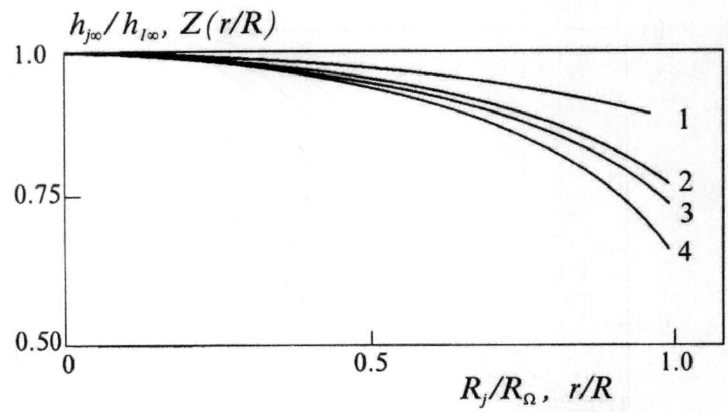

Figure 7.11: Stationary shapes of worn surface for reciprocating motion at different contact density: $a/l = 0.05$ (curve 1), $a/l = 0.25$ (curve 2), $a/l = 0.45$ (curve 3); smooth cylindrical punch of radius R (curve 4).

(7.40) and (7.52) we obtain

$$h_{i\infty} = \frac{(1-\nu^2)P^*}{2aE}\left[\delta V^{*\beta}\right]^{1/\alpha}\left[\frac{1}{V_i^{\beta/\alpha}} + \frac{2}{\pi}\sum_{j=1,\, i\neq j}^{N}\frac{1}{V_j^{\beta/\alpha}}\arcsin\frac{a}{l_{ij}}\right], \qquad (7.59)$$

where the constant δ is specified either by the known displacement velocity of the system of punches in the direction perpendicular to the friction surface, or by the known load P_∞ acting on the system of punches in accordance with Eq. (7.53). Therefore, the relationship between the heights of the punches depends on their arrangement inside the nominal region Ω and on the nature of the system motion, and it is independent of the initial microgeometry of the surface.

Fig. 7.11 illustrates the smoothed stationary shapes of the worn surface for the system of punches considered above, as it reciprocates along the boundary of the elastic half-space (R_Ω is the radius of the region Ω, where punches are arranged). Curves 1–3 were calculated for different values of the parameter a/l. Curve 4 represents a smooth axially symmetric punch whose contact face within the contact area of radius R is of the form:

$$Z\left(\frac{r}{R}\right) = \frac{f\left(\frac{r}{R}\right)}{f(0)}, \quad f\left(\frac{r}{R}\right) = \frac{4(1-\nu^2)}{\pi^2 ER}PE\left(\frac{r}{R}\right), \quad 0 \le r \le R,$$

where $E(x)$ is the complete second-kind elliptical integral. This ensures uniform distribution of pressure over the contact area. This punch is the limiting case of the above model in which the size of each contact spot tends to zero and their density a/l to $1/2$ provided that the contact areas are equal. The results show that

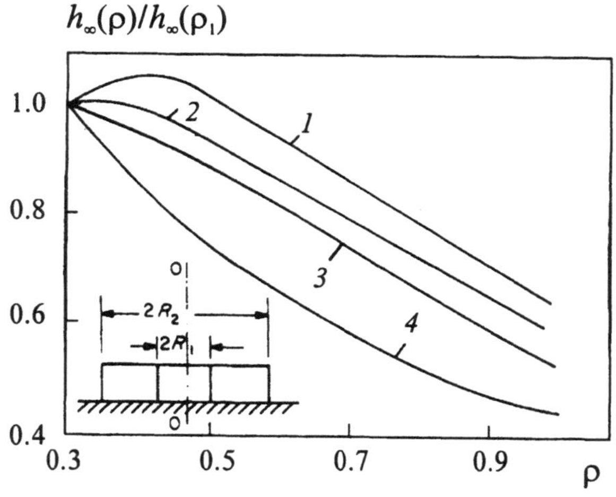

Figure 7.12: Stationary shapes of the worn surface for the system of punches rotating about the axis OO (see scheme), at $R_1/R_2 = 0.3$ and $\Lambda = 0.05$, $a_1 = 0.001$ (curve 1); $\Lambda = 0.1$, $a_1 = 0.01$ (curve 2); $\Lambda = 0.05$, $a_1 = 0.01$ (curve 3); $\Lambda = 0.001$, $a_1 = 0.001$ (curve 4).

the difference in height between the worn punches, located in different distance from the central punch of the model, increases with the punch density.

Because the wear rate depends on sliding velocity as well as load (Eq. (7.38)), the stationary form described by Eq. (7.59), depends critically upon the type of the motion of the system of punches. Calculations were carried out for a system of cylindrical punches which are uniformly located inside the annular region ($R_1 \leq r \leq R_2$), rotating with a constant angular velocity ω about the central point O. Fig. 7.12 illustrates the results. Curves 1 and 3 are constructed for the same values of the relative area of contact Λ ($\Lambda = Na^2/(R_2^2 - R_1^2)$) and for different values of $a_1 = a/R_2$. Curves 1, 4 and 2, 3 are constructed for punches of the same size but for different Λ. The results indicate that, at a constant value of a_1, the difference of the function $h_\infty(\rho)/h_\infty(\rho_1)$ ($\rho = r/R_2$, $\rho_1 = R_1/R_2$) from the function ρ_1/ρ, corresponding to the height distribution of punches without allowance for their interaction, is the greater, the higher the relative area of contact Λ. At the same values of Λ the interaction increases with decreasing size of punches and, hence, with increasing number N, which is proportional to the value of Λ/a_1^2.

Thus, the results show that the punches are worn nonuniformly. Peripheral punches have the largest wear. The shape of the worn surface of the system of punches at the steady-state stage depends essentially on the density of punch arrangement and the type of motion.

7.2.5 Model of equilibrium roughness formation

Analysis of the microgeometry of real surfaces at different stages of the wear process makes possible to conclude that

- during running-in, the surface microgeometry changes and, as a rule, it tends to some stationary microgeometry, the parameters of which do not depend on the initial ones;
- the parameters of the stationary microgeometry depend essentially on the friction conditions (load, type of motion, etc.);
- as a result of the running-in process, the smoothness of the surface can increase or decrease compared to the initial one.

This stationary microgeometry is usually called the ***equilibrium*** or *optimal* roughness. Not the initial, but the equilibrium roughness, together with all other surface properties, determines the wear rate and the friction force in the steady-state stage of wear process.

The system of punches considered above can be used as the simplest mechanical model of a rough surface. Using this model, we can explain the mechanism of equilibrium roughness formation.

The system of equations (7.42) describes the wear kinetics of the model. Parameters of the initial roughness provide the initial conditions for the system (7.42), i.e. the initial load distribution $P_i(0)$ between asperities corresponding to a given surface microgeometry and conditions of loading. The way to calculate the values $P_i(0)$ was described in 2.4.

The parameters of the initial surface microgeometry also determine the number N and the location of the asperities (values of l_{ij}) within the nominal region Ω. The number of contacting asperities varies in the running-in stage because new asperities enter into contact. To take into account this phenomena, we can divide the running-in process into intervals, and assume that within each interval the number of asperities in contact is constant.

From the analysis of the solution of the system of equations (7.42) we can make the following conclusions concerning microgeometry changes in the wear process:

- under a particular loading condition, the wear process consists of running-in and steady-state stages;
- the parameters of the stationary microgeometry corresponding to the steady-state stage of the wear process depend on the asymptotic value of the total load applied to the nominal region Ω, the asperity arrangement, the type of motion, the mechanical properties of contacting bodies, etc., and they are independent of the initial height distribution of the asperities; the initial microgeometry parameters influence the running-in time and the volume of the worn material;
- each contact spot wears uniformly in the steady-state stage of the wear process.

These conclusions are in a good agreement with the experimental observations described above.

The model predicts that the wear rate decreases in the running-in stage for $\alpha > 1$; this is also supported by experimental results (see, for example, Karasik, 1978). To demonstrate this conclusion, we consider the system of punches described in § 7.2.3 which reciprocates on the surface of the elastic half-space so that $V_1 = V_2 = \ldots = V_N$. According to the wear law (7.38), the volume of material Δv_i separating from each contact spot per time interval Δt due to wear, is proportional to P_i^α, i.e.

$$\Delta v_i = K P_i^\alpha \Delta t.$$

Thus the volume of material separated from all contact spots during the time Δt is

$$\Delta v = K \sum_{i=1}^{N} P_i^\alpha \Delta t. \tag{7.60}$$

We can find the extremum of this function using the additional condition

$$\sum_{i=1}^{N} P_i = P_\infty,$$

where P_∞ is the load applied to system of punches.

Using the Lagrange method, we introduce the function Φ:

$$\Phi = K \sum_{i=1}^{N} P_i^\alpha + \lambda \left(\sum_{i=1}^{N} P_i - P_\infty \right),$$

where λ is a Lagrange multiplier.

The extremum point is determined from the condition

$$\frac{\partial \Phi}{\partial P_i} = K\alpha P_i^{\alpha-1} + \lambda = 0$$

or

$$P_i = \left(-\frac{\lambda}{K\alpha} \right)^{1/(\alpha-1)}.$$

The function (7.60) has an extremum if the load is distributed uniformly, and so

$$P_i = P_\infty / N, \quad i = 1, 2, \ldots, N. \tag{7.61}$$

For $\alpha > 1$

$$\frac{\partial^2 \Phi}{\partial P_i^2} = K\alpha\,(\alpha - 1)\, P_i^{\alpha-2} > 0,$$

so the function (7.60) has its minimum value at the point determined by Eq. (7.61) which is the pressure distribution in the steady state stage. So the minimum wear rate $\dfrac{dv}{dt}$ occurs in the steady-state stage for $\alpha > 1$.

Thus, if we model a rough surface as a system of punches, and take their interaction into account, we can explain the existence of the equilibrium roughness, determine its parameters depending on the friction conditions and describe the experimentally observed equilibrium roughness in the wear process. The present mechanical model of the formation of the equilibrium roughness also predicts the minimum wear rate in the steady-state stage which agrees with a number of experimental results.

Note that the method described in this section can also be used to study the wear of an rough elastic body which is in contact with a smooth one. This (inverse) model was investigated in Goryacheva and Dobychin (1988) for asperities of a cylindrical form. A comparison of the results gives

$$H_i - H_j = (h_i - h_j)q,$$

where $h_i - h_j$ is the difference in the heights of the punches in the steady-state stage of the wear process, $H_i - H_j$ is the difference in the heights of asperities of the elastic body in the steady-state stage for the inverse model (the kind of motion, asperity distribution and other conditions are assumed to be the same). The coefficient q is determined by the formula

$$q = \frac{1}{1 - P_\infty / (\pi a^2 EN)}, \quad q > 1.$$

So for the same wear conditions and density of contact spots the model predicts that the difference in asperity heights for the elastic body is larger then for the rigid surface. As was shown in § 7.1.4, this difference is proportional to the load P_∞ for the rough rigid surface, and is proportional to $P_\infty q$ for the rough elastic surface. The value of q is close to 1 for small elastic deformations.

7.2.6 Complex model of wear of a rough surface

The model described above takes into account only the surface continuous wear according to the wear equation (7.38) and so it predicts a monotone character for the wear process in time. However, it is well known that the wear rate in many cases is described by a periodic function, and debris of different scales arise in the wear process. Also contact spot migration occurs at the worn surface.

To describe these experimentally observed results, we include a mechanism for fracture of asperities. We assume that the fracture of punches (asperities) is caused by the damage accumulation process (micropitting). The method of calculation of the damage accumulation is considered in Chapter 5.

We introduce the non-decreasing function $Q_i(t)$ which describes the damage accumulation process within the i-th punch, which is equal to zero for the initial (undamaged) state and is equal to 1 at the instant t_i^* of the punch fracture. So the condition for the fracture of the i-th punch can be written in the form

$$Q_i(t_i^*) = \int_0^{t_i^*} \frac{dt}{T_f\,[P_i(t)]} = 1, \tag{7.62}$$

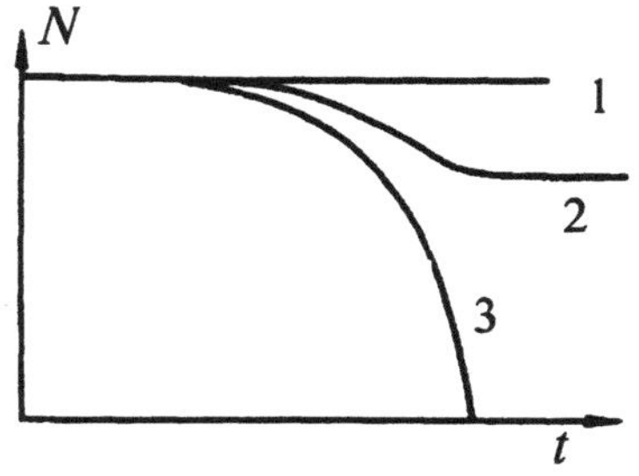

Figure 7.13: The number of contact spots vs. time in wear process.

where $T_f\,[P]$ is the lifetime of the punch acted by the load P, before its fracture. Using Weller's curve, we approximate this function for $P_2 < P < P_1$ by

$$T_f[P] = \frac{1}{K_f} \ln \frac{P_1 - P_2}{P - P_2}, \tag{7.63}$$

where P_1 and P_2 are characteristic values of load, and K_f is a material constant. We assume that the fracture occurs at $t_i^* = 0$ if $P \geq P_1$; fracture does not occur in a finite time if $P \leq P_2$.

At first we examine the model qualitatively and consider the possible ways of wear process development and changes in the surface microgeometry which is modelled by a system of punches. Fig. 7.13 illustrates three different ways of process development.

1. The first is depicted by the curve 1. The fracture of asperities does not occur, and its number does not change in the wear process. The surface continuous wear tends to the steady-state stage. In this case the initial load distribution has to satisfy the condition $P_i(0) < P_1$. In the wear process, there is redistribution of loads between the contact spots. This process leads usually to a stationary load distribution $P_{i\infty}$ at asperities. The case depicted by the curve 1 is realized if the following relations are satisfied simultaneously:

$$Q_i(T) < 1, \quad P_{i\infty} < P_2, \quad i = 1, 2, \ldots, N,$$

where T is a running-in time. This case was investigated in detail in § 7.2.2 based on the model of cylindrical punches in contact with an elastic half-space.

2. The another possible case appears if the process of load redistribution goes slowly compared to the wear accumulation process, and at the initial time there are asperities for which $P_i(0) > P_2$ ($i = 1, 2, \ldots, k$, $k \leq N$). So at an instant t_1 the fracture of some asperities occurs and the loads acting on

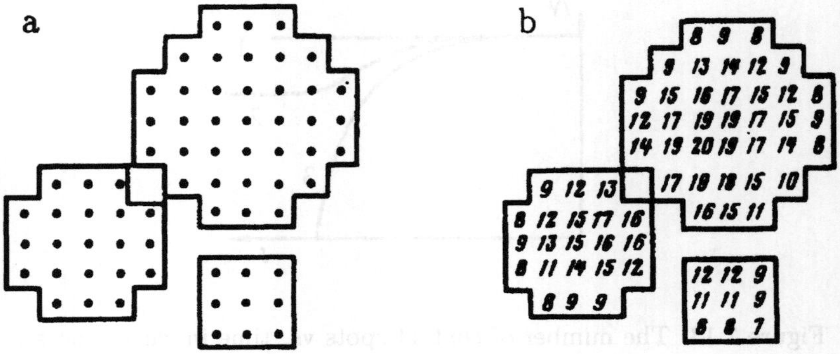

Figure 7.14: Scheme of punch arrangement within three contour regions (a) and the consequence of the punch fracture (b); 7–20 denotes the lifetime of each punch in dimensionless units.

the remainder will increase. This may cause an increasing rate of fracture, leading to the fracture of all asperities (curve 3).

3. The intermediate case occurs if the two competing processes (continuous load redistribution due to the wear of asperities, and discontinuous load increase at some asperities due to the fracture of one or more asperities) proceed so that at some instant t_2, k asperities have failed while the remaining $(N-k)$ asperities are under the condition $P_i(t_2) < P_2$. Then the continuous wear process investigated in § 7.2.2 occurs (curve 2).

Note that the curves 2 and 3 in Fig. 7.13 are smoothed. In the model the number of punches in contact changes step-wise.

Numerical calculations have been carried out for the model schematically represented in Fig. 7.14(a) where the locations of cylindrical punches are denoted by dots. The punches are initially of the same height. To calculate the load redistribution in the wear process, we use the method described in § 7.2.2 which is based on the solution of the system of equations (7.42) and (7.43). Also we calculate the value of $Q_i(\bar{t})$ for each punch. We delete the i-th punch from consideration $(t > t_i^*)$ if the fracture condition (7.62) is satisfied at $t = t_i^*$. To do this we include the coefficient $\mu_i(\bar{t})$ in Eqs. (7.42) and (7.43) which is determined by

$$\mu_i(\bar{t}) = \begin{cases} 0, & \bar{t} \geq t_i^* \dfrac{K_w}{D_0}, \\ 1, & \bar{t} < t_i^* \dfrac{K_w}{D_0}. \end{cases} \tag{7.64}$$

Then Eqs. (7.42) and (7.43) take the form

$$\mu_i(\bar{t})\dot{q}_i(\bar{t})+\sum_{j=1,\ j\neq i}^{N} b_{ij}\mu_j(\bar{t})\dot{q}_j(\bar{t}) = -\dot{D}_1(\bar{t})-\Gamma_i\mu_i(\bar{t})q_i^{\alpha}(\bar{t}),\quad i=1,2,\ldots,N,\quad(7.65)$$

$$\sum_{i=1}^{N}\mu_i(\bar{t})q_i(\bar{t}) = \frac{1-\nu^2}{2aD_0E}P(\bar{t}).\quad(7.66)$$

In the calculations, we considered the reciprocating motion of the system of punches; therefore $V_1 = V_2 = \ldots = V_N = V$ and $\Gamma_1 = \Gamma_2 = \ldots = \Gamma_N = \Gamma$. Varying the dimensionless load which is assumed to be independent of time ($P(\bar{t}) = P_\infty$) and parameter $\frac{K_f l}{K_w}$ (l is a minimum distance between the centers of neighboring cylinders), we obtained three different ways of wear process development described above and represented by curves 1–3 in Fig. 7.13. For small loads $\bar{P}_\infty = \frac{(1-\nu^2)P_\infty}{2aD_0E}$ and small values of ratio $\frac{K_f l}{K_w}$, only the surface wear occurs (curve 1); for larger loads or larger values of ratio $\frac{K_f l}{K_w}$, some punches fracture and the remainder wear. The wear process tends to the steady-state stage (curve 2). Fig. 7.14 illustrates the consequence of punch fracture for the case represented by the curve 3 in Fig. 7.13. The fracture starts at the periphery of the domain and then moves to its center.

Comparing this model to the previous one, we can conclude that it is more realistic because it can explain the following: the migration of contact spots due to the fracture of some group of asperities, the appearance of new contact spots having lower heights, and the periodic character of surface fracture. The periodic behaviour occurs because some time must elapse for the function $Q_i(t)$ to reach 1 when a new group of asperities comes into contact; during that time only the surface wear occurs, and this has a lower rate.

7.3 Control of inhomogeneous surface wear

In §§ 7.1 and 7.2 we showed that the parameters of surface inhomogeneity such as relative size and wear coefficient of hardened and unhardened zones, and density of contact spots in discrete contact influence the shape variation of the surface in the wear process. Based on these solutions we can predict wear if these parameters and other characteristics of wear process are known. But using the wear models considered above we can also solve the inverse problem of finding the parameters of the surface structure which will provide optimal wear.

7.3.1 Problem formulation

In the wear process there is a change of the shape $f(x, y, t)$ of the contacting body surface. For a large class of elements at constant external conditions (load, velocity, etc.) the unsteady wear (running-in) stage is followed by steady-state stage, characterized by stability of all the characteristics, particularly the stationary shape $f_*(x, y)$ of the worn surface. Since operation of the junction in the steady-state stage is most desirable, the problem arises of minimizing the running-in time by making the initial shape $f_0(x, y) = f(x, y, 0)$ as close as possible to the steady-state shape $f_*(x, y)$ (problem 1).

In many cases definite requirements are imposed on the shape of the worn surface. We shall term the shape that most completely satisfies these requirements the *optimal* shape, and denote it by $f_s(x, y)$. This shape must be maintained during almost all the duration of the wear process. So the problem is to create the surface corresponding the given wear conditions for which steady-state shape $f_*(x, y)$ coincides with (or is very close to) the optimal shape $f_s(x, y)$ (problem 2).

We present the mathematical formulations and methods of solution of problems 1 and 2 for some particular cases.

We examine the contact of a punch and an elastic half-space in the presence of wear. We assume that the relative sliding velocity $V(x, y)$ is known and is independent of time, and that the shape and dimensions of the contact region Ω are constant. For definiteness we assume also that the punch wears.

This problem was investigated in Chapter 6 and also in § 7.1 and § 7.2. It was shown that the wear process has an asymptotically stable steady-state stage if the normal load applied to the punch and the rate of its penetration tend to the constant values as $t \to \infty$.

In the steady-state stage the equation, $f_*(x, y)$, of the worn punch surface can be written (apart from a term that is independent of the space coordinates)

$$f_*(x, y) = f_0(x, y) + \lim_{t \to \infty} w(x, y, t) = A\left[\left(\frac{D_\infty (V^*)^\beta (p^*)^\alpha}{K_w(x, y) V^\beta(x, y)}\right)^{1/\alpha}\right], \quad (7.67)$$

where $K_w(x, y) > 0$ is the wear coefficient, α and β are parameters in the wear equation (7.1). A is an operator determined by the characteristics of the wearing body, the half-space properties, the geometry of the contact, the parameters of the surface inhomogeneity, etc.

The constant D_∞ is determined by the value of the given rate of penetration of the punch in the steady-state stage ($\lim_{t \to \infty} \dot{D}(t) = D_\infty$), or by the asymptotic value of the load $P_\infty = \lim_{t \to \infty} P(t)$ in accordance to Eq. (7.6).

It follows from Eq. (7.67) that the steady-state shape $f_*(x, y)$ of the worn surface is influenced by the wear coefficient $K_w(x, y)$, the sliding velocity $V(x, y)$, the value of D_∞ and also the form of the operator A which depends on some parameters. We denote the ensemble of all parameters $\gamma_i(x, y)$ influencing the function $f_*(x, y)$ by Γ.

We will solve the problems 1 and 2 indicated above by variation of the parameters $\gamma_i \in \Gamma$ $(i = 1, 2, \ldots, k)$.

To solve problem 1 we can take the initial surface shape coinciding with the steady-state shape (7.67), i.e. $f_0(x, y) = f_*(x, y)$. Therefore the running-in time is equal to zero, that is the steady-state wear holds for the duration of the wear process.

Problem 2 can be also solved from Eq. (7.67) if we put $f_*(x,y) = f_s(x,y)$ and consider the right-hand side of this equation as a function of the parameters γ_i.

It should be noted that usually the parameters γ_i have practical limitations imposed by the technology used in obtaining the inhomogeneous surfaces. Therefore they belong to a definite class of functions S, i.e. $\Gamma \in S$.

Problem 2 can be formulated as the problem of finding one or more functions $\gamma_i(x,y) \in S$ that minimize the functional F, which is a metric in some space, for example,

$$F = \left[\iint\limits_{\Omega} \left(f_*(x,y) - f_s(x,y) - \frac{1}{\operatorname{mes}\Omega} \iint\limits_{\Omega} (f_*(x',y') - f_s(x',y'))\, dx'dy' \right)^2 dxdy \right]^{1/2} . \tag{7.68}$$

Limitations on the class of functions S arise in many practical problems. They are due to the actual capabilities of the technology used in creating inhomogeneous surfaces, for example, the characteristics of laser hardening, in which treatment is performed by pulses (pointwise) or by strips. Specifically, we can consider S as the class of step functions. If there are no restrictions on Γ, $\min F = 0$.

Note that the function $f_s(x,y)$ obtained from Eq. (7.68) may be complicated, so that the wear process is the only possible way to produce the surface shape determined by this function. In § 7.1 we considered how to obtain a wavy surface with specific properties by wearing of an initially plane locally hardened surface.

Thus, we formulated two problems of wear process optimization:

Problem 1. To decrease the running-in time by making the initial surface shape $f_0(x,y)$ approach the steady-state shape $f_*(x,y)$.

Problem 2. To stabilize the optimal shape $f_s(x,y)$ of the worn surface. The problem may be formulated if the construction and use of the junction allow the parameters $\gamma_i(x,y)$ to be varied within a class S.

We shall examine the solution of problems 1 and 2 for some specific cases.

7.3.2 Hardened surface with variable wear coefficient

We consider that the wearing body is a punch which is a circular cylinder of radius R acted on by a constant load P (see Fig. 3.6). The punch moves with a constant speed V along the surface of an elastic half-space in different directions, so that the half-space surface wears uniformly and remains flat. We assume that the wear coefficient $K_w(r)$ is a function of the radius r $(0 \le r \le R)$. The steady-state shape

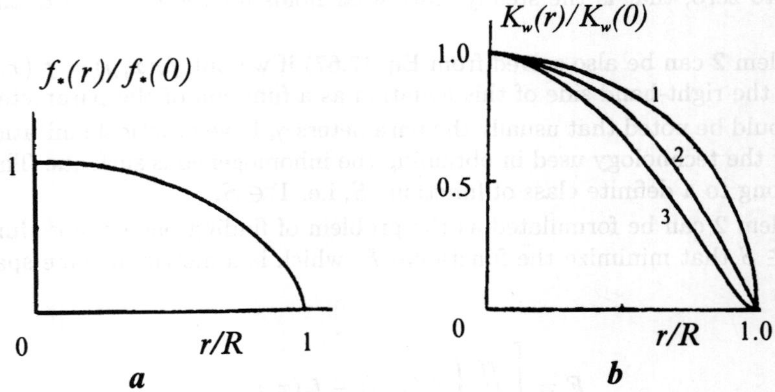

Figure 7.15: Steady-state shape of the worn surface for $K_w(r)$ = const (a) and wear coefficient $K_w(r)$ at $\alpha = 1$ (curve 1), $\alpha = 2$ (curve 2) and $\alpha = 3$ (curve 3) providing the steady-state shape $f_*(r)$ = const (b).

of the worn cylinder surface follows from Eq. (7.67):

$$f_*(r) = \frac{2(1-\nu^2)P}{\pi^2 E}\left(\int_0^R \frac{r\,dr}{K_w^{1/\alpha}(r)}\right)^{-1}\int_0^R K\left(\frac{2\sqrt{rr'}}{r+r'}\right)\frac{r'\,dr'}{K_w^{1/\alpha}(r')(r+r')}, \quad (7.69)$$

where $K(t)$ is the elliptic integral of the first kind.

In the absence of limitations on the shape of the wearing body, the solution of problem 1 is the function $f_0(r) = f_*(r)$, where $f_*(r)$ is determined by Eq. (7.69). If $K_w(r) \equiv K_w$, the initial punch shape $f_0(r)$ providing the steady-state wear throughout the entire time of operation is given by

$$f_0(r) = \frac{4(1-\nu^2)P}{\pi^2 RE}\left(E\left(\frac{r}{R}\right) - 1\right). \quad (7.70)$$

where $E(t)$ is the elliptic integral of the second kind. The plot of the function $f_*(r)/f_*(0)$ is shown in Fig. 7.15(a).

To illustrate the solution of the problem 2, we assume that the optimal shape of the punch surface is flat, i.e. $f_s(r)$ = const, and the wear coefficient $K_w(r)$ admits variation. Then the relation (7.69) is an integral equation for determining the function $K_w(r)$ where the left-hand side is a constant ($f_*(r)$ = const). The solution of this equation is given in Galin (1953), and has the following form:

$$\frac{K_w(r)}{K_w(0)} = \left(1 - \left(\frac{r}{R}\right)^2\right)^{\alpha/2}.$$

The plots of the function $K_w(r)/K_w(0)$ are shown in Fig. 7.15 (b) for various values of α.

Solving the problem 2, we have not considered any restrictions on the wear coefficient variations. As was pointed out in § 7.1 the function $K_w(r)$ can belong to a class of the step functions. The example considered in § 7.1.4 illustrates the solution of the problem in this case. By varying the arrangement of the hardened zones which are the rings of definite thickness, we can satisfy the necessary condition $f(r_i + \rho/2) = \text{const}$ which could be considered there as the optimal surface shape.

The results for surfaces and regions of other shapes, with different natures of the relative motion can be obtained similarly, analytically or numerically, on the basis of the solutions of the contact problems of elasticity theory.

7.3.3 Abrasive tool surface with variable inclusion density

The wear coefficient is not the only parameter influencing the steady-state shape of the worn surface. It was shown in § 7.2 that in discrete contact the relative positions of the individual contacts have a significant influence on the shape of the worn surface. As an example of the problem of optimizing the discrete contact, we shall examine the solution of problem 2 for an abrasive tool, and propose a method for rational design of grinding surfaces to ensure their uniform wear.

The abrasive tool material is a matrix with hard cutting inclusions in it. So as the matrix wear resistance is usually less than that of the abrasive inclusions, the abrasive inclusions during the wear process become practically the only loaded part of the tool surface. In spite of the fact that it is possible for there to be direct contact between the matrix and the treated material, the pressure at those places is much less then on the contact spots of inclusions and treated surface.

This makes it possible to model the tool work surface as a system of punches (inclusions) connected with each other. For the treated body we use the model of an elastic half-space, the surface of which remains flat during the wear process. Each inclusion is modelled by a rigid cylinder of radius a.

The wear of such a system of punches was investigated in § 7.2. It was shown there that the punches wear nonuniformly during the running-in process, and so the steady-state shape of the surface of the system of punches (it is defined by the punch height distribution) differs from the initial one and depends essentially on the punch arrangement inside the nominal contact Ω. The steady-state shape of the surface is given by Eq. (7.59).

The analysis of the wear process in the particular case of a system of punches uniformly distributed inside the circular domain showed that the punches initially distributed at the same level wear nonuniformly; the punches located at the periphery of Ω wear more than those located closer to the center O.

A similar phenomenon occurs in the grinding process when the initially flat tool surface becomes curved due to its nonuniform wear. This causes a decrease in the tool capacity. Usually the tool surface is improved by special treatment that leads to a recovery of its flatness.

We assume here that the tool work surface is a ring with internal and external radii R_1 and R_2 (see Fig. 7.16 (a)). The tool rotates with angular velocity ω on

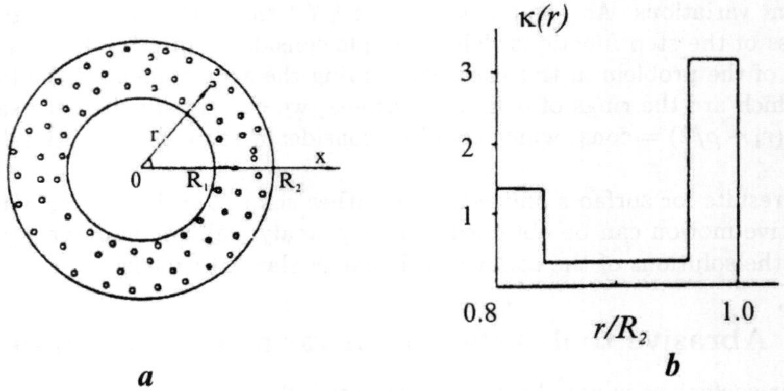

Figure 7.16: Abrasive tool surface with inclusions (a) and variation of the inclusion density vs. radius providing the condition $f_s(r) = \text{const}$

an elastic half-space surface. The system of N punches (inclusions) is distributed inside an annular domain and is acted on by a load P.

If the inclusion size is small and the number of inclusions is large, then it makes sense to speak not of a fixed inclusion position but rather of the function $\kappa(x,y)$, characterizing the contact density of inclusions at the point (x,y) (so $\iint\limits_{\Delta\Omega} \kappa(x,y)dxdy$ is the contact area of inclusions on the subdomain $\Delta\Omega$). The punches are arranged symmetrically with respect to the point O, with relative contact density $\kappa(r)$, which characterizes a variation of contact density with radius r.

To obtain the steady-state shape $f_*(r)$ of the worn surface of the tool, we use Eq. (7.59). For large N the summation in Eq. (7.59) can be replaced by integration, since the additional indentation of the punch resulting from the action of N_r concentrated forces at a distance r ($r \geq 4a$) from this punch inside the annular subdomain Ω_r depends on the overall intensity of these forces and is mostly independent of their arrangement inside Ω_r (see § 2.4). Then for some fixed punch at a distance r from the center O we obtain from Eq. (7.59) and $V(r) = \omega r$,

$$f_*(r) = \frac{(1-\nu^2)Pa}{4E\int\limits_{R_1}^{R_2} r'^{1-\beta/\alpha}\kappa(r')dr'}\left(\frac{1}{r^{\beta/\alpha}}+\right. \tag{7.71}$$

$$\left.\frac{4}{\pi^2 a^2}\int\limits_0^{\pi}\int\limits_{R_1}^{R_2} \kappa(r')\Psi(r,r',\varphi)r'^{1-\beta/\alpha}dr'd\varphi\right),$$

where

$$\Psi(r,\rho,\varphi) = \begin{cases} \arcsin \dfrac{a}{s}, & s > s', \\ 0, & s \leq s', \end{cases} \tag{7.72}$$

$$s^2 = r^2 + r'^2 - 2rr' \cos\varphi, \quad s'^2 = \frac{a^2}{\kappa(r)}.$$

The function $\Psi(r, r', \varphi)$ excludes from the region of integration a circle of radius s' with center at the examined point r, in which there are no contacting inclusions other than the fixed one.

The relation (7.71) can be considered as an integral equation to find the function $\kappa(r)$ which provides the optimal steady-state shape $f_s(r) = f_*(r)$ of the tool surface. Since it is impossible to manufacture the instrument with the density $\kappa(r)$ varying continuously, the solution is sought in the class of step functions $\kappa(r) = \kappa_i$ within the interval (r_{i-1}, r_i), $i = 1, 2, \ldots, n$. The interval size has not to be less than a constant d determined by technological capabilities. Then the optimization problem is to obtain κ_i and r_i which minimize the functional F (see Eq. (7.68)) under the condition $|r_i - r_{i-1}| \geq d$.

The numerical solution for fixed interval dimension $|r_i - r_{i-1}| = d$ was considered for the case when optimal shape is the flat one ($f_s(r) = \text{const}$). The following values of parameters were used:

$$R_1 = 80 \text{ mm}, \; R_2 = 100 \text{ mm}, \; a = 0.09 \text{ mm}, \; N = 10000, \; d = 2 \text{ mm}, \; \alpha/\beta = 1.$$

For these parameters, $\tilde{\kappa} = 0.02$, where $\tilde{\kappa}$ is the density average value ($\tilde{\kappa} = Na^2/(R_2^2 - R_1^2)$). That corresponds to a real abrasive tool inclusion density under the condition that 10% of the inclusions located on the surface are in contact.

The algorithm for the numerical solution is described in Goryacheva and Chekina (1989). The integral equation of the first kind (7.71) was approximately solved by inspection. The problem of constructing the apptoximate solution of Eq. (7.71) in the set of step functions K_n is well-posed in the sence of Tikhonov since K_n is compact in the space L_2 and the integral operator on the right-hand side of Eq. (7.71) is continuous (see Tikhonov and Arsenin, 1974 and Goryacheva, 1987).

The calculation results are presented in Fig. 7.16 (b). The function $\kappa(r)$ guarantees the surface to be practically flat during the wear process. This function consists of three different parts because the densities differing by less than 10% were considered to be indistinguishable for technological reasons.

Thus, for inhomogeneous surfaces it is possible to formulate and solve the problems of wear process optimization by varying the parameters of surface inhomogeneity within the limitations imposed by practice.

Chapter 8

Wear of Components

In this chapter we give some applications of the methods presented in Chapter 6 to the analysis of the wear kinetics of some components. Study of wear kinetics makes it possible to predict the durability of moving parts of machines during operation; this is one of the most important problems in tribology.

The first junction investigated in this chapter is the plain journal bearing. Recently considerable success has been gained in the calculation of the wear kinetics of journal bearings of different types. An algorithm accounting for wear of the journal only was developed in Blyumen, Kharach and Efros (1976) for a plain journal bearing with thick-wall sleeve. This study is based on Hertzian contact and a power-law dependence of the wear rate on the contact pressure. (This is the wear equation that is used in most studies.) A more complete solution of this problem was given by Usov, Drozdov and Nikolashev (1979), where journal and sleeve wear were both taken into account.

Journal bearings with antifriction coatings were studied by Bogatin and Kanibolotsky (1980), Kuzmenko (1981), Kovalenko (1982), Goryacheva and Dobychin (1984a, 1984b), Soldatenkov (1985). The design of a sliding pair with a protective coating which prevents severe wear and decreases the friction losses is of interest for engineering. The wear of journal bearings depends on the coating location, either at the bush or at the shaft surface. In the calculation of wear of the journal bearing with coated bush the simplifying assumption is usually made that the thickness of the coating remains constant in the process of wear. Some researches ignore the coating when calculating the contact characteristics of bearings.

In § 8.1–§ 8.3 we discuss the wear of a thin antifriction coating in plain journal bearings when coating is either at the bush or at the shaft surface. In calculation of the wear kinetics we do not use the assumptions we just noted; this allows us to obtain a better model of journal bearings with antifriction coatings.

We also discuss the important rail-wheel contact problem in this chapter. In his monograph devoted to the mechanics of rolling contact Kalker (1990) stated: "The motion that rail and wheel perform with respect to each other is very complicated and varied, yet it is found that the worn form of wheel and rail converge to standard forms. It wóuld be interesting if such standards could arise from theoretical studies and simulations." Some approaches to rail and wheel wear analysis are presented in § 8.4.

In § 8.5 we discuss a model of the wear of a tool in rock cutting. This problem was investigated in a set of theoretical and experimental works. Some new approaches to the solution have been recently proposed by Hough and Das (1985) and Appl, Wilson and Landsman (1993). The model presented in this chapter was developed by Checkina, Goryacheva and Krasnik (1996). It is based on the analysis of worn tool profiles obtained experimentally. It takes into account the shape variation of both contacting bodies caused by the wear or cutting process. The model is used for calculation of the pressure distribution in a contact zone, and of the variation of forces during cutting process. The influence of tool wear on contact characteristics is also investigated.

We note that the wear kinetics of such widely used moving components as piston rings, slides and guides can be calculated by using the solutions of the wear contact problems described in § 6.2 and § 6.8.

8.1 Plain journal bearing with coating at the bush

8.1.1 Model assumptions

We consider the plain journal bearing with an antifriction element (coating) located at the bush (direct sliding pair, DSP). Fig. 8.1 illustrates the scheme of contact in the plain journal bearing consisting of the shaft S_1, the bush S_2 and the coating S_0. The shaft S_1 is loaded uniformly along its directrix with a load P per unit length.The shaft rotates with angular velocity ω about the axis Oz which is perpendicular to the scheme plane. The wear occurs in the sliding process.

Before investigating the wear kinetics of this junction we make some assumptions. Usually the wear resistance of the shaft is greater than the wear resistance of the antifriction coating. So we neglect the wear of the shaft and assume that only the coating S_0 wears.

It is typical for journal bearings that the elasticity modulus of the antifriction coating is 2-3 orders less than the moduli of the bush and shaft materials. Because of this we will assume that the bodies S_1 and S_2 are rigid and S_0 is elastic.

Antifriction coatings, as a rule, have a thickness of 10–100 μ. Such small thickness of coatings can be explained by their low heat conductivity. The thinner the coating, the less is its size instability due to heat expansion and swelling and the greater is the stiffness of the junction. For this reason we will assume in what follows that the initial thickness of the antifriction coating $h_0 = h(0)$ is small, i.e. $h_0/R_0 \ll 1$, where R_0 is the inner radius of the coating.

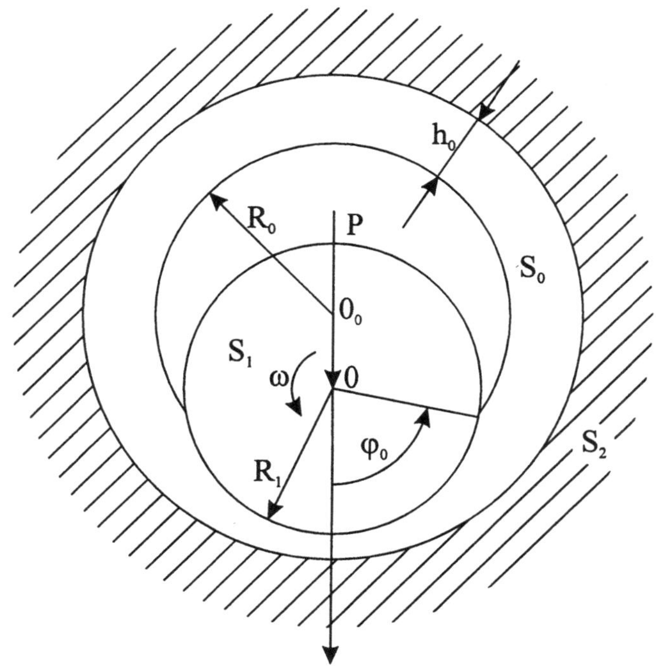

Figure 8.1: Scheme of plain bearing with coating applied on the bush (direct sliding pair, DSP)

8.1.2 Problem formulation

Under the assumptions of § 8.1.1 the wear kinetics of the journal bearing is reduced to a study of the wear of a thin coating S_0 of initial thickness h_0 applied on a rigid bush S_2. The coating wears by contact interaction with a rigid shaft S_1 (Fig. 8.1), loaded by the linear load P and rotating with the angular velocity ω.

We assume that the wear rate of the coating $\partial h/\partial t$ depends on the contact pressure $p(\varphi, t)$ and the linear velocity $V = \omega R_1$ (R_1 is the radius of the shaft) according to the relation

$$\frac{\partial h(\varphi, t)}{\partial t} = -K_w V p^{\alpha}(\varphi, t), \tag{8.1}$$

where K_w is a wear coefficient, $K_w = c p_0^{-\alpha}$, c, p_0 and α are characteristics which depend on the mechanical properties of the contacting pair, roughness parameters and a friction coefficient, and can be determined theoretically from wear models or experimentally.

It was shown by Aleksandrov and Mhitaryan (1983) that for a thin elastic layer it is possible to neglect the influence of the tangential contact stress on the normal one and to consider the thin elastic layer as a Winkler foundation for which the normal elastic displacement $u(x)$ is proportional to the contact pressure $p(x)$

$$u(x) = khp(x),$$

where h is the layer thickness, k is a coefficient characterizing the layer compliance; for the layer bonded to the rigid foundation it was determined by Aleksandrova (1973) as

$$k = \frac{1-2\nu}{2G(1-\nu)}.$$

Here G and ν are the shear modulus and the Poisson ratio for the layer, respectively ($G = \frac{E}{2(1+\nu)}$, E is the Young modulus).

It should be noted that due to nonuniform wear the layer thickness h varies along the contact region $|\varphi| \leq \varphi_0$, i.e $h = h(\varphi, t)$. In previous studies of the wear of plain journal bearing these changes were neglected (see, for example, Kovalenko, 1982). Based on the method of § 6.8 we generalize the Winkler model and use the following relation to describe the layer compliance at an arbitrary instant of time:

$$u_r(\varphi, t) = kh(\varphi, t)p(\varphi, t), \tag{8.2}$$

where $u_r(\varphi, t)$ is the radial displacement of the boundary points of S_0.

In the process of wear, not only the layer thickness changes, but the contact angle φ_0 varies with a certain rate $v = d\varphi_0/dt$. Assuming that the rate v is positive, we will use the magnitude φ_0 as a time parameter. In this case a real time t is determined by the formula:

$$t(\varphi_0) = \int\limits_{\varphi_{0,0}}^{\varphi_0} \frac{d\varphi}{v(\varphi)}, \tag{8.3}$$

where $\varphi_{0,0} = \varphi_0(0)$.

Substituting the real time t by the parameter φ_0 in Eqs. (8.2) and (8.1), we obtain

$$u_r(\varphi, \varphi_0) = kh(\varphi, \varphi_0)p(\varphi, \varphi_0), \tag{8.4}$$

$$\dot{h}(\varphi, \varphi_0) = -\frac{K_w V}{v(\varphi_0)} p^{\alpha}(\varphi, \varphi_0), \tag{8.5}$$

where

$$\dot{h}(\varphi, \varphi_0) = \frac{\partial h(\varphi, \varphi_0)}{\partial \varphi_0}, \quad h(\pm\varphi_0, \varphi_0) = h_0.$$

To Eqs. (8.4) and (8.5) we add the condition of contact of the bodies S_1 and S_0 within the region $|\varphi| \leq \varphi_0$

$$h_0 - h(\varphi, \varphi_0) + u_r(\varphi, \varphi_0) = d(\varphi, \varphi_0), \tag{8.6}$$

where

$$d(\varphi,\varphi_0) = \Delta\left(\frac{\cos\varphi}{\cos\varphi_0} - 1\right), \tag{8.7}$$

Δ is the initial clearance ($\Delta = R_0 - R_1$).

Also we take into account the equilibrium equation

$$R_1 \int_{-\varphi_0}^{\varphi_0} p(\varphi,\varphi_0)\cos\varphi\, d\varphi = P. \tag{8.8}$$

Eqs. (8.4), (8.5), (8.6) and (8.8) comprise the basic system of equations of the problem.

8.1.3 Method of solution

We give the solution developed by Soldatenkov (1985).

From Eqs. (8.4) and (8.6) we derive the relationship for the contact pressure

$$p(\varphi,\varphi_0) = \frac{d(\varphi,\varphi_0) - h_0 + h(\varphi,\varphi_0)}{kh(\varphi,\varphi_0)}. \tag{8.9}$$

Substituting Eq. (8.9) in Eq. (8.8), we transform the equilibrium condition to the form

$$P = \frac{R_1}{k}\int_{-\varphi_0}^{\varphi_0} \frac{d(\varphi,\varphi_0) - h_0 + h(\varphi,\varphi_0)}{h(\varphi,\varphi_0)}\cos\varphi\, d\varphi. \tag{8.10}$$

Substituting $\varphi_0 = \varphi_{0,0}$ and Eq. (8.7) in Eq. (8.10), and taking into account, that $h(\varphi,\varphi_{0,0}) = h_0$, we find the following relation between the problem characteristics

$$\frac{kPh_0}{R_1\Delta} = \frac{\varphi_{0,0}}{\cos\varphi_{0,0}} - \sin\varphi_{0,0}. \tag{8.11}$$

It should be remarked that the elastic displacement at any point is always less than the layer thickness, i.e. $u_r(\varphi,\varphi_0) < h(\varphi,\varphi_0)$. From this condition it follows

$$\cos\varphi_0 > \frac{\Delta}{h_0 + \Delta}.$$

This is a restriction on the initial characteristics of the bearing.

Differentiating Eq. (8.10) with respect to the parameter φ_0 and taking into account Eq. (8.5), we derive the following relationship to determine the rate of change of the contact angle φ_0

$$v(\varphi_0) = \frac{VK_w \displaystyle\int_{-\varphi_0}^{\varphi_0} \frac{p^\alpha(\varphi,\varphi_0)\,[1 - kp(\varphi,\varphi_0)]}{h(\varphi,\varphi_0)}\cos\varphi\, d\varphi}{\Delta\cdot\dfrac{\sin\varphi_0}{\cos^2\varphi_0}\displaystyle\int_{-\varphi_0}^{\varphi_0}\frac{\cos^2\varphi\, d\varphi}{h(\varphi,\varphi_0)}}. \tag{8.12}$$

So we have the system of equations (8.5), (8.9) and (8.12) to calculate the functions $p(\varphi, \varphi_0)$, $h(\varphi, \varphi_0)$ and $v(\varphi_0)$. The real time t can be calculated from Eq. (8.3) where the initial contact angle $\varphi_{0,0}$ is found from Eq. (8.11).

We introduce the dimensionless coordinate $\bar{\varphi} = \varphi/\varphi_0$ and corresponding functions $\bar{h}(\bar{\varphi}, \varphi_0) = h(\bar{\varphi}\varphi_0, \varphi_0)/h_0$, $\bar{p}(\bar{\varphi}, \varphi_0) = p(\bar{\varphi}\varphi_0, \varphi_0)K_w^{1/\alpha}$. Then Eq. (8.5) is transformed to the following

$$\frac{\partial \bar{h}(\bar{\varphi}, \varphi_0)}{\partial \varphi_0} - \frac{\bar{\varphi}}{\varphi_0}\frac{\partial \bar{h}(\bar{\varphi}, \varphi_0)}{\partial \bar{\varphi}} = -\frac{V}{h_0 v(\varphi_0)}\bar{p}^{\alpha}(\bar{\varphi}, \varphi_0). \tag{8.13}$$

The boundary conditions for Eq. (8.13) are $\bar{h}(\bar{\varphi}, \varphi_{0,0}) = 1$, $\bar{h}(\pm 1, \varphi_0) = 1$.

The numerical calculation is based on step by step integration of the partial differential equation (8.13) along characteristics, taking into account the boundary conditions and Eqs. (8.3), (8.9), (8.10) and (8.12). That is, using the known values (on the first step - from initial conditions) φ_0, $\bar{h}(\bar{\varphi}, \varphi_0)$, $\bar{p}(\bar{\varphi}, \varphi_0)$ and, consequently, $v(\varphi_0)$ and $d\varphi_0$, we determine the increment of the function $\bar{h}(\bar{\varphi}, \varphi_0)$ along the characteristics of Eq. (8.13). The characteristics are the family of hyperbolas $\varphi_0 = C/\bar{\varphi}$; C is a parameter of the family. Then we determine the increment of the time dt from Eq. (8.3) and the new value of the pressure $\bar{p}(\bar{\varphi}, \varphi_0 + d\varphi_0)$ from Eq. (8.9). Values of $\varphi_0 + d\varphi_0$, $t + dt$, $\bar{h}(\bar{\varphi}, \varphi_0 + d\varphi_0)$, $\bar{p}(\bar{\varphi}, \varphi_0 + d\varphi_0)$ are initial data for the next step in respect to the angle φ_0.

Based on this procedure we calculate the changes of the contact angle, contact pressure and thickness of the coating in time.

Note that Soldatenkov (1987) used a similar procedure to calculate the wear of a thin coating applied on the bush of a plain journal bearing, taking into account the elastic properties of the bush and the shaft.

8.1.4 Wear kinetics

Calculations were carried out for the following values of the parameters:

$$\frac{kP}{R_1} = 0.9 \cdot 10^{-2}, \quad \frac{\Delta}{h_0} = 10^{-1}, \quad \alpha = 1, \quad \frac{k}{K_w} = 9 \cdot 10^5.$$

Fig. 8.2 illustrates the contact pressure distribution for times $\bar{t} = \dfrac{tV}{h_0}$. The dependence of the maximum contact pressure $p_{\max}$, the minimum value of the coating thickness $h_{\min}$ and the contact angle φ_0 on time are presented in Fig. 8.3. The results show that the maximum contact pressure and the average contact pressure decrease during the wear process.

Based on the results, we can divide the wear process of this type of journal bearing into two stages: the running-in $(0 \leq t \leq T)$, and steady-state stage $(t > T)$. In the running-in stage, the values of $p_{\max}$ change considerably according to a non-linear law. It is evident that the running-in time T has to satisfy the condition $T \ll T^*$, where T^* is the bearing lifetime determined from the condition $h(0, \varphi_0(T^*)) = 0$.

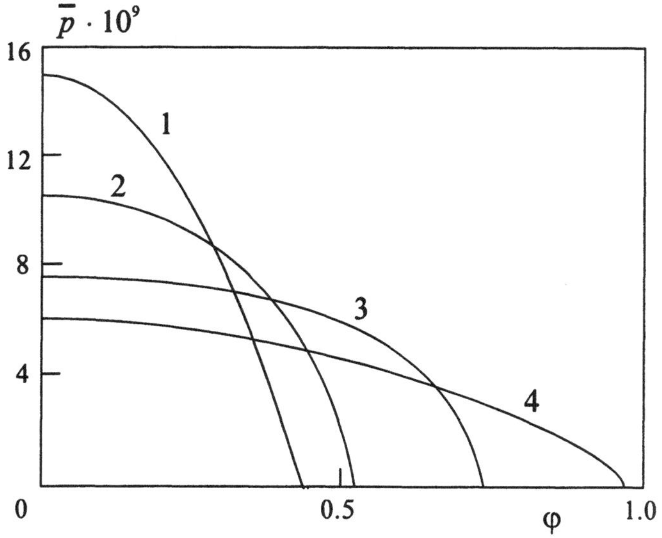

Figure 8.2: Pressure distribution within the contact region $\varphi \leq \varphi_0$ (φ is expressed in radians) for the journal bearing (DSP) at different instants of time: $\bar{t} = 0$ (curve 1); $\bar{t} = 0.2 \cdot 10^7$ (curve 2); $\bar{t} = 0.9 \cdot 10^7$ (curve 3); $\bar{t} = 5.6 \cdot 10^7$ (curve 4).

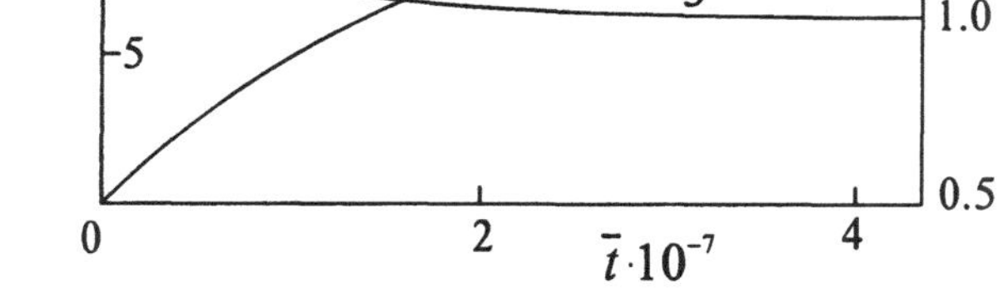

Figure 8.3: Dependence of the minimum value of the coating thickness $\bar{h}_{\min}$ (curve 1), the contact angle φ_0 (curve 2) and the maximum contact pressure $\bar{p}_{\max}$ (curve 3) on time for the plain journal bearing (DSP).

The near-linear dependence of the value of $h_{\min}$ on time makes it possible to calculate the lifetime T^* using linear interpolation of the function $h_{\min}(t)$. For the case under consideration $\bar{T}^* = 1.5 \cdot 10^8$.

In the steady-state stage the values of the maximum contact pressure $p_{\max}$ change with an approximately constant rate. Due to this fact we can suggest some simplifications to the steady-state analysis.

8.1.5 Steady-state stage of wear process

For the bearing under consideration, the contact pressure cannot be stationary in the steady-state stage because the contact angle varies due to coating wear. However, the analysis of the numerical results shows that the contact pressure in the steady-state stage can be characterized by the stationary function $p_s(\varphi, \varphi_0)$, including the contact angle φ_0 as parameter. This function can be determined from the following equation which is obtained by differentiating Eq. (8.6) with respect to φ_0 and taking into account Eq. (8.5)

$$K_w V \left[p_S(\varphi, \varphi_0)\right]^\alpha \frac{1 - kp_S(\varphi, \varphi_0)}{\cos\varphi} = \Omega(\varphi_0), \tag{8.14}$$

where

$$\Omega(\varphi_0) = v(\varphi_0)\Delta \frac{\sin\varphi_0}{\cos^2\varphi_0}. \tag{8.15}$$

We introduce the function

$$\varepsilon(\varphi_0) = \frac{\displaystyle\int_{-\varphi_0}^{\varphi_0} \left| K_w V \left[p(\varphi, \varphi_0)\right]^\alpha \frac{\left[1 - kp(\varphi, \varphi_0)\right]}{\cos\varphi} - \Omega(\varphi_0) \right| d\varphi}{\Omega(\varphi_0)} \tag{8.16}$$

which characterizes the deviation of the wear process from the steady-state stage. We assume that the steady-state stage begins at $t(\tilde{\varphi}_0) = T$ if $\varepsilon(\tilde{\varphi}_0) < 0.05$.

The approximate formula and tables for calculation of the running-in time T and the contact angle $\tilde{\varphi}_0$ corresponding to the time T are in Goryacheva and Dobychin (1988).

We will obtain here the characteristics of the steady-state stage of the wear process ($t > T$), indicating these characteristics by the index S. To simplify the analysis we consider the case $kp \ll 1$ which is most common in practice. From Eq. (8.14) we obtain

$$p_S(\varphi, \varphi_0) = \left(\frac{\Omega_S(\varphi_0)}{K_w V} \cos\varphi \right)^{\frac{1}{\alpha}}. \tag{8.17}$$

Substituting Eq. (8.17) into equilibrium condition (8.8), we find

$$P = R_1 \left(\frac{\Omega_S(\varphi_0)}{K_w V} \right)^{\frac{1}{\alpha}} \int_{-\varphi_0}^{\varphi_0} (\cos\varphi)^{\frac{1+\alpha}{\alpha}}\, d\varphi.$$

From this relationship we obtain

$$\Omega_S(\varphi_0) = K_w V \left(\frac{P}{R_1 C_\alpha(\varphi_0)}\right)^\alpha, \tag{8.18}$$

where $C_\alpha(\varphi_0) = \int\limits_{-\varphi_0}^{\varphi_0} (\cos\varphi)^{\frac{1+\alpha}{\alpha}}\, d\varphi$.

Substituting Eq. (8.18) into Eq. (8.17) we obtain

$$p_S(\varphi, \varphi_0) = \frac{P}{R_1 C_\alpha(\varphi_0)} (\cos\varphi)^{\frac{1}{\alpha}}. \tag{8.19}$$

From Eqs. (8.15) and (8.18) we find the relationship for the rate $v_S(\varphi_0)$

$$v_S(\varphi_0) = K_w V \left[\frac{P}{R_1 C_\alpha(\varphi_0)}\right]^\alpha \frac{\cos^2\varphi_0}{\Delta \cdot \sin\varphi_0}. \tag{8.20}$$

Then the real time t can be calculated from the following relationship obtained from Eq. (8.3)

$$t_S(\varphi_0) = \int\limits_{\tilde{\varphi}_0}^{\varphi_0} \frac{d\varphi}{v_S(\varphi)} + T. \tag{8.21}$$

Eqs. (8.18)-(8.21) completely describe the steady-state stage of the wear process.

For $\alpha = 1$ these equations take a simple form. In this case the function $C_\alpha(\varphi_0) = C_1(\varphi_0)$ is

$$C_1(\varphi_0) = \varphi_0 + \frac{1}{2}\sin 2\varphi_0.$$

From Eqs. (8.19), (8.20) and (8.21) we obtain the contact pressure $p_S(\varphi, \varphi_0)$, the angle rate $v_S(\varphi_0)$ and the time $t_S(\varphi_0)$ in this case as

$$p_S(\varphi, \varphi_0) = \frac{P\cos\varphi}{R_1\left(\varphi_0 + \sin\varphi_0\cos\varphi_0\right)}, \tag{8.22}$$

$$v_S(\varphi_0) = \frac{K_w V P \cos^2\varphi_0}{C_1(\varphi_0) R_1 \cdot \Delta \cdot \sin\varphi_0}, \tag{8.23}$$

$$t_S(\varphi_0) = \frac{R_1\Delta}{K_w V P\left(\dfrac{\varphi_0}{\cos\varphi_0} - \sin\varphi_0 - \dfrac{\tilde{\varphi}_0}{\cos\tilde{\varphi}_0} + \sin\tilde{\varphi}_0\right)} + T. \tag{8.24}$$

From Eq. (8.6) we can obtain the relationship for the limit contact angle φ_0^* which is found from the condition $h(0, \varphi_0^*) = 0$:

$$\varphi_0^* = \arccos\frac{\Delta}{h_0 + \Delta}. \tag{8.25}$$

Thus if we know the load P applied to the shaft, the geometric characteristics of the bearing (R_1, Δ, h_0), the shaft linear velocity V, the mechanical properties of the coating (k) and the wear characteristics (K_w and α), we can calculate the lifetime of the bearing and the characteristics of the wear process using the method described above.

8.2 Plain journal bearing with coating at the shaft

Journal bearings in which the thin antifriction coating is located on the shaft are finding more and more applications.

The scheme of such a junction (inverse sliding pair, ISP) is presented in Fig. 8.4. As in the previous case, we assume that the coating wears, i.e. the wear of the bush is negligibly small compared to the wear of the soft antifriction coating. We assume also that in operation the coated shaft (journal) remains a circular cylinder with decreasing radius due to wear of the coating.

Thus, the geometry of the contact remains the same for any instant of time. So in the wear kinetics calculation we can use the solution of the same contact problem in which the thickness of the coating is determined from the wear equation at each step of the wear process. This distinguishes the problem from the contact problem for the coated bush and shaft described in the previous section where the equations (8.5), (8.9) and (8.12) were solved simultaneously.

That is why we will first describe the contact problem for the coated shaft and the bush, and then will study the wear kinetics of the junction taking into account the relationship between the contact characteristics and the magnitude of the wear.

8.2.1 Contact problem formulation

The plain journal bearing relates to a cylindrical joint with conforming surfaces. Such joints are widespread in engineering (journal bearing, hinges, piston liner assemblies etc.).

In this study we take into account elastic properties of a shaft and a bush. We consider an elastic infinite plate S_2 (Fig. 8.4) with a round hole of radius R_2 and an elastic disk S_1 of radius R_1 inserted into it. A thin layer S_0 of initial thickness h_0, whose elastic properties differ from those of the disk, is applied on the disk surface. It is supposed that the radii R_2 and $R_0 = R_1 + h_0$ are close, i.e. $(R_2 - R_0)/R_2 \ll 1$ and the layer thickness is small, $h_0/R_1 \ll 1$. In this joint the cylinder S_1 with coating S_0 is an analogue of the journal, the cylinder itself is an analogue of shaft, and the elastic body S_2 is the model of the bush. The journal is loaded by a normal force uniformly distributed over its length, so that at each section perpendicular to the journal axis there is a linear load P. The journal rotates with angular velocity ω.

Hertz theory applied to the calculation of the contact characteristics of this junction may lead to considerable errors, since in this case the condition that the dimension of the contact region must be small compared with the dimensions of each body is not always satisfied.

To solve the contact problem we use the method suggested by Kalandiya (1975). We introduce the system of coordinates (XOY) related to the center O of the disk. Simultaneously we consider the plane of the complex variable $z = x + iy$ where $x = X/R_2$ and $y = Y/R_2$. In the x, y-plane the radii of the disk with coating and

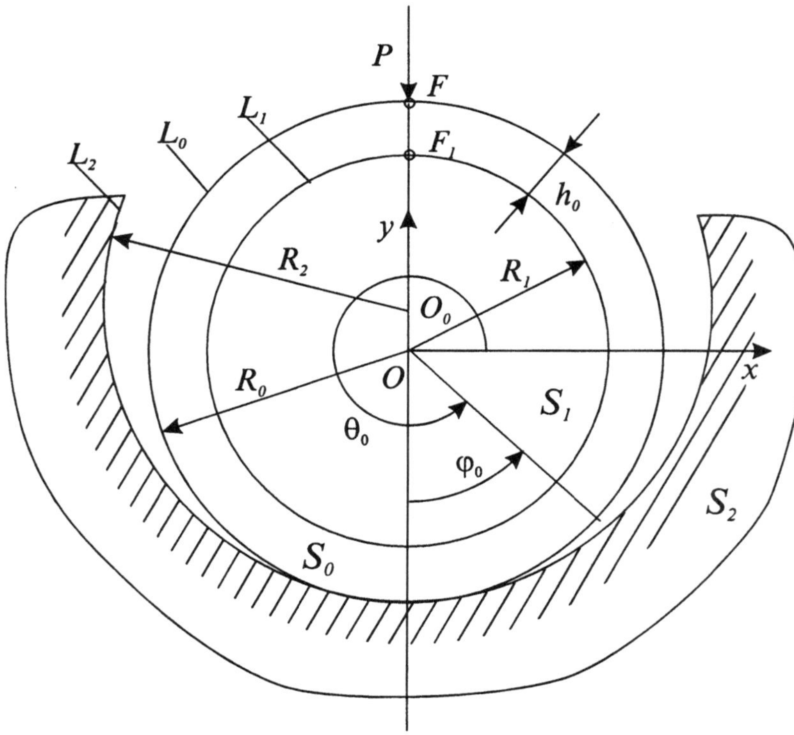

Figure 8.4: Scheme of a plain journal bearing with coating applied on the shaft (inverse sliding pair, ISP).

the hole are $\rho = R_0/R_2$ ($\rho < 1$) and 1, respectively. The center of the hole in the undeformed state is at the point $z_0 = i(1-\rho)$. At the point F $(z_F = i\rho)$ the load P is applied. The load P direction passes through the point of initial contact of the bodies S_2 and S_0 opposite to the y-axis. The load presses the disk against the elastic plate S_2 and as a result of elastic deformations they come into contact along the contact arc γ characterized by the angle $2\varphi_0$.

We denote the contours of the disk, the hole and the external contour of the layer by L_1, L_2 and L_0, respectively. Points on the contours L_0 and L_2 have coordinates $t_0 = \rho e^{i\theta}$ and $t_2 = e^{i\theta} + z_0$, respectively; θ is the polar angle calculated from the OX axis (see Fig. 8.4). To provide the contact of the bodies S_0 and S_2 along the contact arc γ, the dimensionless radial displacements $u_r^{(0)}$, $u_r^{(1)}$ and $u_r^{(2)}$ of points on the contours L_0, L_1 and L_2, respectively, have to satisfy the following relationship which reflects the equality of curvatures within the contact region

$$u_r^{(0)}(\theta) + u_r^{(1)}(\theta) - u_r^{(2)}(\theta) = (1-\rho)(1+\sin\theta). \qquad (8.26)$$

For a thin layer ($h_0 \ll 2R_0\varphi_0$, where φ_0 is semi-contact angle) for which the modulus of elasticity of the layer S_0 is smaller than that of the shaft, the radial displacements $u_r^{(0)}(\theta)$ are proportional to the layer thickness h_0 and the normal contact stress $\sigma_r(\theta)$, i.e.

$$u_r^{(0)}(\theta) = k\delta_0\sigma_r(\theta), \tag{8.27}$$

where $\delta_0 = \dfrac{h_0}{R_2}$. It was shown by Aleksandrov and Mhitaryan (1983) that $k = \dfrac{1-2\nu}{2G(1-\nu)}$, where G and ν are the shear modulus and the Poisson ratio for the layer S_0, respectively. The relation (8.27) corresponds to the Winkler model.

It should be noted here that alongside the normal stress $\sigma_r(\theta)$ within the contact region γ there is a tangential stress $\tau_{r\theta} = \mu\sigma_r(\theta)$ where μ is the coefficient of friction, caused by friction of the surfaces. But due to the small value of the coefficient of friction μ for the junction under consideration, it is possible to ignore the influence of the tangential stress on the normal stress within the contact region, i.e. to find the normal contact stress by neglecting the tangential one.

Then the following boundary conditions are satisfied on the contours L_0 and L_2

$$\begin{aligned}
&\tau_{r\theta}^{(0)}(t_0) = \tau_{r\theta}^{(2)}(t_2) = 0, \quad t_0 \in L_0, \quad t_2 \in L_2, \\
&\sigma_r^{(0)}(t_0) = \sigma_r^{(2)}(t_2), \quad t_0 \in \gamma_0, \quad t_2 \in \gamma_2, \\
&\sigma_r^{(0)}(t_0) = 0, \quad t_0 \in L_0 \setminus \gamma_0, \quad t_0 \neq i\rho, \\
&\sigma_r^{(2)}(t_2) = 0, \quad t_2 \in L_2 \setminus \gamma_2,
\end{aligned} \tag{8.28}$$

where γ_0 and γ_2 are the parts of the contours L_0 and L_2, respectively, which are in contact after deformation; $\sigma_r^{(i)}$ and $\tau_{r\theta}^{(i)}$ are the normal and tangential stresses on the contour L_i $(i = 0, 1, 2)$.

Taking into account the boundary condition at L_0 and the small thickness of the layer S_0, we obtain the following boundary condition on L_1

$$\begin{aligned}
&\tau_{r\theta}^{(1)}(t_1) = 0, \\
&\sigma_r^{(1)}(t_1) = \sigma_r^{(0)}(t_0).
\end{aligned} \tag{8.29}$$

The equilibrium condition takes the form

$$\int_{\gamma_0} \sigma_r^{(0)}(t_0)\, dt_0 = -\frac{P}{R_2}. \tag{8.30}$$

8.2.2 The main integro-differential equation

To solve the problem we use the method suggested by Kalandiya (1975). Differentiating two times Eq. (8.26) and adding the result to the initial one, we obtain

$$\begin{aligned}
&\frac{d^2u_r^{(0)}(\theta)}{d\theta^2} + u_r^{(0)}(\theta) + \frac{d^2u_r^{(1)}(\theta)}{d\theta^2} + u_r^{(1)}(\theta) = \\
&\frac{d^2u_r^{(2)}(\theta)}{d\theta^2} + u_r^{(2)}(\theta) + 1 - \rho.
\end{aligned} \tag{8.31}$$

Taking into account Eqs. (8.27)-(8.30), we can reduce Eq. (8.31) to the following integro-differential equation for the unknown function $\sigma_r(t_1) = \sigma_r^{(1)}(t_1)$, $t_1 \in \gamma'$ (γ' is a part of the contour L_1 corresponding to the contact arc γ)

$$\begin{aligned} &\eta\sigma_r(t_1) + \frac{t_1}{\pi i}\int\limits_{\gamma'} \frac{\sigma_r'(t)\,dt}{t-t_1} - \frac{\lambda}{2\pi i}\int\limits_{\gamma'} \frac{\sigma_r(t)\,dt}{t} + \\ &u\left[t_1\sigma_r'(t_1) + t_1^2\sigma_r''(t_1)\right] = pH_1(t_1) + \\ &\frac{qiP}{2\pi}\left(\frac{\rho-\delta_0}{t_1} - \frac{t_1}{\rho-\delta_0}\right) - m(1-\rho), \end{aligned} \tag{8.32}$$

where

$$\begin{aligned} \tau &= \frac{2}{\mu_2(\kappa_1+1)(\rho-\delta_0)+\mu_1(\kappa_2+1)}, \\ \eta &= \frac{1}{2}\tau\left[\mu_1(\kappa_2-1)-\mu_2(\kappa_1-1)(\rho-\delta_0)-4\mu_1\mu_2 k\delta_0\right], \\ \lambda &= \frac{1}{2}\tau\mu_1(\kappa_2+1), \quad u = 2\tau\mu_1\mu_2 k\delta_0, \\ m &= 2\tau\mu_1\mu_2, \quad p = \tau\mu_2(\kappa_1+1)(\rho-\delta_0), \\ q &= \tau\kappa_2\mu_1, \quad \kappa_i = 3-4\nu_i, \quad \mu_i = \frac{E_i}{2\left(1-\nu_i^2\right)}, \\ \sigma_r'(t) &= \frac{d\sigma_r}{dt}, \quad \sigma_r''(t) = \frac{d^2\sigma_r}{dt^2}. \end{aligned} \tag{8.33}$$

Here E_i and ν_i are the Young's moduli and Poisson's ratios for the bodies S_1 ($i = 1$) and S_2 ($i = 2$).

The points t_1 and t in Eq. (8.32) are on the contour L_1. However, it follows from Eq. (8.29) that the normal stresses found from Eq. (8.32) coincide with the stresses $\sigma_r^{(0)}(t_0)$ occurring within the contact region γ at L_0 for $t_0 = \rho t_1/(\rho - \delta_0)$.

Note that if $\delta_0 = 0$, Eq. (8.32) coincides with that obtained by Kalandiya (1975) for the contact problem for two elastic cylinders.

The function $H_1(t_1)$ in Eq. (8.32) is determined by the load applied to the body S_1 and has the form

$$\begin{aligned} H_1(t_1) &= \Re\left[F_1(t_1) - t_1F_1'(t_1) - \frac{1}{2}F_1(0)\right], \\ F_1(t_1) &= \lim_{z\to t_1}\frac{1}{2\pi i}\int\limits_{L_1'}\frac{\sigma_r(t)\,dt}{t-z}, \end{aligned} \tag{8.34}$$

where L_1' is the part of the contour L_1 where the load is applied.

If the load P is applied to the body S_1 at the point F_1 with coordinate $t_1 = i(\rho - \delta_0)$, the function $F_1(t_1)$ has the form

$$F_1(t_1) = -\frac{P}{2\pi R_2 i\left[(\rho - \delta_0) - t_1\right]}. \tag{8.35}$$

Eq. (8.32) and the equilibrium condition (8.30), which can be written in the form

$$\int\limits_{\gamma'} \sigma_r(t_1)\, dt_1 = -\frac{P}{R_2}, \tag{8.36}$$

are the complete system of equations to determine the normal pressure $\sigma_r(t_1)$ within γ' and the contact angle θ_0.

We map the circumference $|z| = \rho - \delta_0$ onto the real axis using the following function

$$\begin{aligned} i(\rho - \delta_0)\frac{\xi - i\beta}{\xi + i\beta} &= t_1, \\ \beta &= \frac{\cos\theta_0}{1 + \sin\theta_0}. \end{aligned} \tag{8.37}$$

Then the contact arc transforms into the segment $[-1, 1]$, and the function $H_1(t_1)$ becomes ($F_1(t_1)$ is determined by Eq. (8.35))

$$H_1(t_1) = H(\xi) = \frac{P}{8\pi R_2(\rho - \delta_0)} \cdot \frac{\xi^2 + \beta^2}{\beta^2}.$$

Eqs. (8.32) and (8.36) take the following forms, respectively

$$\begin{aligned} &\frac{\eta\beta}{\xi^2 + \beta^2}\sigma_r(\xi) - \frac{1}{2\pi}\int\limits_{-1}^{1}\frac{\sigma_r'(\zeta)\, d\zeta}{\zeta - \xi} - \frac{\lambda\beta^2}{\pi\,(\xi^2 + \beta^2)}\int\limits_{-1}^{1}\frac{\sigma_r(\zeta)\, d\zeta}{\zeta^2 + \beta^2} - \\ &\frac{u\xi}{2\beta}\sigma_r'(\xi) - \frac{u\,(\xi^2 + \beta^2)}{4\beta}\sigma_r''(\xi) = -\frac{m\beta}{\xi^2 + \beta^2}(1 - \rho) + \\ &\frac{P}{\pi R_2} \cdot \frac{\beta}{\xi^2 + \beta^2}\left[q\frac{\xi^2 - \beta^2}{\xi^2 + \beta^2} + p\frac{\xi^2 - \beta^2}{8(\rho - \delta_0)\beta^2}\right], \end{aligned} \tag{8.38}$$

$$\int\limits_{-1}^{1}\frac{\xi^2 - \beta^2}{(\xi^2 + \beta^2)^2}\sigma_r(\xi)\, d\xi = \frac{P}{2\beta R_2(\rho - \delta_0)}. \tag{8.39}$$

8.2.3 Method of solution

The system of equations (8.38) and (8.39) was solved approximately by Multhopp. His method was used by Kalandiya (1975) to solve Eq. (8.38) when $\delta_0 = 0$. We describe here the main idea of the method.

Introducing the new variable ϑ by equation $\xi = \cos\vartheta$, we rewrite Eqs. (8.38) and (8.39) in the form

$$\frac{\eta\beta}{\cos^2\vartheta+\beta^2}\sigma_r(\vartheta) + u\frac{\cos\vartheta}{2\beta\sin\vartheta}\left(1+\frac{\cos^2\vartheta+\beta^2}{2\sin^2\vartheta}\right)\sigma_r'(\vartheta) -$$

$$u\frac{\cos^2\vartheta+\beta^2}{4\beta\sin^2\vartheta}\sigma_r''(\vartheta) + \frac{1}{2\pi}\int_0^\pi \frac{\sigma_r'(\vartheta')\,d\vartheta'}{\cos\vartheta'-\cos\vartheta} -$$

$$\frac{\lambda\beta^2}{\pi\left(\cos^2\vartheta+\beta^2\right)}\int_0^\pi \frac{\sigma_r(\vartheta')\sin\vartheta'\,d\vartheta'}{\cos^2\vartheta'+\beta^2} = \tag{8.40}$$

$$\frac{P\beta}{\pi R_2\left(\cos^2\vartheta+\beta^2\right)}\left[q\frac{\cos^2\vartheta-\beta^2}{\cos^2\vartheta+\beta^2}+p\frac{\cos^2\vartheta+\beta^2}{8(\rho-\delta_0)\beta^2}\right] -$$

$$\frac{m\beta}{\cos^2\vartheta+\beta^2}(1-\rho),$$

$$\int_0^\pi \frac{\left(\cos^2\vartheta-\beta^2\right)\sin\vartheta}{\left(\cos^2\vartheta+\beta^2\right)^2}\sigma_r(\vartheta)\,d\vartheta = \frac{P}{2\beta R_2(\rho-\delta_0)}. \tag{8.41}$$

We construct the Lagrange interpolation polynomial for an unknown function $\sigma_r(\vartheta)$ choosing interpolation nodes within the segment $[-1,1]$ as the roots of the Chebyshev polynomial of the second kind of degree n, i.e. the points

$$x_k = \cos\vartheta_k, \quad \vartheta_k = \frac{k\pi}{n+1}, \quad k = 1,2,\ldots,n.$$

Then the Lagrange polynomial which coincides with the function $\sigma_r(\vartheta)$ in the points $\vartheta = \vartheta_k$,i.e. $\sigma_k = \sigma_r(\vartheta_k)$ has the form

$$L\left[\sigma_k, x\right] = \frac{2}{n+1}\sum_{k=1}^{n}\sigma_k\sum_{m=1}^{n}\sin m\vartheta_k \sin m\vartheta. \tag{8.42}$$

Replacing integrals on the left-sides of Eqs. (8.40) and (8.41) by finite sums, and giving ϑ the values ϑ_k $(k = 1,2,\ldots,n)$, we obtain the system of equations for the unknown function at the nodes of interpolation:

$$A_k\sigma_k + \sum_{j=1}^{n} a_{kj}\sigma_j + b_{1k}\sigma_k' + b_{2k}\sigma_k'' = B_k, \quad k = 1,2,\ldots,n, \tag{8.43}$$

where

$$A_k = \frac{\eta\beta}{\cos^2\vartheta_k+\beta^2}, \quad B_k = -\frac{m\beta}{\cos^2\vartheta_k+\beta^2}(1-\rho),$$

$$b_{1k} = \frac{u\cos\vartheta_k}{2\beta\sin\vartheta_k}\left(1+\frac{\cos^2\vartheta_k+\beta^2}{2\sin^2\vartheta_k}\right), \quad b_{2k} = -u\frac{\cos^2\vartheta_k+\beta^2}{4\beta\sin^2\vartheta_k},$$

$$a_{kj} = \frac{1}{(n+1)\sin\vartheta_k}\sum_{m=1}^{n} m\sin m\vartheta_j \sin m\vartheta_k -$$

$$\frac{\lambda\beta^2}{(n+1)\left(\cos^2\vartheta_k+\beta^2\right)}\cdot\frac{\sin\vartheta_j}{\left(\cos^2\vartheta_j+\beta^2\right)}-$$

$$\frac{2\beta^2\left(\rho-\delta_0\right)}{(n+1)\left(\cos^2\vartheta_k+\beta^2\right)}\left[q\frac{\cos^2\vartheta_k-\beta^2}{\cos^2\vartheta_k+\beta^2}+\right.$$

$$\left.p\frac{\cos^2\vartheta_k+\beta^2}{8\left(\rho-\delta_0\right)\beta^2}\right]\frac{\sin\vartheta_j\left(\cos^2\vartheta_j-\beta^2\right)}{\left(\cos^2\vartheta_j+\beta^2\right)^2}.$$

It should be noted that the system (8.43) includes not only the values of σ_k but also the first and the second derivatives of the normal pressure. The polynomial (8.42) does not provide the Hermitian interpolation of the function $\sigma_r(\vartheta)$, i.e. the values of the first and the second derivatives of the interpolation polynomial (8.42) calculated at the points x_i, do not coincide with the values of the corresponding derivatives of the function $\sigma_k(x)$ at the same points. Because of this, we calculated the values σ_k' and σ_k'' following the standard procedure by using the values of the function (8.42) at the k-th and at the nearby nodes. Then the system (8.43) is reduced to the system of n linear algebraic equations to determine the values σ_k.

To evaluate the influence of the number n of nodes on the solution of the system (8.43), we solved this system for $n = 7$, $n = 15$ and $n = 31$ (the first, second and the third approach by Multhopp, respectively). The results showed that for all values of β the second approach differs from the first one by less than 0.1%.

After the calculation of the values σ_k at the points ϑ_k we find the load P from the following equation

$$P = R_2\frac{2\pi\beta(\rho-\delta_0)}{n+1}\sum_{k=1}^{n}\frac{\sin\vartheta_k\left(\cos^2\vartheta_k-\beta^2\right)}{\left(\cos^2\vartheta_k+\beta^2\right)^2}\sigma_k. \tag{8.44}$$

8.2.4 Contact characteristics analysis

In calculations we assume that we are given the geometric characteristics (R_2,ρ,δ_0), the elastic characteristics of the contacting bodies ($\mu_1, \kappa_1, \mu_2, \kappa_2, k$), and the parameter β which is determined by the angle θ_0 from Eq. (8.37).

The results of numerical calculations are shown in Fig. 8.5 where the dependence of the contact angle $\phi_0 = \theta_0 - \frac{3}{2}\pi$ on the load P is presented. The curves 1–3 are plotted for the following parameters: $E_1 = E_2 = 2\cdot 10^5$ MPa, $\nu_1 = \nu_2 = 0.3$; $R_2 = 10^{-2}$ m, $\rho = 0.995$, $k = 0.5\cdot 10^{-3}$ MPa^{-1}. Curve 1 corresponds to $\delta = 0$ (disk without coating), curve 2 – to $\delta = 2\cdot 10^{-3}$, curve 3 – to $\delta = 5\cdot 10^{-3}$. Curve 4 is plotted for the rigid bodies S_1 and S_2 and the elastic ring S_0 ($\delta = 2\cdot 10^{-3}$). With this combination of properties of contacting bodies there is some limitation in increasing of the contact arc due to increasing of the load. This process is stopped when a displacement at any point of contact will reach the value δ. Such

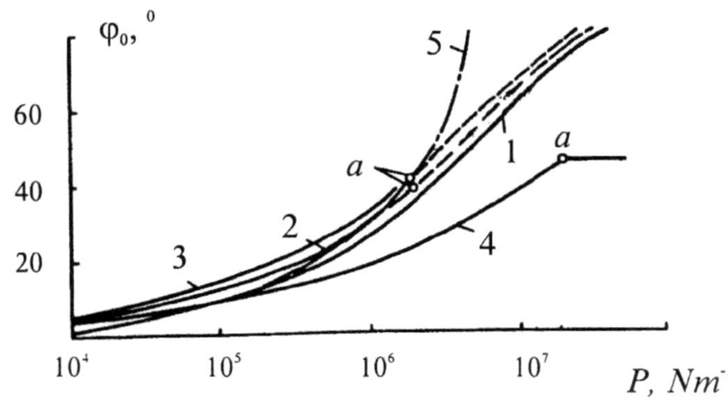

Figure 8.5: Dependence of the contact angle upon the load for a plain bearing (ISP) at different coating thickness: $\delta = 0$ (curve 1), $\delta = 2 \cdot 10^{-3}$ (curve 2) and $\delta = 5 \cdot 10^{-3}$ (curve 3), at $E_1 = E_2 = 2 \cdot 10^5$ MPa. Curve 4 is calculated neglecting the elasticity of bodies S_1 and S_2; curve 5 is calculated from the Hertz theory of contact of elastic bodies S_1 and S_2.

a situation is marked on the curves by the point a and the load corresponding to this point is P_a. The parts of the curves for $P > P_a$ can be considered as unrealistic.

Curve 5 corresponds to the Hertz theory of contact of the bodies S_1 and S_2, neglecting of the coating existence.

The principal conclusions of this study are the following:

1. It is expedient to distinguish three regions for the value of the parameter δ/φ_0 .

   If $\delta/\varphi_0 > 5 \cdot 10^{-2}$ it is possible to consider the bodies S_1 and S_2 as rigid, and S_0 as elastic. In this case the relation between the load and the size of contact arc obeys a simple analytical expression

$$P = R_2 \frac{1-\rho}{k\delta_0} \left(\frac{\phi_0}{\cos\phi_0} - \sin\phi_0 \right), \tag{8.45}$$

   which follows from the solution of the differential equation

$$\sigma_r'' + \sigma_r = \frac{1-\rho}{k\delta_0}$$

   taking the form

$$\sigma_r = \frac{1-\rho}{k\delta_0} \left(1 - \frac{\cos\phi}{\cos\phi_0} \right).$$

   The results calculated from Eq. (8.45) and from Eq. (8.43) for $\delta/\varphi_0 > 5 \cdot 10^{-2}$ are in a good agreement.

If $\delta/\varphi_0 < 5 \cdot 10^{-3}$ it is possible to ignore the coating S_0 in calculations.

If $5 \cdot 10^{-3} \leq \delta/\varphi_0 \leq 5 \cdot 10^{-2}$ we must take into consideration the elastic properties of the three bodies S_0, S_1 and S_2.

2. The soft coating decreases the contact pressure and increases the size of contact arc compared to the characteristics of the journal bearing without coating.

3. Hertz theory gives a good approximation to the contact characteristics of the plain journal bearing with a small contact angle (low loading), but does not agree with experiment for the bearing with coating.

8.2.5 Wear analysis

The results were used to study the wear kinetics of the plain journal bearing with a journal coated by a thin solid lubricant.

In calculations we used the wear law in the form of Eq. (8.1). The operating time was measured by the number N of journal revolutions. During the wear of the junction such characteristics as the contact pressure $p(\varphi, N) = -\sigma_r(\varphi, N)$, the contact angle $\varphi_0(N)$, the thickness of coating $\delta_0(N)$, and the journal radius $R_0(N)$ depend on N.

Modelling the wear process, we calculate the contact characteristics after each revolution assuming that they are constant during each revolution and are changed step-wise at the instant that a new revolution begins. The wear at any fixed revolution is determined by the contact characteristics at the previous revolution.

We introduce the wear at the $(N+1)$-th revolution as

$$\Delta w(N+1) = K_w R_0(N) \int_{-\varphi_0(N)}^{\varphi_0(N)} p^\alpha(\varphi, N) d\varphi, \tag{8.46}$$

where $\varphi = \theta - \frac{3}{2}\pi$.

We used the following procedure for calculating the wear kinetics of the junction. From the contact problem analysis (see §§8.2.1–8.2.3) we determine the initial values of $\varphi_0(0)$ and $p(\varphi_0(0))$. Then, using the relation (8.46) for $N = 0$ we estimate the wear throughout the first revolution ($N = 0$) of a journal, and then we calculate the radius $R_0(1) = R_0(0) - \Delta w(1)$ and the new coating dimensionless thickness $\delta(1) = \dfrac{h(1)}{R_2}$ where $h(1) = h_0 - \Delta w(1)$. This completes one sequence of steps. In order to study the wear kinetics we have to repeat such a sequence as many times as necessary.

Fig. 8.6 illustrates the dependence of the coating wear $\bar{w}(N) = 1 - \dfrac{h(N)}{h_0}$ (curves 1 and 2) and the contact angle φ_0 (curves 1′ and 2′) on the parameter

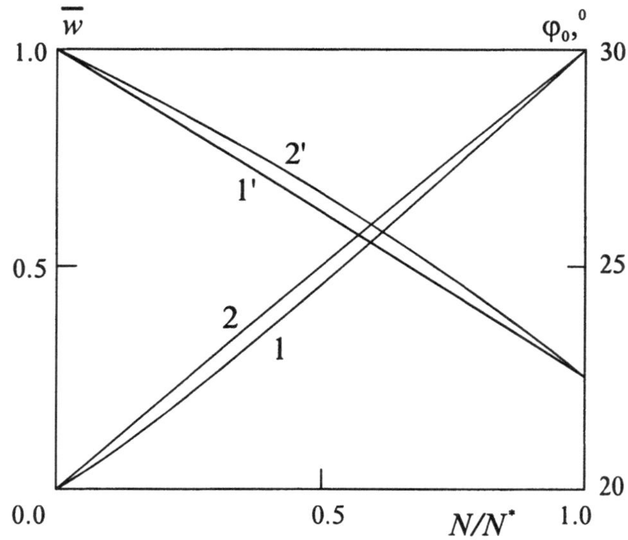

Figure 8.6: Variation of the coating wear $\bar{w}$ (curves 1 and 2) and contact angle φ_0 (curves $1'$ and $2'$) in wear process of the plain bearing with ISP for $K_w = 10^{-14}\text{Pa}^{-1}$ and $\alpha = 1$ (curves 1, $1'$) and $K_w = 10^{-19}\text{Pa}^{-2}$ and $\alpha = 2$ (curves 2, $2'$).

N/N^* which is the ratio of the current number of revolution to the number of revolutions N^* corresponding to the complete wear of the coating ($h(N^*) = 0$).

The results were calculated for

$$R_2 = 10^{-2}\text{m}, \quad \rho(0) = 0.995, \quad \delta_0 = 2 \cdot 10^{-3},$$

$$\mu_1 = \mu_2 = 8.46 \cdot 10^4\text{MPa}, \quad \kappa_1 = \kappa_2 = 1.8,$$

$$k = 0.5 \cdot 10^{-3}\text{MPa}^{-1}, \quad P = 1.08 \cdot 10^6\text{N/m}.$$

For this case $\theta_0(0) = \dfrac{5\pi}{3}$. Curves 1 and $1'$ are calculated for $K_w = 10^{-14}\text{Pa}^{-1}$ and $\alpha = 1$; curves 2 and $2'$ correspond to $K_w = 10^{-19}\text{Pa}^{-2}$ and $\alpha = 2$.

From the results we conclude that if the coating wear rate is a power function of the pressure, the wear of the coating is proportional to the number of revolutions. The contact angle decreases nearly linearly in the wear process. To understand this we may use the following simple argument. Since the journal radius (and consequently the contact angle) decreases during wear, the sliding distance per revolution also decreases. Simultaneously the contact pressure increases, resulting in increase of the wear intensity in accordance with the wear equation (8.1). Considering that the wear per revolution is the product of the wear intensity by sliding distance, it is clear that by virtue of the competing influences of the operating time on these quantities the wear per revolution will change very little. We can consider that this is a characteristic feature of the wear of such sliding bearings.

This result can be used to calculate the lifetime of a junction within a range of operation conditions. These conditions are usually specified by the limiting value of some parameter. We often take this to be the bearing radial clearance, with the magnitude of which the secondary dynamic loads in the machine assemblies and the accuracy are associated. We shall consider that the magnitude of this clearance Δ^* is specified in advance. Since junction wear takes place only at the expense of the journal coating, then $\Delta^* = R_2 - R_0(N^*)$, where $R_0(N^*)$ is the critical value of the journal radius achieved for N^* revolutions. Because of Eq. (8.46) and the initial value of the radial clearance $\Delta_0 = R_2 - R_0(0)$ the limiting wear can be written in the form

$$\Delta^* - \Delta_0 = \sum_{N=1}^{N^*} \Delta w(N) =$$

$$K_w \sum_{N=0}^{N^*-1} R_0(N) \int_{-\varphi_0(N)}^{\varphi_0(N)} p^\alpha(\varphi, N)\, d\varphi, \tag{8.47}$$

$$\Delta^* - \Delta_0 \leq h_0.$$

Thus, determination of the junction service life reduces to determining N^*, satisfying the conditions (8.47).

Considering that the journal bearing wear is nearly proportional to the number of revolutions, we can find a more effective and highly accurate calculation technique by partitioning the limiting wear magnitude $\Delta^* - \Delta_0$ into M uniform intervals $\Delta h = (\Delta^* - \Delta_0)/M$ and calculating the average wear per revolution on each interval. In fact, determining the junction geometry at the end of the m-th interval ($m = l, 2, \ldots, M$) and finding from Eqs. (8.38) and (8.39) the corresponding contact characteristics $p(\varphi, N_m)$, $\varphi_0(N_m)$, we can use Eq. (8.46) to calculate the average wear $\Delta\tilde{w}_m$ per revolution

$$\Delta\tilde{w}_m = \frac{1}{2}\left(\Delta w_{(m-1)} + \Delta w_m\right),$$

where

$$\Delta w_m = K_w \left(R_2(0) - m\Delta h\right) \int_{-\varphi_0(N_m)}^{\varphi_0(N_m)} p^\alpha\left(\varphi, N_m\right) d\varphi.$$

Then the approximate value of N^* is determined as follows:

$$N^* = \Delta h \sum_{m=1}^{M} \Delta\tilde{w}_m^{-1}.$$

Note that a very good approximation to this result can be obtained if we determine the average wear per revolution at the beginning and at the end of assembly operation, i.e. $\Delta\tilde{w}_* = \left(\Delta w_*(0) + \Delta w_*(N^*)\right)/2$. This is explained by a characteristic

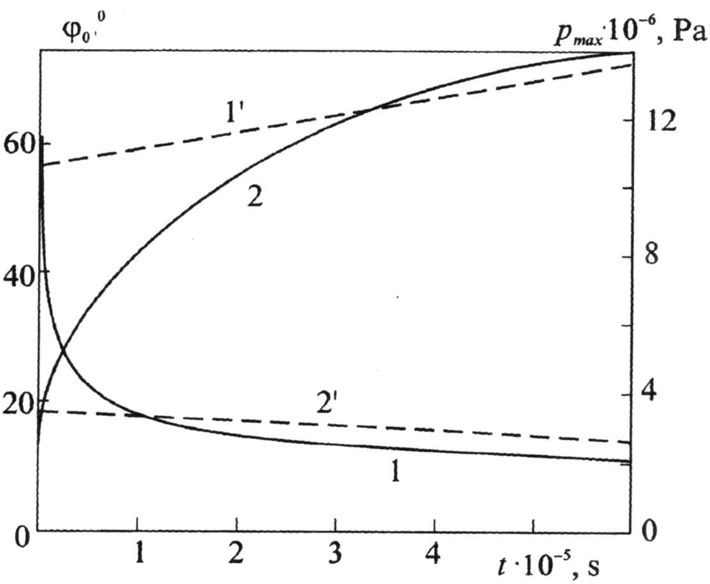

Figure 8.7: Changes of the maximum contact pressure $p_{\max}$ (curves 1 and 1′) and the contact angle φ_0 (curves 2 and 2′) in time for the plain journal bearings with DSP (solid lines) and ISP (dashed lines).

of sliding bearing wear kinetics, noted previously and amounting to the fact that the wear per revolution remains practically constant during operation. Thus, we can calculate the approximate value $\tilde{N}^*$ of N^* as

$$\tilde{N}^* = \frac{\Delta^* - \Delta_0}{\Delta \tilde{w}_*}.$$

This method makes it possible to simplify the calculations considerably and at the same time ensure high accuracy.

The results show also that failure to account for coating properties in calculating the journal bearing service life leads to underestimation of the junction service life, which is due to the errors in evaluating the contact zone dimensions and the pressure distribution.

8.3 Comparison of two types of bearings

The results for the previous problems make it possible to compare kinetics of changes of contact and tribotechnical characteristics for two types of plain bearings, which are DSP and ISP described in § 8.1 and in § 8.2, respectively.

Fig. 8.7 illustrates the dependence of contact angle and maximum contact pressure on operating time for the plain journal bearings with DSP (solid lines) and ISP (dashed lines). Calculations were performed for the following initial data:

$$P = 10^5\text{N/m}, \quad R_2 = 2.505 \cdot 10^{-2}\text{m}, \quad R_1 = 2.5 \cdot 10^{-2}\text{m},$$
$$k = 8.35 \cdot 10^{-4}\text{MPa}^{-1}, \quad \alpha = 1.4,$$
$$K_w = 1.2 \cdot 10^{-19} \cdot \text{Pa}^{-1.4}, \quad \omega = 10\text{s}^{-1}, h_0 = 3 \cdot 10^{-4}\text{m},$$
$$\mu_1 = \mu_2 = 7.7 \cdot 10^4\text{MPa}, \quad \nu_1 = \nu_2 = 0.3.$$

The kinetics of changes of parameters for DSP and ISP differ in principle: for the DSP the contact angle increases and maximal pressure diminishes in the process of wear; for the ISP the contact angle diminishes and maximal pressure increases. The evolution of contact characteristics for DSP looks more favorable than for ISP. The difference in the initial values for p_m and θ for these types of junctions can be explained by the fact that for DSP the bodies S_0 and S_1 are considered as rigid, and for ISP as being elastic.

There is a second significant discrepancy between the two kinds of wear processes. For DSP the shape of a bush changes during the wear process. This feature leads to a difference between the running-in stage of wear process and its steady-state stage. The first stage is characterized by intense changing of parameters and non-linear dependence of the contact pressure, contact angle and the wear rate on the operating time; over the second stage these relations are very close to the linear ones.

For ISP, there is no shape variation. Consequently for this junction the steady-state conditions are valid over the whole operating time, the dependences of $p_{\max}(t)$, $\varphi_0(t)$ and $h_{\min}(t)$ are always slightly different from linear ones. This considerably simplifies calculations of contact and tribotechnical characteristics of such joints.

In the special case ($\alpha = 1$) it can be strictly proved that the ISP lifetime is higher than DSP, all other things being equal. Let us examine the case $\alpha = 1$ and a small contact angle φ_0. The wear for the N-th revolution for ISP is calculated from Eq. (8.46) as

$$\Delta w^{(2)}(N+1) = K_w \int\limits_{-\varphi_0(N)}^{\varphi_0(N)} p(\varphi, N) R_0(N) d\varphi \approx K_w P.$$

The lifetime of DSP is determined by the wear at the point where the maximum contact pressure occurs. The friction distance during one revolution for this point is $2\pi R_1$. The wear for the N-th revolution for this scheme is determined by the formula

$$\Delta w^{(1)}(N+1) = 2\pi R_1 K_w p(0, N).$$

By virtue of the fact that $2\pi R_1 p(0,N) > P$, we obtain $\Delta w^{(1)}(N+1) > \Delta w^{(2)}(N+1)$. From this relationship, it follows that for the equal limiting wear $(\Delta^* - \Delta_0)$ the lifetime of ISP is always higher than DSP.

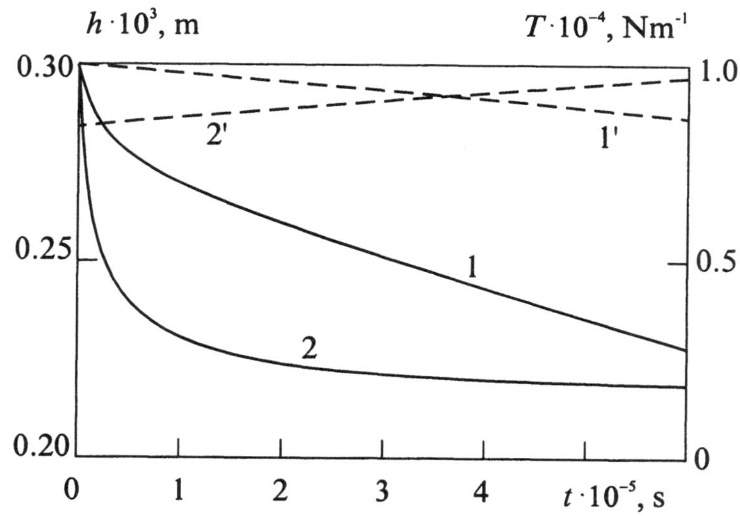

Figure 8.8: Changes of the coating thickness h (curves 1 and 1$'$) and the friction force T (curves 2 and 2$'$) in time for the plain bearing with DSP (solid lines) and ISP (dashed lines).

The another important tribological characteristic of the journal bearing is the friction force. It must be noted that the character of the dependence of the friction force on time for both types of bearings depends on the friction law. Particularly, if the tangential stress τ is a power function of contact pressure

$$\tau = \mu^* p^m,$$

with a power $m > 1$, then DSP is more favorable than ISP in respect to the friction force. Fig. 8.8 illustrates the dependence of the friction force on time for DSP (solid lines) and ISP (dashed lines) for $\mu^* = 10^{-8}\text{Pa}^{-1}$ and $m = 2$. In this case the friction force decreases in the wear process for DSP and it increases for ISP. It should be noted that the results depend essentially on the parameter m.

From these results it is evident that the kinetics of changes of contact and friction characteristics of plain bearings with direct and inverse pairs differ considerably. So one should pay attention to their functional properties when choosing the type of configuration for a plain bearing.

8.4 Wheel/rail interaction

In general, rail and wheel profiles are chosen to satisfy simultaneously the following conditions:

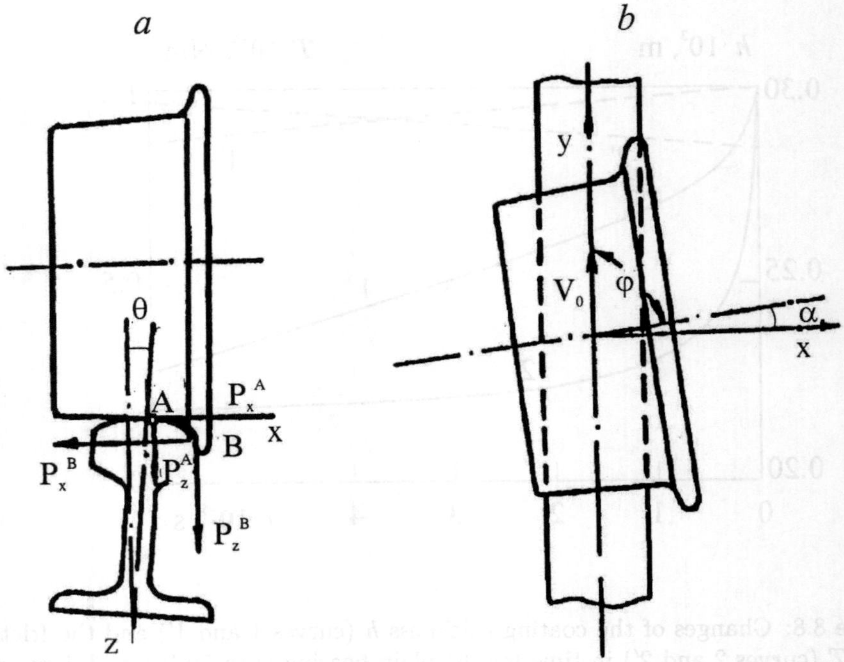

Figure 8.9: Relative position of a rail and a wheel at the planes $y = 0$ (a) and $z = 0$ (b).

- provision of wheel stability in contact with rail;
- reduction of contact fatigue defects;
- reduction of wear of rails and wheels.

Excessive wear and damage of rails and wheels are great problems for heavy haul railways. It should be noted that the most wear occurs at the side of the rail and at the crest of the wheel travelling on curved track.

In what follows we present the model for evaluating the tribological aspects of wheel/rail curve interaction developed in Bogdanov et al. (1996) . The results could be used for selection of the rail and wheel profiles, which provide decreasing wear rate and rate of fatigue damage accumulation in rails.

8.4.1 Parameters and the structure of the model

We consider a contact of a rail and a wheel travelling on a curved track. Fig. 8.9 illustrates the relative position of rail and wheel in contact. The geometry of contact is described by the angle θ of the rail inclination about the vertical axis

Oz (rail inclination angle), the attack angle $\alpha = 90^0 - \varphi$, where φ is the angle between the axis of rotation of the wheel and the longitudinal axis Oy of the rail. We assume that the angles θ and α are random variables.

The profiles of the rail and the wheel are given and can be changed in the wear process.

Actually there is two-point contact between the wheel and the rail (the first is on the running part of the rail head and the second is on the side of the rail). At the contact spots on the top and on the side of the rail characterized by the points A and B of initial contact, the vertical P_z^A and P_z^B and lateral P_x^A and P_x^B forces of interaction between the rail and wheel are applied. These forces are also considered as random variables. They are obtained from the dynamic model of track and rolling stock interaction described by Verigo and Kogan (1986).

We consider the cyclic interaction of wheels with the fixed part of the rail on a curved track. As the result of this process the rail and the wheels are worn and damage accumulates inside the contacting bodies.

The problem may be split into several stages which are shown schematically in Fig. 8.10.

At first we solve the contact problem for rail and wheel to find the shape, size and the position of the contact zones and the contact stresses.

Then, using the contact stress distribution, we calculate the internal stresses in the rail and wheel, and the damage accumulation function. With this we determine the areas where the fatigue damage is concentrated. These problems are indicated in the left column of Fig. 8.10.

The results of the contact problem analyses are also used to calculate the wear rate of the rail and wheel surfaces and to determine the worn shapes of the rail and wheel. These problems are indicated in the right column of Fig. 8.10.

We now discuss each problem in detail.

8.4.2 Contact characteristics analysis

Due to the deformation of the bodies, the contact of the rail and wheel occurs within the contact zones, including the points of initial contact. Determination of the initial points of contact is geometric problem which is described in detail by Bogdanov et al. (1996).

The initial data are rail and wheel profiles (both bodies are cylindrical) and the angles θ and α. The wheel and rail profiles are given pointwise, and then third order spline-approximations are used to produce twice continuously differentiable functions describing the profiles. After that these functions are rewritten for a common system of coordinates.

The system of equations for determination of the initial contact points A and B contains the conditions that the shape functions coincide and the normals are collinear at these points. We used an iterative method to solve the equations.

The analysis of the contact problem for the rail and wheel is based on various assumptions. The deformations of the bodies in contact are considered to be elastic. Determination of the stresses within the contact zone of elastic bodies

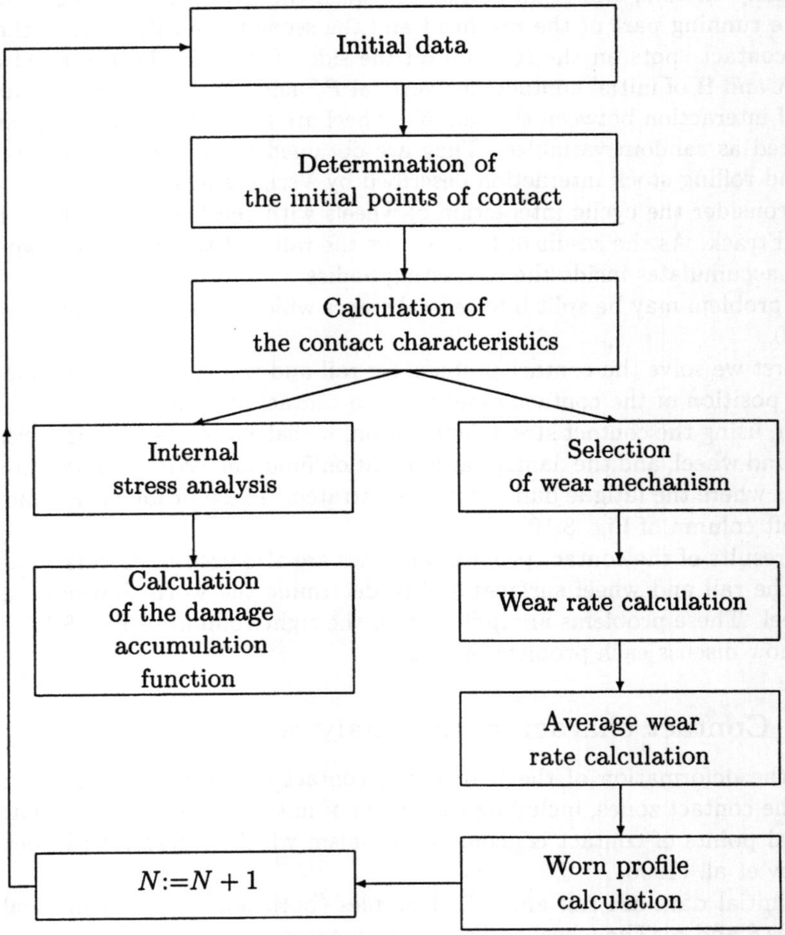

Figure 8.10: The stages of calculation of the wear and damage accumulation processes in a wheel and in a rail (N is the cycle number).

with profiles which cannot be represented adequately by their curvature radii at the initial contact point is a severe problem. In order to avoid some difficulties we model the contacting bodies by a simple Winkler elastic foundation.

The second simplification of the contact problem is connected with neglecting the tangential stress in the contact region when we calculate the contact pressure. It is well known that the tangential contact stress does not influence the normal contact stress in the contact of bodies characterized by the same elastic moduli. If the elastic properties of contacting bodies are different, there is some influence, but it is still small.

We consider some initial point (x_0, y_0, z_0) of contact of rail and wheel and place the origin O of the system of coordinates $O\xi\eta\zeta$ there. The axis $O\zeta$ coincides with the common normal to the contacting surfaces at the point (x_0, y_0, z_0), the axis $O\eta$ is aligned with a rail generatrix, and the axis $O\xi$, which is in the tangential plane, is determined by the condition that the axes $O\xi$, $O\eta$ and $O\zeta$ form a right handed triple.

Undeformed surfaces of the rail and the wheel in this system of coordinates are described by equations $\zeta_1 = f_1(\xi)$ and $\zeta_2 = f_2(\xi, \eta)$, respectively. The separation between the two surfaces near the initial point of contact is given by

$$h = |\zeta_1 - \zeta_2| = f(\xi, \eta) = |f_1(\xi) - f_2(\xi, \eta)|.$$

Under the normal load P the surfaces of the rail and wheel have the displacements $w_1(\xi, \eta)$ and $w_2(\xi, \eta)$, respectively. The boundary condition for displacements within the contact region Ω can be written

$$w_1(\xi, \eta) + w_2(\xi, \eta) + h = D, \tag{8.48}$$

where D is the approach of the bodies under the load.

According to the Winkler model, the contact pressure $p(\xi, \eta)$ at any point depends only on the displacement at that point, thus

$$w_1(\xi, \eta) = K_1 p(\xi, \eta), \quad w_2(\xi, \eta) = K_2 p(\xi, \eta), \tag{8.49}$$

where K_1 and K_2 are the coefficients which characterize the elastic compliances of the rail and wheel, respectively. Assuming $K = K_1 = K_2$, from Eqs. (8.48) and (8.49) we obtain the following relationship within the contact region Ω

$$2Kp(\xi, \eta) = D - f(\xi, \eta). \tag{8.50}$$

Outside the contact region for the model under consideration the normal displacements satisfy the conditions

$$w_1(\xi, \eta) = 0, \quad w_2(\xi, \eta) = 0, \quad D \leq h. \tag{8.51}$$

Adding to Eqs. (8.50) and (8.51) the equilibrium condition

$$\iint\limits_{\Omega} p(\xi, \eta) d\xi d\eta = P, \tag{8.52}$$

we obtain the complete system of equations for determination of the contact pressure $p(\xi, \eta)$, approach D and the contact region Ω.

The normal load $P = P^A$ (or $P = P^B$) acting on each contact region $\Omega = \Omega^A$ (or $\Omega = \Omega^B$) is equal to the sum of the projections of the forces P_z^A and P_x^A (P_z^B and P_x^B) on the axis $O\zeta$.

8.4.3 Wear analysis

We consider a cyclic interaction of the wheel moving along the fixed part of a curved track, as a result of which the wheel and the rail wear.

To calculate the wear rate of the rail and the wheel at the N-th cycle characterized by the given shape of the rail and wheel and the given probability density function $\rho(\theta, \alpha, \vec{P}^A, \vec{P}^B)$ ($\vec{P}^A$ and $\vec{P}^B$ are the vectors of forces acted at the contact regions Ω^A and Ω^B, respectively), we represent the process of contact interaction as a number of elementary interactions. We can treat the elementary.interaction as a single passage of the wheel along the fixed part of the rail. For each elementary interaction the external contact parameters (θ, α, $\vec{P}^A$, $\vec{P}^B$, etc.) are assumed to be given and fixed. Using the wear rates calculated for each elementary interactions and averaging them over the ensemble of external parameters, we obtain the desired rail and wheel wear rates.

Let us consider this procedure in more detail. The wheel moves along the rail with a constant speed V_0. The mutual position of the rail and wheel is described by the angle of inclination θ and the attack angle α. From the solution of the contact problem we know the contact pressure distribution at the contact zones Ω^A and Ω^B:

$$\begin{aligned} p^A(\xi, \eta) &= \Pi^A(\xi, \eta, \theta, \alpha, P^A), \\ p^B(\xi, \eta) &= \Pi^B(\xi, \eta, \theta, \alpha, P^B), \end{aligned} \tag{8.53}$$

where ξ and η are the local coordinates in the vicinity of the initial contact points A and B, and Π^A and Π^B are the functions obtained from the contact problem analysis.

The contact pressure in the presence of the relative sliding produces the wear of the contacting surfaces. We assume that the wear rates of the rail $\partial W_r^i/\partial t$ and the wheel $\partial W_w^i/\partial t$ are described by the equations

$$\begin{aligned} \frac{\partial W_r^i}{\partial t} &= F_r(p^i, V^i), \\ \frac{\partial W_w^i}{\partial t} &= F_w(p^i, V^i), \end{aligned} \tag{8.54}$$

where W_r^i and W_w^i are the wear of the rail and the wheel at the fixed point (ξ, η), V^i is the sliding speed, F_r and F_w are the known functions, $i = A, B$ depending on the contact point under consideration.

The sliding speed V^A for the wheels mounted on a common axle while traveling on curved track is determined by the difference of lengths of their trajectories

$$V^A = \frac{D_r V_0}{2R_c}, \tag{8.55}$$

where R_c is the radius of the track curvature, D_r is the distance between the wheels at a common axle.

The sliding speed $V^B(\xi,\eta)$ at the contact zone Ω^B located at the lateral edges of the rail head and the wheel depends on the distance of Ω^B from the instantaneous center of rotation of the wheel of the radius R. The function $V^B(\xi,\eta)$ can be determined from the following relationship

$$V^B(\xi,\eta) = \frac{V_0}{R}\left[\left(z^B_{rc} - \Delta + \xi\cos\beta\right)^2 + \left(y^B_{rc} + \eta\right)^2\right]^{1/2}, \qquad (8.56)$$

where $(x^B_{rc}, y^B_{rc}, z^B_{rc})$ are the coordinates of the initial contact point B at the system of coordinates $(Ox_r y_r z_r)$ coupled to the rail (the axis Oz_r coincides with the axis of symmetry of the rail and the axis Oy_r is collinear to the rail generatrix; the origin O is at the top of the rail); Δ is the displacement of the instantaneous axis of rotation from the point O, β is the angle between the axis Oz_r and the tangential plane to the rail surface at the point B.

Note that we neglect the real speed distribution within the contact zone Ω^A assuming it to be constant, because the characteristic size of the contact region is significantly less than the distance D_r.

In contrast, we take into account the speed distribution $V^B(\xi,\eta)$ because the values of z^B_{rc} and y^B_{rc} on the one hand, and ξ and η on the other, can be commensurable.

From Eq. (8.54) we can find the wear of the rail $\delta W^i_r(\lambda_r)$ and of the wheel $\delta W^i_w(\lambda_w)$ in the elementary interaction (λ_r and λ_w are curvilinear coordinates at the rail and wheel profiles, respectively)

$$\begin{aligned} \delta W^i_r(\lambda_r) &= \frac{1}{V_0}\int\limits_{a^i(\lambda_r)}^{b^i(\lambda_r)} F^i_r\left[p^i\left(\lambda_r - \lambda^i_r, \eta\right), V^i\right] d\eta, \\ \delta W^i_w(\lambda_w) &= \frac{1}{V_0}\int\limits_{a^i(\lambda_w)}^{b^i(\lambda_w)} F^i_w\left[p^i\left(\lambda_w - \lambda^i_w, \eta\right), V^i\right] d\eta. \end{aligned} \qquad (8.57)$$

where λ^i_r and λ^i_w are the curvilinear coordinates of the initial contact points at the rail and wheel, respectively, $a^i(\lambda_r)$ $(a^i(\lambda_w))$ and $b^i(\lambda_r)$ $(b^i(\lambda_w))$ are the functions describing the boundaries of the contact zones at the rail (wheel) surface. The contact pressure $p^i(\xi,\eta)$ and the sliding speed $V^i(\xi,\eta)$ are determined by Eqs. (8.53), (8.55) and (8.56).

The elementary wear δW^i_r (δW^i_w) can be represented as a function of the external parameters θ, α and $\vec{P}^A$ and $\vec{P}^B$, i.e.

$$\begin{aligned} \delta W^i_r(\lambda_r) &= \Phi^i_r\left(\lambda_r, \theta, \alpha, \vec{P}^i\right), \\ \delta W^i_w(\lambda_w) &= \Phi^i_w\left(\lambda_w, \theta, \alpha, \vec{P}^i\right). \end{aligned} \qquad (8.58)$$

We recall that the external contact parameters are constant during one elementary interaction and are random variables described by the probability density function $\rho\left(\theta,\alpha,\vec{P}^A,\vec{P}^B\right)$ for the full process of the contact interaction.

Averaging Eq. (8.58) over the set of the external parameters, we obtain the average wear $\delta\overline{W}_r(\lambda_r)$ at the point λ_r ($\delta\overline{W}_w(\lambda_w)$ at the point λ_w) at the N-th cycle as

$$\delta\overline{W}_r(\lambda_r) \equiv \overline{W}_r^A(\lambda_r) + \delta\overline{W}_r^B(\lambda_r) =$$
$$\int\limits_{\Sigma}\left[\Phi_r^A\left(\lambda_r,\theta,\alpha,\vec{P}^A\right) + \Phi_r^B\left(\lambda_r,\theta,\alpha,\vec{P}^B\right)\right]\rho\left(\theta,\alpha,\vec{P}^A,\vec{P}^B\right)d\Omega, \tag{8.59}$$
$$\delta\overline{W}_w(\lambda_w) \equiv \overline{W}_w^A(\lambda_w) + \delta\overline{W}_w^B(\lambda_w) =$$
$$\int\limits_{\Sigma}\left[\Phi_w^A\left(\lambda_w,\theta,\alpha,\vec{P}^A\right) + \Phi_w^B\left(\lambda_w,\theta,\alpha,\vec{P}^B\right)\right]\rho\left(\theta,\alpha,\vec{P}^A,\vec{P}^B\right)d\Omega,$$

where Σ is the range of admissible values of the external parameters.

The values $\delta\overline{W}_r(\lambda_r)$ and $\delta\overline{W}_w(\lambda_w)$ determined from Eq. (8.59) make it possible to analyze the wear kinetics of the rail and wheel. For this aim we change the rail and wheel profile in accordance with the wear functions $\delta\overline{W}_r(\lambda_r)$ and $\delta\overline{W}_w(\lambda_w)$ and the given step in time of the N-th cycle, and repeat the procedure of calculations described above with the new rail and wheel profiles. Using the necessary number of cycles N, we can study the profile evolution.

8.4.4 Fatigue damage accumulation process

The solution of the contact problem described in § 8.4.2 makes it possible to find the internal stresses in the rail and wheel and to study the fatigue damage accumulation process.

In this study we will use the phenomenological approach which was described in details in Chapter 5 to analyse the fatigue damage accumulation process. It is based on the linear summation theory of damage. The model can be used to determine the possible places of the fatigue crack initiation.

For definiteness we describe the process of damage accumulation inside the rail.

We suppose that the damage Δd accumulated at a fixed point of the rail cross-section for each elementary interaction with the moving wheel is determined by the maximum value $\tau_{\max}^*$ of the principal shear stress $\tau_{\max}$ at this point, and is calculated by the formula

$$\Delta d = k_d(\tau_{\max}^*)^n, \tag{8.60}$$

where k_d and n are coefficients characterizing the material properties ($n > 1$). We assume that the internal stresses do not depend on the level of the damage of the contacting bodies. Since the minimum value of the function $\tau_{\max}$ for one interaction is equal to zero, the value $\tau_{\max}^*$ coincides with the amplitude of the function $\tau_{\max}$.

Averaging the value Δd over the set of the external parameters according to the probability density function in the m-th cycle, we calculate the average damage Δd_m accumulated at the fixed point for the m-th cycle.

The damage D accumulated at some point for N cycles is calculated as

$$D(x,z) = D_0(x,z) + \sum_{m=1}^{N} \Delta d_m,$$

where $D_0(x,z)$ is the initial damage at the point (x,z).

The most probable region of a fracture is identified with the region having the maximum value of the function $D(x,z)$.

8.4.5 Analysis of the results

Contact characteristics

We studied the influence of the inclination angle θ and the attack angle α on the characteristics of the contact interaction of the rail and wheel (the size and location of contact zones, the pressure distribution within each contact zone, etc.).

Three kinds of rail profiles (new, moderately worn and severely worn profiles) were considered in contact interaction with a new wheel. These profiles are shown in Fig. 8.11.

Fig. 8.12 illustrates the location of the contact regions on the rail surface for the contact of the new rail (Fig. 8.11 (a)) and the new wheel (Fig. 8.11 (d)). The results of calculations show that the shapes of the contact regions of the low worn rails and wheels on the running part of the rail and on its lateral edge are close to elliptical. The eccentricity of the elliptic region on the running part is nearly zero, i.e. the region is nearly a circle, but the ellipse at the lateral edge of the rail is stretched along the rail generatrix.

For the contact of new rails and wheels, the values of the angles θ and α have considerable influence on the contact pressure at the region located on the lateral edge of the rail. The maximum and average values of contact pressure increase as the angles θ or α increase.

In contrast, in the contact between the severely worn rail and the new wheel, the maximum and average pressure are essentially independent of the inclination and attack angles. In addition, the comparison of the contact characteristics within the region located on the running part of the rail for the new and the severely worn rails show that the contact area for the severely worn rail is $5-6$ times less than for new one, and the contact pressure increases considerably. So for the worn rail the contact pressure at the running part of the rail can reach the yield stress. It can give rise to the specific configuration on the external edge of the worn rail shown in Fig. 8.13.

It was established by Bogdanov et al. (1996) that the attack angle α has a considerable influence on the location of the contact region on the lateral edge of the rail, and the distance between this region and the instantaneous axis of rotation, and in turn affects the sliding velocity and the wear rate (see Eqs. (8.54) and

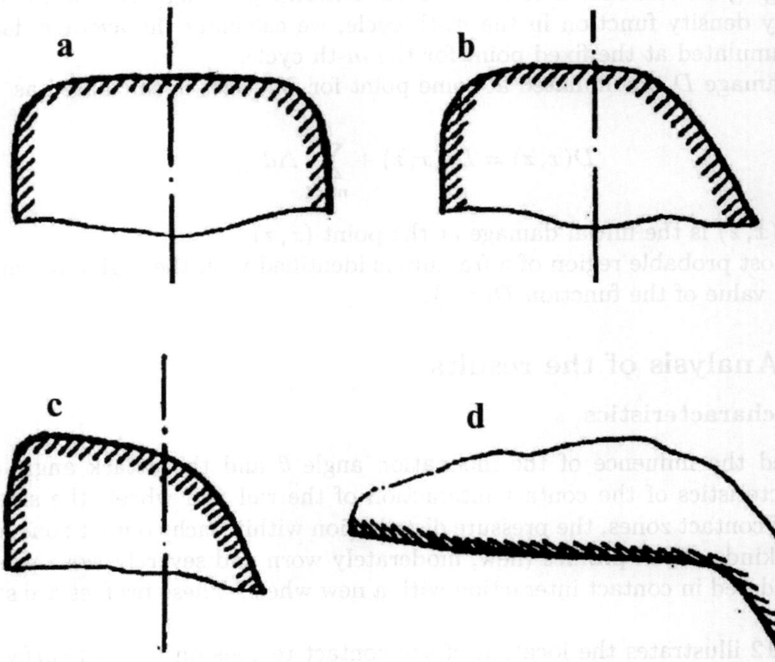

Figure 8.11: Profiles of a rail and a wheel used in the analysis of the contact characteristics: new rail (a), moderately worn rail (b), severely worn rail (c), new wheel (d).

(8.56)). The fact that the attack angle is the important characteristic determining the rail wear is supported by the experimental results discussed by Xia-Qiu Wang (1994).

Damage accumulation process

The analysis of the damage accumulation process from the model described in §8.4.4 makes it possible to differentiate two main groups of parameters determining the damage accumulation rate and the points where the damage accumulation function reaches its maximum value.

The first group includes the parameters which have considerable influence on contact characteristics (size and location of the contact region, maximum contact pressure, etc.) during the elementary interaction. They are the profiles of the rail and wheel, the loads applied to the contact regions, attack and inclination angles. This group of parameters also includes the parameter n in the damage rate equation (8.60) which largely determines the depth where the damage accumulation

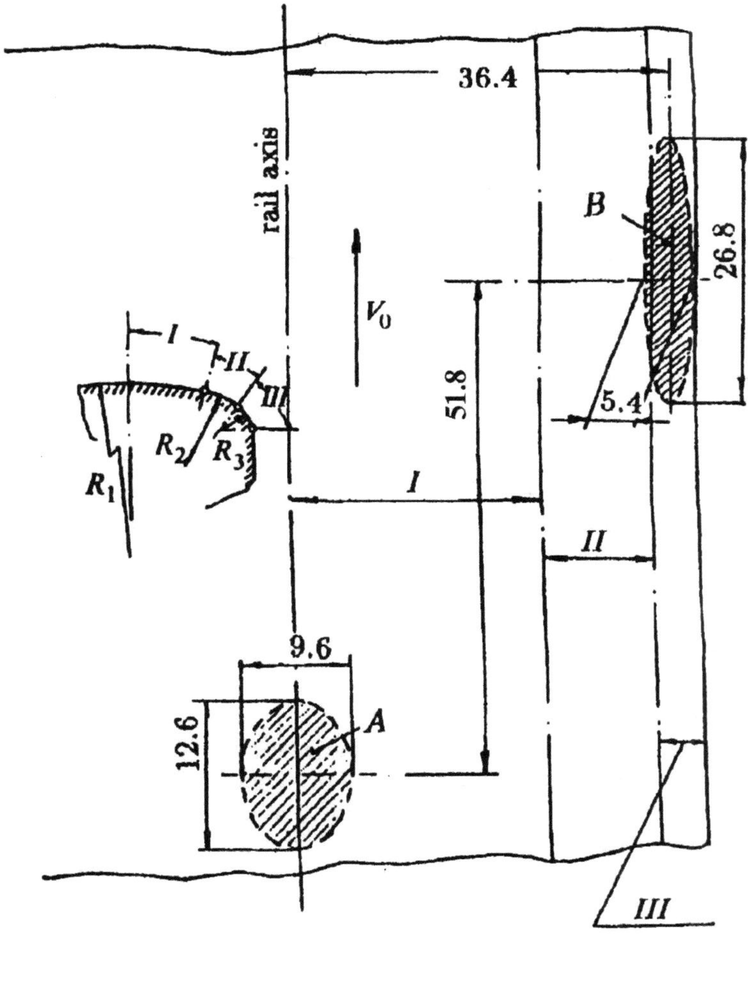

Figure 8.12: Location of the contact zones on the rail surface for the contact of a new rail and a new wheel for $\alpha = 0.06$ rad, $\theta = 0$, $P^A = 6.6 \cdot 10^4$ N, $P^B = 1.1 \cdot 10^5$ N (all sizes are given in millimeters).

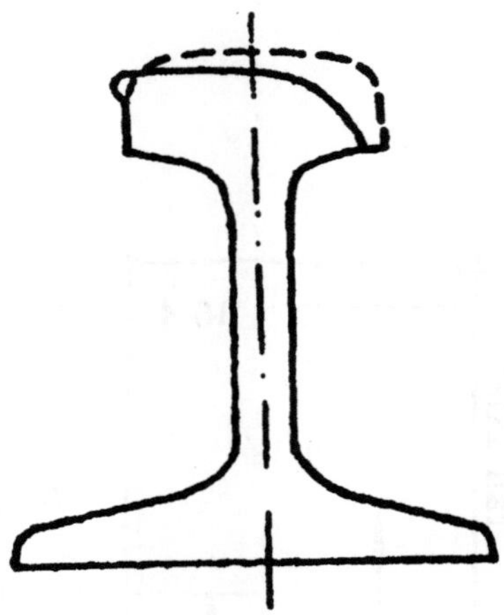

Figure 8.13: The worn rail profile in curve track ($R_c = 303$ m) after 2 years (solid line) and the new one (dashed line).

function reaches its maximum value. Since the contact region at the lateral edge of the rail is more extended than that at the running part, the maximum value of the damage function is localized closer to the surface at the lateral edge of the rail.

The second group includes the parameters which determine the statistical characteristics of the elementary interaction ensemble. For instance, the greater the range of location of the initial contact points at the rail profile, the less is the damage concentration, and the greater is the time needed to achieve the critical value of the damage function at some point.

In calculations we found the ratio of the damage to k_d. The parameter n which influences the location of the point of maximum damage was chosen between the limits from 5.8 to 9.5 that correspond to different structures of the rail steel. Fig. 8.14 illustrates the damage accumulation function distribution within the new rail head in contact with the new wheel.

Wear kinetics

We used Eq. (8.59) to calculate the values of $\delta\overline{W}_r(\lambda_r)$ and $\delta\overline{W}_w(\lambda_w)$. The functions $F_r(p,V)$ and $F_w(p,V)$ in the wear equations (8.54) were taken in the form given by Specht (1987):

$$F(p,V) = \kappa\frac{\mu}{\gamma}pV, \tag{8.61}$$

where

$$\kappa = \begin{cases} \kappa_m, & \text{if} \quad \mu pV \leq Q^*, \\ \kappa_s, & \text{if} \quad \mu pV > Q^*, \end{cases}$$

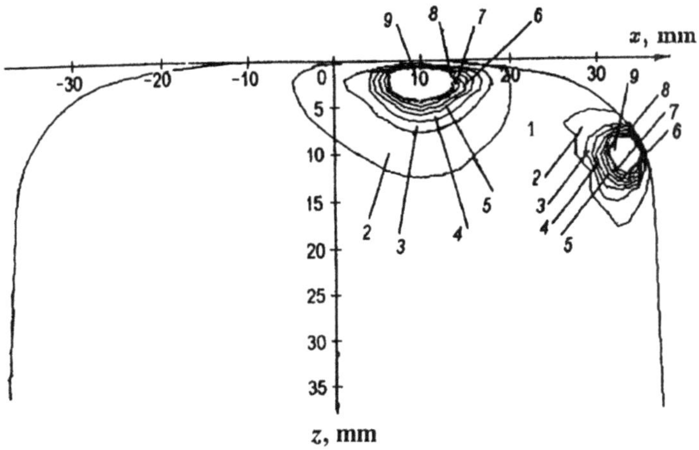

Figure 8.14: Damage distribution within the rail head for D in the intervals: (1) $(0, 140]$; (2) $(140, 1500]$; (3) $(1500, 3000]$; (4) $(3000, 4300]$; (5) $(4300, 5800]$; (6) $(5800, 7200]$; (7) $(7200, 8600]$; (8) $(8600, 10000)$; (9) $D = 10000$ (D is measured in some conventional units).

μ is the friction coefficient, γ is the density of material, κ_m and κ_s are the wear coefficients, Q^* is the critical value of the specific capacity of friction. Eq. (8.61) reflects the jump in wear rate corresponding to the transition from the mild to the severe wear regime for large values of the specific capacity of friction ($\mu pV > Q^*$). The values of γ, κ_m, κ_s and Q^* can be different for the rail and the wheel, but the results presented here were calculated under the assumption that these values are the same for both contacting bodies.

The function $\rho(\theta, \alpha, \vec{P}^A, \vec{P}^B)$ was taken from Romen (1969) where the solution of the dynamic model of the contact interaction of a carriage and a railway was obtained. This function corresponds to a track with radius of curvature $R_c =$ 350m, and the speed $V_0 = 20\text{ms}^{-1}$.

Fig. 8.15 illustrates the wear rate distribution along the rail (a) and wheel (b) profiles. The maximum wear rate occurs at the lateral sides of the rail and wheel.

This model makes it possible to calculate the evolution of the rail and wheel profiles in the wear process. Fig. 8.16 illustrates the rail profiles occurring after different number of cycles in contact interaction of an initially new rail with a new wheel. The results show that the worn profile calculated from the model is very close to the shape presented in Fig. 8.13. This suggests that the model can be used to predict the wear of rails and wheels in contact interaction and to evaluate the influence of different parameters on the wear and damage accumulation processes.

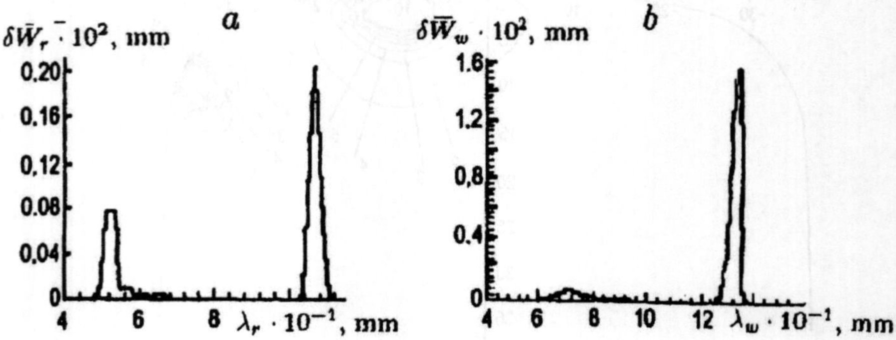

Figure 8.15: Wear rate distribution along the rail (a) and wheel (b) profiles.

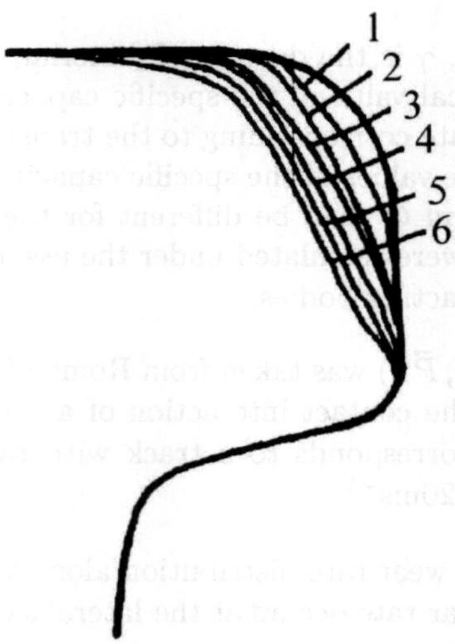

Figure 8.16: Evolution of the rail profile in wear process for $N = 1.34 \cdot 10^6$ cycles (curve 1), $N = 2.68 \cdot 10^6$ cycles (curve 2), $N = 4.02 \cdot 10^6$ cycles (curve 3), $N = 5.37 \cdot 10^6$ cycles (curve 4), $N = 6.71 \cdot 10^6$ cycles (curve 5), $N = 8.05 \cdot 10^6$ cycles (curve 6).

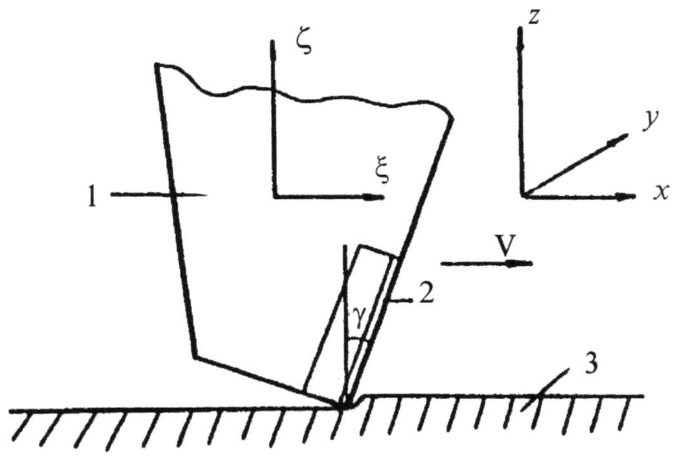

Figure 8.17: Scheme of the tool/rock contact.

8.5 A model for tool wear in rock cutting

A specific feature of the cutting tool – worked material (rock) pair is the variation of shape of both elements caused by fracture or wear. Thus, the problem of cutting tool operation modelling is significantly different from the traditional wear contact problems described in Chapter 6, where shape variation of only one body is taken into account. The shape variation leads to pressure redistribution in the contact zone; this in turn influences the rock fracture and tool wear. The interconnected non-stationary contact problem including wear and fracture must be studied.

To solve this problem it is necessary to develop a model of worked material fracture in cutting. Rock fracture has been studied deeply by Cherepanov (1987) and Atkinson (1987). However, a general model of rock cutting has not yet been developed. This can be considered as an obstacle for modelling of cutting tool wear. However, since the processes of rock fracture and tool wear are interconnected, information on tool wear process (shape variation, size and position of wear land) can be used for modelling the process in the contact zone.

The experimental data obtained for a tool with a diamond-hard alloy insert have been used as the basis of the model. Fig. 8.17 illustrates a schematic of the cutter (1) with the insert (2) in contact with rock (3). The (x, y, z) coordinate system is fixed on the rock. The cutter is moving along the x-axis with the speed V; γ is the rake angle.

Experiments have shown the following:

1. The wear area is inclined relative to the horizontal axis. Fig. 8.18 illustrates worn tool profiles presented by Checkina, Goryacheva and Krasnik (1996). The profiles of the worn tool were obtained for cutting sand-cement blocks; the cutting depth was 10 mm, velocity of the tool displacement (cutting speed) V=1.25 ms^{-1}, $\gamma = 15°$.

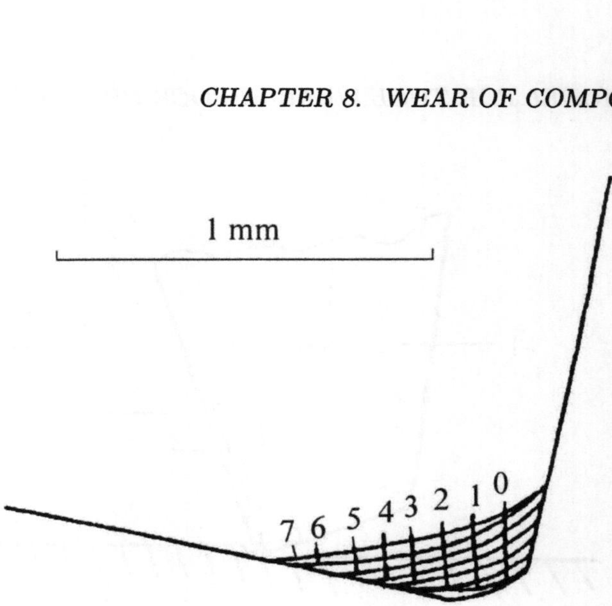

Figure 8.18: Experimental worn tool profiles. Curve number corresponds to the path of the tool in contact with rock; measurements are in kilometers.

2. When the cutting depth is equal to several millimeters, the size of the face wear area is fraction of 1 mm (see Fig. 8.18).

3. The cutting force components oscillate during the process of cutting.

In what follows we describe a model of the tool wear in cutting which was developed by Checkina, Goryacheva and Krasnik (1996). This model reproduces the features revealed in the experiments, and investigates the influence of the model parameters on tool wear and also the influence of the tool shape variation on the characteristics of the cutting process.

8.5.1 The model description

We treat the problem as two-dimensional, considering the tool width along the y-axis to be much greater than the size of contact zone in the x-direction. We introduce the (ξ, ζ) coordinate system moving with the tool. The tool shape in this coordinate system is described by the function $f(\xi, t)$. Shape variation with time is caused by the tool wear; its initial shape is $f(\xi, 0) = f_0(\xi)$.

The following relationships hold between the coordinate systems (x, z) and (ξ, ζ):

$$x = \xi + Vt, \quad z = \zeta - c(t), \tag{8.62}$$

where $c(t)$ is the cutting depth.

The model of rock deformation

We consider two types of rock boundary displacement taking place in the contact zone simultaneously. They are elastic displacement $u_z(x,t)$ along the z-axis described by the equation

$$u_z(x,t) = k\tilde{p}(x,t) \tag{8.63}$$

($\tilde{p}(x,t) \geq 0$ is the contact pressure at the point x of the rock surface, k is a coefficient) and irreversible displacement $w(x,t)$ along the z-axis governed by the relationship

$$\frac{dw(x,t)}{dt} = F(V)\tilde{p}(x,t). \tag{8.64}$$

Irreversible displacement is caused by the rock fracture (crushing) under the tool.

It should be mentioned that Eq. (8.64) can describe different types of process, depending on the function $F(V)$. For $F(V) \sim V^{\beta}$ this equation is equivalent to the one used for the calculation of wear. In each case the type of the function $F(V)$ should be chosen in accordance with the mechanical characteristics of the fractured rock. As it will be shown below, simultaneous consideration of the two mechanisms for the rock boundary displacement in the contact zone allows us to obtain a wear area shape similar to that obtained experimentally (Fig. 8.18).

Contact conditions

The following relationship between the shape of rock boundary $z_0(x)$, the shape of the tool, the cutting depth and rock displacement due to elastic deformation and crushing is satisfied in the contact zone

$$z_0(x) - w(x,t) - u_z(x,t) = f(x - Vt, t) - c(t). \tag{8.65}$$

This equation can be written in differential form by taking into account Eqs. (8.63) and (8.64)

$$F(V)\tilde{p}(x,t) + k\frac{d\tilde{p}(x,t)}{dt} = \frac{dc(t)}{dt} - \frac{df(x - Vt, t)}{dt}. \tag{8.66}$$

In (ξ, ζ) coordinate system, Eq. (8.66) has the form

$$F(V)p(\xi,t) + k\left(\frac{\partial p(\xi,t)}{\partial t} - \frac{\partial p(\xi,t)}{\partial \xi}V\right) = \frac{dc(t)}{dt} + \frac{\partial f(\xi,t)}{\partial \xi}V - \frac{\partial f(\xi,t)}{\partial t}, \tag{8.67}$$

where $p(\xi,t) = \tilde{p}(\xi + Vt, t)$ and the following relationship obtained from Eq. (8.62) is taken into account

$$\frac{dp(\xi,t)}{dt} = \frac{\partial p(\xi,t)}{\partial t} - \frac{\partial p(\xi,t)}{\partial \xi}V. \tag{8.68}$$

A similar relationship for the tool shape is

$$\frac{df(\xi,t)}{dt} = \frac{\partial f(\xi,t)}{\partial t} - \frac{\partial f(\xi,t)}{\partial \xi}V. \tag{8.69}$$

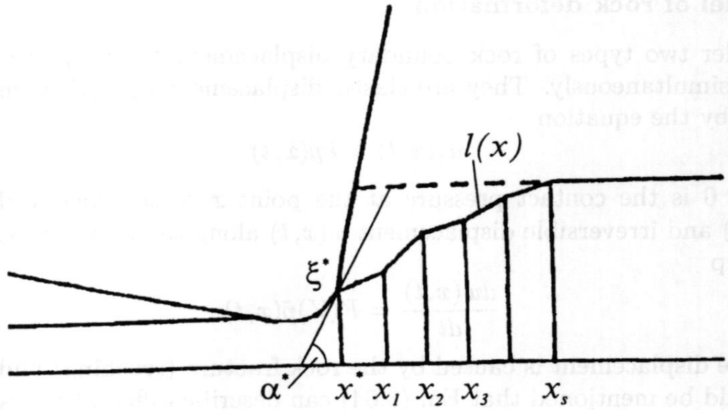

Figure 8.19: Scheme of the crack propagation.

Here $\frac{\partial f(\xi,t)}{\partial t}$ is the tool shape variation caused by wear.

The pressure at the ends $a(t)$ and $b(t)$ of the contact zone is equal to zero, that is

$$p(a(t),t) = p(b(t),t) = 0. \tag{8.70}$$

The coordinate $a(t)$ of the leading point of contact zone is obtained from the formula

$$z_0(a+Vt) = -c(t) + f(a,t) \tag{8.71}$$

which is based on Eqs. (8.63)-(8.65) and Eq. (8.70). Initial conditions for the differential equation (8.67) depend on the type of the tool motion. Eq. (8.67) in combination with Eqs. (8.70) and (8.71) can be used for the contact pressure calculation.

Chip formation

Rock brittle fracture leading to chip formation occurs in parallel with rock elastic deformation and crushing when the tool penetration depth is considerable. Chip formation is one of the causes of cutting force oscillation in tool operation. The fragment separation is caused by propagation of the crack which originates near the cutter nose. This statement can be verified by noting that the front face of the tool is worn only near the nose (see Fig. 8.18).

We assume that separation of rock fragment occurs at the instant t^* when pressure at the point ξ^* is equal to a critical value p^*. Thus

$$p(\xi^*,t^*) = p^* \tag{8.72}$$

then crack propagation originates from the point $x^* = \xi^* + Vt^*$. The crack is supposed to be a polygonal line $l(x)$ shown in Fig. 8.19. Its inclination angle δ_i at each segment $[x_i, x_{i+1}]$ of the length $\Delta x = |x_{i+1} - x_i|$ is a random value, uniformly distributed at $[0; \alpha^*]$, α^* is the inclination angle of the tool profile to the x-axis at the point x^*.

The crack propagates up to a point x_s of the rock boundary. The shape of the rock boundary ahead of the tool is changed as the result of chip fragment separation.

$$z_0(x)|_{t=t^*+0} = \begin{cases} l(x) & \text{if } x^* \le x \le x_s, \\ z_0(x)|_{t=t^*-0} & \text{if } x > x_s. \end{cases} \tag{8.73}$$

Tool wear model

The following relationship is used to model the tool shape variation due to wear

$$\frac{\partial f_n(\xi,t)}{\partial t} = K_w p_n(\xi,t) v \tag{8.74}$$

Here $\partial f_n(\xi,t)/\partial t$, $p_n(\xi,t)$ are the wear rate and contact stress in the direction normal to the friction surface, v is the relative velocity of the worn body and the abrasive medium (rock surface) in the tangential direction, K_w is the wear coefficient.

From geometrical consideration we have the following relations:

$$v = V/\cos\alpha,$$

$$\frac{\partial f_n(\xi,t)}{\partial t} = \cos\alpha \frac{\partial f(\xi,t)}{\partial t},$$

$$\cos^{-2}\alpha = 1 + \left(\frac{\partial f(\xi,t)}{\partial \xi}\right)^2,$$

where α is the inclination angle of the tool profile to the x-axis at each point ξ.

Tool shape variation caused by wear can be described by the formula following from Eq. (8.74)

$$\frac{\partial f(\xi,t)}{\partial t} = K_w p_n(\xi,t) \left[1 + \left(\frac{\partial f(\xi,t)}{\partial \xi}\right)^2\right] V. \tag{8.75}$$

Thus, we propose a mathematical description for the following main processes taken into account in this model:

- elastic deformation of rock, Eq. (8.63);
- rock crushing, Eq. (8.64);
- chip formation, Eqs. (8.72) and (8.73);
- tool wear, Eq. (8.75).

The contact condition written in differential form (8.67), together with the boundary condition (8.71), gives the possibility of calculating contact characteristics (the value of the pressure $\tilde{p}(\xi, t)$ and coordinates of the ends of contact zone $a(t)$ and $b(t)$), and hence of modelling the development of the entire process. Numerical procedure and results of the modelling are described below.

To reveal the role of separate mechanisms in the process of tool operation, we first consider the simplified situation when only some of them occur.

8.5.2 Stationary process without chip formation and tool wear

We analyse the pressure distribution in the contact zone when only elastic deformation and crushing described by Eqs. (8.63) and (8.64) are taken into account. The tool shape is assumed to be a wedge with angle 90°, rake angle $\gamma = 15°$, and cutting edge roundness is equal to zero, that is the ***absolutely sharp*** cutter is considered. The shape of the cutter does not change: $f(\xi, t) = f_0(\xi)$, where

$$f_0(\xi) = \begin{cases} A\xi & \text{if } \xi \geq 0, \\ -\xi/A & \text{if } \xi < 0, \end{cases}$$

where $A = \cot \gamma$.

We study the stationary motion of the tool with constant cutting depth, assuming $z_0(x) = 0$, that is the rock surface is originally flat:

$$\frac{\partial p(\xi, t)}{\partial t} = 0; \qquad \frac{dc(t)}{dt} = 0.$$

In this case Eq. (8.67) turns into

$$F(V)p(\xi, t) - k\frac{\partial p(\xi, t)}{\partial \xi}V = \begin{cases} AV & \text{if } \xi \geq 0, \\ -V/A & \text{if } \xi < 0. \end{cases} \tag{8.76}$$

Eq. (8.76) has the stationary solution

$$p(\xi) = \begin{cases} \dfrac{AV}{F(V)}\left(1 - \exp\dfrac{F(V)(\xi - a)}{kV}\right) & \text{if } \xi \geq 0, \\ \dfrac{V}{AF(V)}\left(-1 + \exp\dfrac{F(V)(\xi - b)}{kV}\right) & \text{if } \xi < 0. \end{cases} \tag{8.77}$$

Coordinate a of the leading point of contact is given, the coordinate b is obtained from the condition

$$b = -\frac{kV}{F(V)}\ln\left[A^2\left(1 - \exp\frac{-F(V)a}{kV}\right) + 1\right].$$

Eq. (8.70) and the condition of pressure continuity at the point $\xi = 0$ have been used to construct these relationships.

Fig. 8.20 illustrates the functions $p(\xi)/p(0)$ for different values of the param-

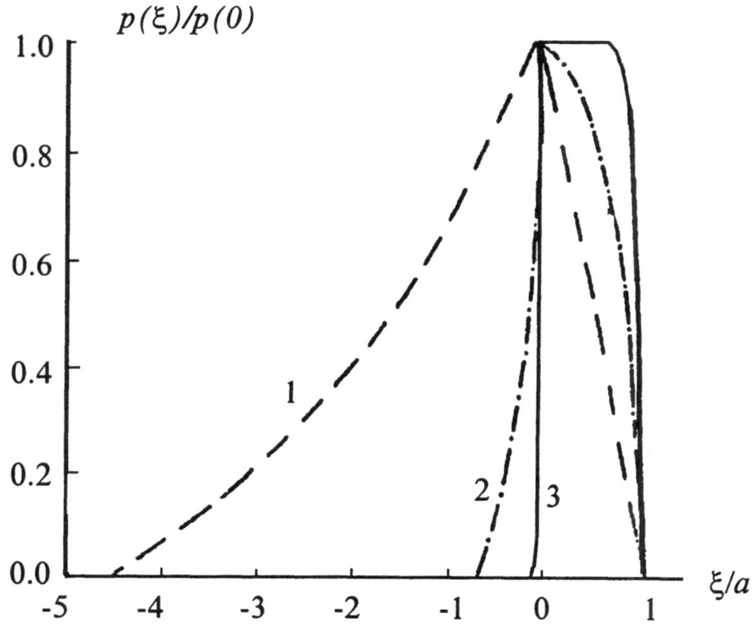

Figure 8.20: Contact pressure distribution for various values of β: $\beta = 0.3$ (curve 1), $\beta = 3$ (curve 2), $\beta = 30$ (curve 3) (tool operation without chip formation).

eter $\beta = \dfrac{F(V)a}{kV}$. The results show that the contact zone size b/a decreases as β increases (that is when the role of crushing increases), the pressure distribution on the front face tending to a constant (curve 3). Increasing the effect of elastic deformation causes an increase of the contact zone size on the rear face.

We can conclude from Eq. (8.77) that the pressure $p(\xi)$ is independent of the velocity V in the stationary stage, if $F(V)$ is a linear function. As is shown in a set of experimental investigations by Vorozhtsov et al. (1989), the components of the cutting force depend only slightly on the velocity V; in future we shall suppose $F(V) = \lambda V$.

8.5.3 Analysis of the cutting process

To analyse the model behaviour, we developed a numerical procedure. It includes a step-by-step in time solution of the differential equation (8.67) in the process of the cutter displacement in the x- and z-direction; instant changes of the rock shape occur ahead of the tool in accordance with Eq. (8.73) when condition (8.72) holds. Permanent wear of the tool is calculated on the basis of Eq. (8.76). Since the variation of cutting forces due to rock fragment separation, and the tool wear are processes with different time scales, the time-averaged value of the contact pressure was used for calculation of wear. This procedure significantly reduced the calculation time.

The system of Eqs. (8.67), (8.71)–(8.73) was solved in dimensionless form. The system depends on the dimensionless parameter $\lambda p^* = \bar{p}^*$.

The calculation has been carried out for a tool, which has a wedge shape with angle 90° , $\gamma = 15°$, and cutting edge roundness is 0.2 mm. It is supposed that the tool and rock are out of contact originally, initial conditions being $p(\xi, 0) = 0$, $a(t) - b(t) = 0$. At first tool penetration is $c(t) = c_0 t$ with constant rate c_0 ($c_0/V = 0.2$) , then cutting with constant depth takes place. The dependence of cutting depth on time is illustrated by Fig. 8.21(a).

The vertical (P_v) and horizontal (P_h) components of cutting force are calculated from

$$P_v = \int_b^a p(\xi, t) d\xi,$$

$$P_h = \int_b^a p(\xi, t) \frac{\partial f(\xi, t)}{d\xi} d\xi.$$

Note that the force P_h is caused by the rock crushing under the tool. It is only a part of the cutting force horizontal component. The other part of the force which is caused by the chip formation is not considered here.

Cutting process without tool wear

First we analysed the cutting process without tool wear, and assumed that the cutter shape is independent of time $f(\xi, t) = f_0(\xi)$. Fig. 8.21 (b)-(d) illustrates the variations of $a(t)$ and $b(t)$, $P_v(t)$ and $P_h(t)$, respectively. The calculation has been carried out for $\Delta x = 0.4$ mm. It should be mentioned that the cutting depth and the size of contact zone are shown in dimensional units (mm) to make the comparison with experimental results easier.

Fig. 8.21 demonstrates that initially rock crushing without chip formation occurs, and cutting force components and size of contact zone increase monotonously. Then after the beginning of chip formation, essential oscillations of cutting force components occur, and the size of contact zone does not increase appreciably, in spite of the growth of the cutting depth. After transition to operation with constant cutting depth, the process quickly becomes quasistationary.

It should be mentioned that the characteristics of the cutting process turn out to be sensitive to change of penetration speed dc/dt. When the speed is changed abruptly from 0.2 V to zero (constant cutting depth) the value of vertical force P_v (Fig. 8.21 (c)), as well as the frequency of contact parameter oscillation (Fig. 8.21 (b)–(d)) and size of contact zone $a(t) - b(t)$ (Fig. 8.21 (b)), diminish.

Fig. 8.22 gives a typical view of cracks arising successively in penetration and horizontal displacement of the cutter. This figure shows that fragments of different

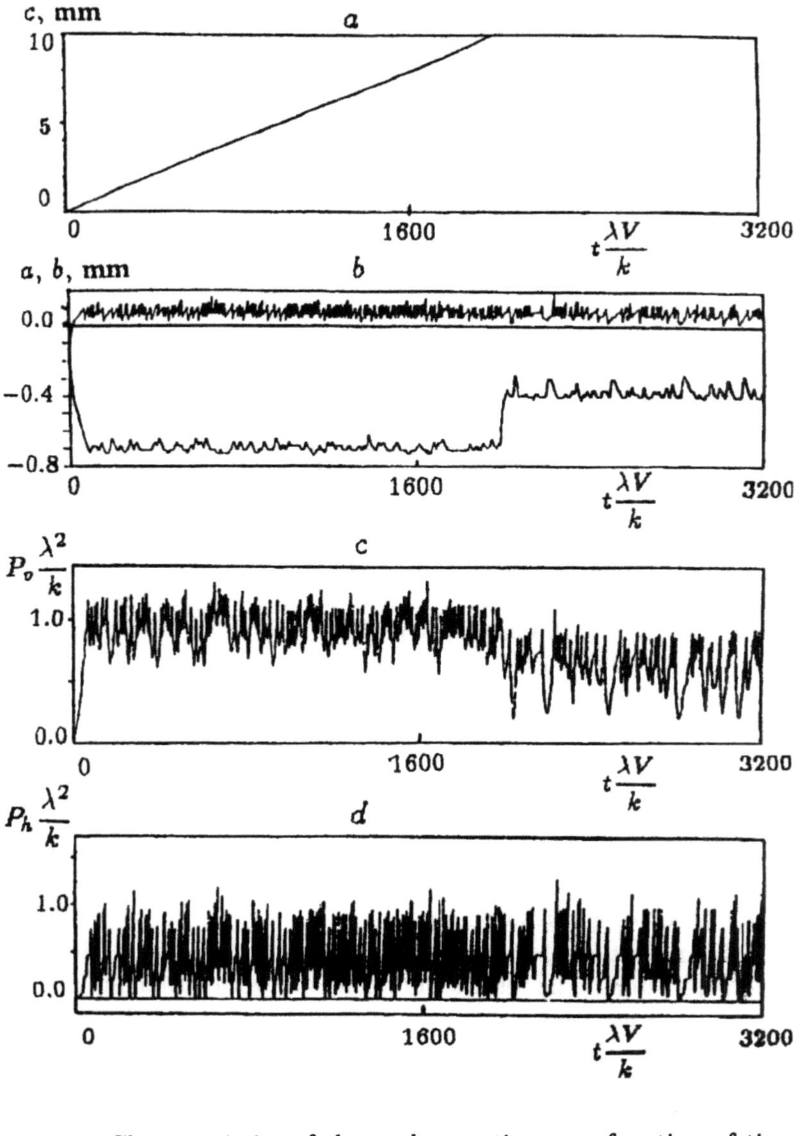

Figure 8.21: Characteristics of the tool operation as a function of time (cutting process without tool wear): (a) the tool penetration $c(t)$; (b) the coordinates of the edge points $a(t)$ and $b(t)$ of contact zone; vertical P_v (c) and horizontal P_h (d) components of cutting force at $\bar{p}^* = 0.84$, $\lambda/k = 40$ mm^{-1}.

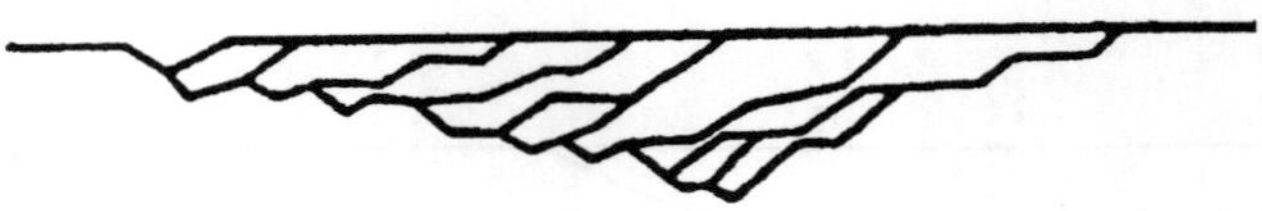

Figure 8.22: Typical view of cracks arising successively in cutter penetration and horizontal displacement obtained from calculations.

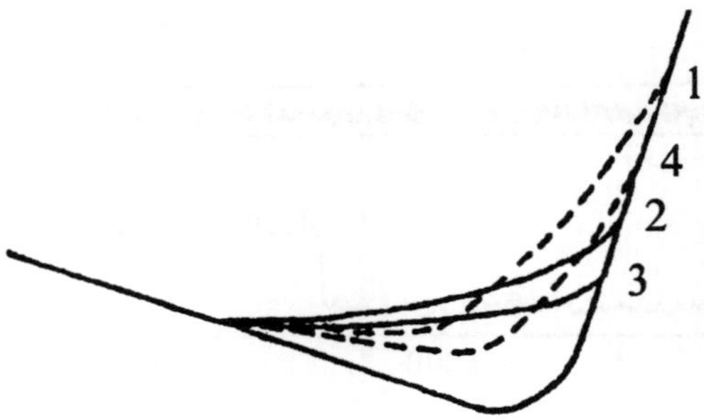

Figure 8.23: Profiles of worn cutter calculated at $\bar{p}^* = 0.84$, $\lambda/k = 40$ mm^{-1} (curve 1); $\bar{p}^* = 0.28$, $\lambda/k = 40$ mm^{-1} (curve 2), $\bar{p}^* = 0.14$, $\lambda/k = 20$ mm^{-1} (curve 3); $\bar{p}^* = 0.28$, $\lambda/k = 10$ mm^{-1} (curve 4).

sizes are separated in cutting.

8.5.4 Influence of tool wear on the cutting process

Tool shape variation in the wear process leads to gradual variation of contact characteristics, unlike the case analysed above when, in the absence of wear, the cutting process becomes quasistationary.

Fig. 8.23 illustrates typical profiles of the worn cutter calculated for different sets of model parameters. The results show that the worn cutter profile depends essentially on the rock mechanical characteristics that describe elastic deformation, crushing, and brittle fracture of rock.

The parameters that correspond to curve 2 in Fig. 8.23 give a shape for the wear area extremely close to that obtained experimentally (Fig. 8.18). In this case the calculations accurately reproduce the details of the wear process for a real tool.

Fig. 8.24 illustrates the function $P_v(t)$ at different stages of the tool wear. The calculation has been carried out for the parameters that correspond to curve 1 in Fig. 8.23. The sizes of wear area S in Fig. 8.24 are determined as

$$S(t) = \max_{\tau<t} a(\tau) - \min_{\tau<t} b(\tau).$$

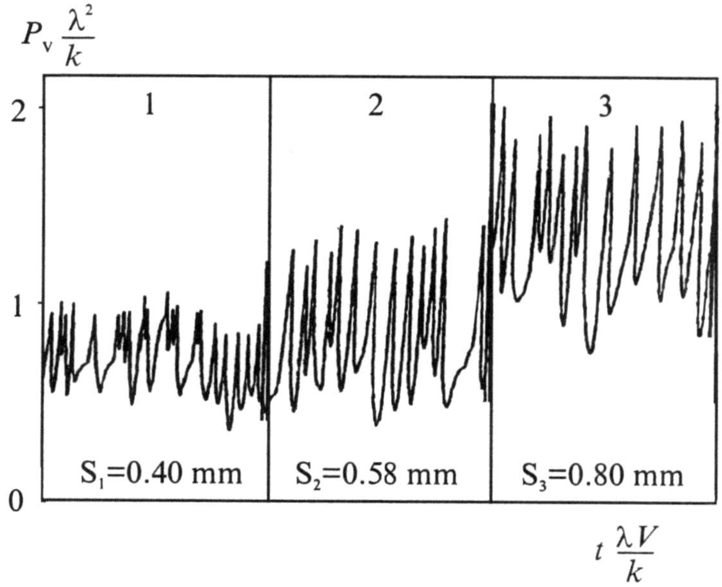

Figure 8.24: Dependence of the vertical component of the cutting force on time at different stages of the tool wear for $\bar{p}^* = 0.84$ and $\lambda/k = 40\ \mathrm{mm}^{-1}$.

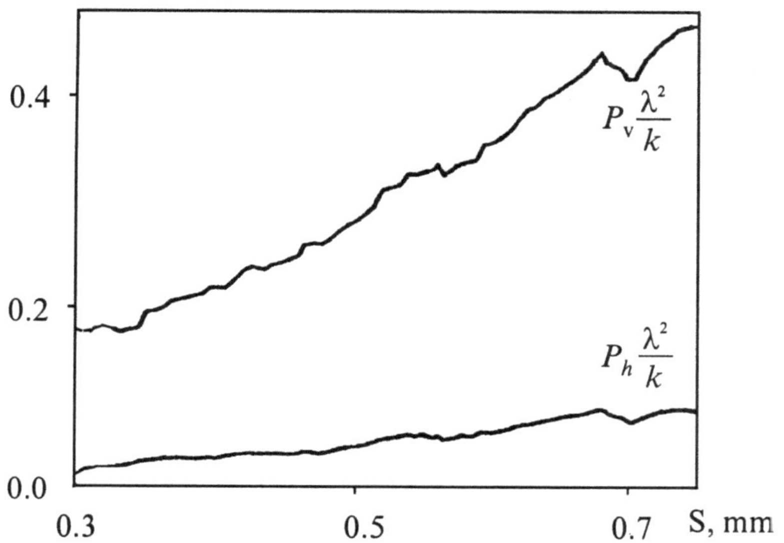

Figure 8.25: Dependence of the averaged cutting force components P_v and P_h on the wear area size.

The time interval $\Delta t = 400k/(\lambda V)$ is the same for regions 1, 2, 3. The mean values of the vertical force $\bar{P}_v$ are $0.69k/\lambda^2$; $0.77k/\lambda^2$; $1.31k/\lambda^2$ for the regions 1, 2, and 3, respectively. As follows from Fig. 8.24, wear causes growth of the oscillation amplitudes and the value of $\bar{P}_v$; it causes the oscillation frequency to diminish. The behaviour of the P_h component is similar. In practice, growth of the oscillation amplitude and the mean value of the forces can cause the tool to break-down.

Cutting force components P_v and P_h averaged over a large number of time steps are shown in Fig. 8.25 as a function of the wear area size S. The calculation parameters here are the same as for the curve 2 in Fig. 8.23. It is interesting to mention that variations of the vertical and horizontal components of cutting force with the size of wear area obtained in the experiment on cutting of cement-sand blocks described above are also close to linear ones; they coincide qualitatively with the results of calculation. This is one more confirmation of the idea that the tool profile variation in wear can be an indicator of the processes occurring in rock fracture.

The investigation of the proposed model allows us to conclude that the analysis of tool shape variation caused by wear provides important information that can be used for modelling of the processes in the contact zone.

The model is based on simultaneous consideration of tool wear, rock elastic deformation, crushing, and brittle fracture leading to rock fragment separation. Numerous phenomena observed in tests confirm the adequacy of the theory.

The investigation revealed the influence of the tool wear on various characteristics of the cutting process, and also the influence of the rock mechanical characteristics on tool shape variation caused by wear.

This model allows us to predict the cutting process characteristics for tools with different initial geometrical parameters (cutter shape, rake angle, etc.) and could be used for the optimal choice of these parameters.

Chapter 9

Conclusion

In this book we have considered various contact problems which reproduce the peculiarities of friction interaction. The solutions of these problems have two main applications.

Some of them can be used *to explain the friction and wear processes*, i.e. to solve some fundamental problems of tribology. We may include in this set the problems of discrete contact (Chapter 2), the problems of sliding and rolling contact (Chapter 3), the contact problems for inhomogeneous bodies (Chapter 4), the models of fatigue wear of surfaces in contact with rough body (Chapter 5), and so on.

The main idea of the approach used in the book to investigate the discrete contact of rough surfaces, is to take into account the interaction between contact spots. This approach was a basis for analysis of contact characteristics and internal stresses and for modelling the wear process of rough surfaces. It allowed us to explain some important features of the process known from experiments, such as the fact that the process of surface fracture can have a stationary, a periodic, or a catastrophic type; the effect of saturation of the real contact area; the equilibrium roughness formation, and so on.

In some models, we took into account simultaneously the effects of contact discreteness and mechanical inhomogeneity of contacting bodies. This allowed us to analyze the stresses within the coatings, the thickness of which is commensurable with the typical size and the distance between asperities, and to determine the type of the coating fracture for different loading conditions. Other models were used to analyze the effect of thin surface films in sliding and rolling friction in regimes of elasto-hydrodynamic or boundary lubrication. All these models help us to understand the mechanical aspects of the processes occuring in contact interaction.

The contact problems described in Chapters 6 – 8 and partly in Chapter 3 can be used for *calculation of contact characteristics of different junctions* taking into account friction and wear. This applied problem is one of the most important tasks of tribology.

Some of the models are of both fundamental and applied use. For instance, we used the model of wear in discrete contact to analyze the fundamental problem of wear of rough bodies, and also to calculate the worn shape of abrasive tools with various inclusion density (Chapter 7).

Sometimes, the models considered in the book can be used at different scales. Thus, the problem of sliding contact of viscoelastic bodies can model the macrocontact of bodies with smooth surfaces and also the microcontact of an asperity of the rough surface. Using this model at the microlevel, we calculated the mechanical component of the friction force in Chapter 3.

There is one important feature of most of the models. They allow us *to predict the characteristics of the process* under given loading and friction conditions. This is one of the main tasks of the wear contact problems. The evolution in time of the pressure distribution, the shape of the worn surface, and the approach of the elements of junction is predicted from the wear contact problem solution. Based on the solutions, we can also to calculate the life time of junctions and the duration of the running-in stage.

The approaches developed in the book can also be used *to optimize friction and wear process*. Among the optimization parameters under consideration there are the thickness and mechanical properties of coatings, the parameters of local hardening of surfaces, etc. In Chapter 7, we formulated some problems of optimization of the wear process and gave their solutions.

Finally, the problems with complicated boundary conditions considered in this book allow us *to evaluate the accuracy of simplified models*, which are widely used in tribology. We can now answer the following questions: "For what values of the roughness parameters and loading conditions can we neglect the interaction of contact spots and calculate the real contact pressure and real contact area based on Hertz theory? What are the contact conditions which allow us to neglect the influence of the thin surface film in calculating the friction force in sliding contact? Is it possible without significant loss of accuracy to ignore the deformation of the substrate (to consider it as rigid) for given properties of the coating?", etc.

Of course, by their nature, contact problems are an idealization of real processes in contact interaction. The formulations of the problems include only some mechanisms of the processes. To carry out the idealization correctly, experimentally obtained results should be thoroughly analyzed. The comparison of the model prediction and experimental data proves whether the governing mechanisms of the process are chosen correctly or not.

It should be noted that some very important questions concerning the effects of residual stresses and plastic deformations of surface layers, heating in friction interaction, changes of surface structure and the mechanical properties in the processes of friction and wear are beyond the scope of this book. These questions pose new formulations of contact problems. Some of them have already been investigated, other problems are waiting for their solutions.

We hope that this book will be useful for specialists in contact mechanics and tribology, and it will stimulate new research of the complicated processes occuring in friction interaction.

Chapter 10

References

Pages on which references occur are indicated by the numbers in square brackets. The abbreviation I.R. means In Russian.

Aleksandrov, V.M. (1967) Axisymmetric problem of the effect of an annular die on an elastic half-space, (I.R.) *Izv. AN SSSR, Mechanics of Solids*, **4**, 108-116. [216]

Aleksandrov, V.M. (1968) Asymptotic methods in contact problems of the theory of elasticity, *J. Appl. Math. Mech.*, **32**, 4, 672-683. [221]

Aleksandrov, V.M., and Kovalenko, E.V. (1978) Axisymmetric wear contact problem for the linear deformable foundation, *Izv. AN SSSR, Mechanics of Solids*, **5**, 58-66. [10]

Aleksandrov, V.M., and Kovalenko, E.V. (1984) Mathematical methods in the wear contact problems , in *Nonlinear Models and Problems in Mechanics of Solids*, (I.R.) Nauka, Moscow, pp. 77-89. [10]

Aleksandrov, V.M., and Kudish, I.I. (1979) Asymptotic analysis of plain and axisymmetric contact problems taking into account a surface structure of contacting bodies, *Izv. AN SSSR, Mechanics of Solids*, **1**, 58-70. [50]

Aleksandrov, V.M., and Mhitaryan, S.M. (1983) *Contact Problems for Bodies with Thin Coatings and Stringers*, (I.R.) Nauka, Moscow. [4, 8, 279, 288]

Aleksandrova, G.P. (1973) Contact problems of the bending of plates lying on an elastic foundation, (I.R.) *Izv. AN SSSR, Mechanics of Solids*, **1**, 97-106. [229, 230, 280]

Alekseev, N.M., and Dobychin, M.N. (1994) Wear models, in Belyi, V.A., Ludema, K.C., and Myshkin, N.K. (Eds.) *Tribology in the USA and the Former Soviet Union: Studies and Application*, Allerton Press, pp. 67-88. [61]

Amontons, G. (1699) De la résistance cause dans les machines, *Mémoire de l'Académie Royale*, **A**, 275-282. [66, 82]

Appl, F.C., Wilson, C.C., and Landsman, I. (1993) Measurement of forces, temperatures and wear of PDC cutters in rock cutting, *Wear*, **169**, 9-24. [278]

Archard, J.F. (1951) Elastic deformation and the contact of surfaces, *Nature*, **172**, 918-919. [5]

Archard, J.F. (1953) Contact and rubbing of flat surfaces, *J. Appl. Phys.*, **24**, 981-988. [197]

Atkinson, B.K., Ed. *Fracture Mechanics of Rock*, Academic Press Geology Series, Academic Press. [313]

Barus, C. (1893) Isothermal, isopiestics, and isometrics relative viscosity, *Am. Journal Sci.*, **45**, 87-96. [154]

Batra, S.K., and Ling, F.F. (1967) On deformation friction and interface shear stress in viscoelastic elastic layered system under a moving load, *Am. Soc. Lubr. Eng. Trans.*, **10**, 294-301. [122]

Bartenev, G.M., and Lavrentiev, V.V. (1972) *Friction and Wear of Polymers*, (I.R.) Khimiya, Moscow. [14]

Bershadsky, L.I. (1981) *Self-Organization and Reliability of Tribosystems*, (I.R.) Znanie, Kiev. [194]

Bhushan, B., and Gupta, B.K., Eds. (1991) *Handbook of Tribology. Materials, Coatings, and Surface Treatments*, McGraw-Hill, New York. [1]

Blyumen, A.V., Kharach, G.M., and Efros, D.G. (1976) Analytic evaluation of wear intensity and service life of shaft-sleeve contact with inverse friction pair, (I.R.) *Vestnik Mashinostroeniya*, **2**, 29-32. [277]

Bogatin, O.V., and Kanibolotsky, M. A. (1980) Study and identification of sliding bearing sleeve wear process, *Soviet Journal of Friction and Wear*, **1**, 3, 533-542. [277]

Bogatin, O.V., Morov, V.A., and Chersky, I.N. (1983) *Calculation of Polymer Friction Components*, (I.R.) Nauka, Novosibirsk. [10]

Bogdanov, V.M., Goryachev, A.P., Goryacheva, I.G., Dobychin, M.N., Zakharov, S.M., Krivonogov, V.G., Soldatenkov, I.A., and Checkina, O.G. (1996) A study of wheel-rail contact, wear and cumulative damage, *Soviet Journal of Friction and Wear*, **17**, 1, 12-26. [300, 301, 307]

Boussinesq, J. (1885) *Application des Potentiels à l'Étude de l'Équilibre et du Mouvement des Solides Élastique*, Gauthier-Villars, Paris. [20]

Bowden, F.P., and Tabor, D. (1950) *The Friction and Lubrication of Solids*, Part 1, Clarendon, Oxford. [1, 6]

Bowden, F.P., and Tabor, D. (1964) *The Friction and Lubrication of Solids*, Part 2, Clarendon, Oxford. [1]

Braat, G.F.M., and Kalker, J.J. (1993) Theoretical and experimental analysis of the rolling contact between two cylinders coated with multilayered viscoelastic rubber, in Aliabadi, M.H., and Brebbia, C.A. (Eds.) *Contact Mechanics, Computational Techniques*, Contact Mechanics Publication, pp. 119-126. [7, 108, 109, 123]

Bushe, N.A. (1994) Evaluation of metallic materials in tribosystem compatibility, in Belyi, V.A., Ludema, K.C., and Myshkin, N.K. (Eds.) *Tribology in the USA and the Former Soviet Union: Studies and Application*, Allerton Press, pp. 139-155. [194]

Carslaw, H.S., and Jaeger, J.C. (1960) *Conduction of Heat in Solids*, Clarendon Press, Oxford. [185]

Carter, F.W. (1926) On the action of a locomotive driving wheel, *Proc. Roy. Soc., A*, **112**, 151. [87]

Cesari, L. (1959) *Asymptotic Behavior and Stability Problems in Ordinary Differential Equations*, Springer–Verlag, Berlin. [257]

Challen, J.M., and Oxley, P.L.B. (1986) Prediction of Archard's wear coefficient for metallic sliding friction assuming a low cycle fatigue wear mechanism, *Wear*, **111**, 275-288. [197]

Checkina, O.G. (1996) The modelling of fracture of surface layers in contact of rough bodies, (I.R.) *9 Conf. on Strength and Plasticity, Moscow*, **1**, 186-191. [182]

Checkina, O.G., Goryacheva, I.G., and Krasnik, V.G. (1996) The model for tool wear in rock cutting, *Wear*, **198**, 33-38. [278, 313]

Cherepanov, G.P. (1987) *Mechanics of Rock Fracture in Drilling Process*, (I.R.) Nedra, Moscow. [313]

Chichinadze, A.V., Ed. (1995) *Fundamentals of Tribology*, (I.R.) Nauka i Tekhnika, Moscow. [1]

Cole, S.J., and Sayles, R.S. (1991) A numerical model for the contact of layered elastic bodies with real rough surfaces, *Journal of Tribology*, **113**, 2, 334-340. [13, 112]

Collins, W.D. (1963) Potential problems for a circular annulus, *Edinburgh Mathematical Society. Proceedings*, Ser. 2, **13**, 235-246. [216]

Collins, J.A. (1981) *Failure of Materials in Mechanical Design*, Wiley-Int. Publ., New York. [169, 173]

Conry, T.F., Wang, S., and Cusano, C. (1987) A Reynolds–Eyring equation for elastohydrodynamic lubrication in line contacts, *Journal of Tribology*, **109**, 648-658. [152]

Cooper, J.R., Dowson, D., and Fisher, J. (1991) Birefringent studies of polyethylene wear specimens and acetabularcups, *Wear*, **151**, 391-401. [182]

Coulomb, Ch.A. (1785) Théorie des machines simples, *Mémoire de Mathématiques et de Physique de l'Académie Royale*, 161-331. [61]

Czichos, H. (1980) System approach to wear problems, *Wear Control Handbook, ASME*, 17-34. [193]

Demkin, N.B. (1963) *Real Contact Area of Solid Surfaces*, (I.R.) Acad. Sci. USSR, Moscow. [165]

Demkin, N.B. (1970) *Contact of Rough Surfaces*, (I.R.) Nauka, Moscow. [5, 14]

Deryagin, B.V. (1934) Molecular theory of friction and sliding, (I.R.) *J. Physical Chemistry*, **5**, 9, 1165-1172. [6]

Dowson, D. (1978) *History of Tribology*, Longman, London. [1]

Dowson, D., and Higginson, G.R. (1966) *Elastohydrodynamic Lubrication. The Fundamentals of Roller and Gear Lubrication*, Pergamon, Oxford. [152]

Dupuit, J. (1837) *Essai et Expériences sur le Tirage de Voitures et sur le Frottement de Second Espèce*, Carilian–Gœury, Paris. [7]

Evans, A.G., and Marshall, D.B. (1981), by Rigney, D.A. (Ed.) Fundamentals of friction and wear materials, *Am. Soc. for Metals*, Metals Park, OH, 439. [197]

Elsharkawy, A.A., and Hamrock, B.J. (1994) EHL of coated surfaces, part 1: Newtonian results, *Journal of Tribology*, **116**, 29-36. [153]

Flom, D.G. (1962) Dynamic mechanical losses in rolling contacts, in *Rolling Contact Phenomena*, Bidwell, Elsevier, London, pp. 97-112. [62]

Flom, D.G., and Bueche, A.M. (1959) Theory of rolling friction for spheres, *J. Appl. Phys.*, **30**, **11**, 1725-1730. [62]

Fleischer, G. (1973) Energetische Methode der Bestimnung des Verschleißes, *Schmierungstechnik*, **4**, 9-16. [197]

Fromm, H. (1927) Berechnung des Schlupfes beim Rollen deformierbaren Scheiben, *ZAMM*, **7**, 27-58. [87]

Galakhov, M.A., and Usov, P.P. (1990) *Differential and Integral Equations of the Mathematical Theory of Friction*, (I.R.) Nauka, Moscow. [10]

Galanov, B.A. (1984) 3-D contact problems for rough elastic bodies under elastic-plastic deformations of asperities, *J. Appl. Math. Mech.*, **48**, 6, 1020-1029. [50]

Galin, L.A. (1945) Pressure of punch with friction and adhesion domains, *J. Appl. Math. Mech.*, **9**, 5, 413-424. [6]

Galin, L.A. (1953) *Contact Problem of the Theory of Elasticity*, (I.R.) Gostekhizdat, Moscow. (English translation by Moss, H., North Carolina State College, Department of Mathematics, 1961). [4, 5, 17, 32, 37, 66, 70, 76, 82, 214, 217, 222, 226, 228, 272]

Galin, L.A. (1976a) Contact problems of the theory of elasticity in the presence of wear, *J. Appl. Math. Mech.*, **40**, 6, 981-986. [10]

Galin, L.A., Ed. (1976b) *The Advancement of Elasticity Theory in USSR*, (I.R.) Nauka, Moscow. [4, 8, 55, 216, 220]

Galin, L.A. (1980) *Contact Problems of Elasticity and Viscoelasticity Theories*, (I.R.) Nauka, Moscow. [4, 10, 64, 87, 88, 169]

Galin, L.A. (1982) *Elastoplastic Problem*, (I.R.) Nauka, Moscow. [4, 169]

Galin, L.A., and Goryacheva, I.G. (1977) Axisymmetric contact problem of the theory of elasticity in the presence of wear, *J. Appl. Math. Mech.*, **41**, N 5, 826-831. [10]

Galin, L.A., and Goryacheva, I.G. (1983) Three-dimensional contact problem of the motion of a stamp with friction, *J. Appl. Math. Mech.*, **46**, 6, 819-824. [6, 76, 77]

Galin, L.A., and Shmatkova, A.A. (1968) Motion of a punch along a viscoelastic half-plane, *J. Appl. Math. Mech.*, **32**, 3, 445-453. [7]

Garkunov, D.P. (1985) *Tribotechniques*, (I.R.) Mashinostroenie, Moscow. [239]

Gladwell, G.M.L. (1980) *Contact Problems in the Classical Theory of Elasticity*, Sijthoff and Noordhoff, Alphen aan den Rijn. [4, 6, 8, 33, 55, 104, 108, 125, 169, 216]

Gladwell, G.M.L., and Gupta, O.P. (1979) On the approximate solution of elastic contact problems for a circular annulus, *J. Elast.*, **9**, 335-348. [216]

Glagolev, N.I. (1945) Resistance of cylindrical bodies in rolling, *J. Appl. Math. Mech.*, **9**, 4, 318-333. [87]

Godet, M. (1984) The third-body approach: a mechanical view of wear. *Wear*, **100**, 437. [61]

Goldstein, R.V., and Spector, A.A. (1986) Variational methods of analysis of space contact and mixed problems with friction, in *Mechanics of Deformable Bodies*, (I.R.) Nauka, pp. 52-73. [6, 87]

Goodman, L.E. (1954) Contact stress analysis of normally loaded rough spheres, *J. Appl. Mech.*, **29**, 515. [5]

Goryacheva, I.G. (1973) Contact problem of rolling of a viscoelastic cylinder on a base of the same material. *J.Appl. Math. Mech.*, **37**, 5, 877-885. [82, 90, 91]

Goryacheva, I.G. (1974) *Investigation of Rolling Friction Taking into Account Viscoelasticity and Partial Slip in Contact Region*, Ph.D. thesis, (I.R.) Moscow State University. [87]

Goryacheva, I.G. (1979a) A method of solving the contact problem of elasticity theory in the presence of wear, in *Current Problems of Mechanics of Deformable Bodies*, (I.R.) Dnepropetrovsk State University, Dnepropetrovsk, pp. 79-84 [10]

Goryacheva, I.G. (1979b) Plane and axisymmetric contact problems for rough elastic bodies, *J. Appl. Math. Mech.*, **43**, 1, 99-105. [50, 53]

Goryacheva, I.G. (1980) Wear contact problem for the ring inserted into cylinder, *J. Appl. Math. Mech.*, **44**, 2, 363-367. [10]

Goryacheva, I.G. (1987) Wear contact problem for the system of punches, *Izv. AN SSSR, Mechanics of Solids*, **6**, 62-68. [10, 275]

Goryacheva, I.G. (1988) *Contact Problems in Tribology*, D.Sci. thesis, (I.R.) Institute for Problems in Mechanics of Russian Academy of Sciences. [217, 220, 259]

Goryacheva, I.G. (1989) Changes in the contact characteristics of junctions during the wear, in *Proceedings of the 5-th International Congress on Tribology*, **3**. Helsinki, 360-366. [10]

Goryacheva, I.G., and Chekina, O.G. (1989) Control of surface deformation during wear, *Soviet Journal of Friction and Wear*, **10**, 1, 1-7. [275]

Goryacheva, I.G., and Checkina, O.G. (1990) Surface fatigue damage model, *Soviet Journal of Friction and Wear*, **11**, 3, 1-11. [172]

Goryacheva, I.G., and Dobychin, M.N. (1980) Estimation of accuracy of the method of calculation of the contact stiffness of rough bodies with taking into account the interaction of contact spots, (I.R.) *Mashinovedenie*, **1**, 70-77. [40]

Goryacheva, I.G., and Dobychin, M.N. (1984a) Influence of coating on contact characteristics of plain journal bearings, *Soviet Journal of Friction and Wear*, **5**, 3, 66-72. [277]

Goryacheva, I.G., and Dobychin, M.N. (1984b) Wear kinetics of plain journal bearings with solid lubricant, *Soviet Journal of Friction and Wear*, **5**, 4, 581-588. [277]

Goryacheva, I.G., and Dobychin, M.N. (1988) *Contact Problems in Tribology*, (I.R.) Mashinostroenie, Moscow. [4, 36, 37, 38, 260, 266, 284]

Goryacheva, I.G., and Feldstein, I.V. (1995) The method of investigation of defects system influence on stress-strain state of elastic bodies, *ZAMM*, **76**, 173-174. [105]

Goryacheva, I.G., and Feldstein, I.V. (1996) Analysis of the influence of the internal system of defects on the stress state of elastic bodies, *Izv. AN SSSR, Mechanics of Solids*, **5**, 55-61. [105]

Goryacheva, I.G., and Soldatenkov, I.A. (1983) Theoretical study of running-in and the steady-state wear regime of solid lubricant coatings, *Soviet Journal of Friction and Wear*, **4**, 3, 26-35. [234]

Goryacheva, I.G., and Torskaya, E.V. (1992) Contact problems for bodies with wear resistance variable on the surface, *Soviet Journal of Friction and Wear*, **13**, 1, 151-158. [253, 254]

Goryacheva, I.G., and Torskaya, E.V. (1994) Stress state of coating under discrete loading, *Soviet Journal of Friction and Wear*, **15**, 3, 1-7. [116, 122]

Goryacheva, I.G., and Torskaya, E.V. (1995) Periodic contact problem for a system of punches and an elastic layer bounded to an elastic substrate, *Soviet Journal of Friction and Wear*, **16**, 4, 642-652. [117, 122]

Gradshteyn, I.S., and Ryzhik, I.M. (1963) *Tables of Integrals, Series and Products*, 4th Edn. (I.R.) Nauka, Moscow. (English translation, Jeffrey, A. (Ed.), Academic Press Inc., New York, 1965). [66, 76, 83, 85, 90, 244]

Greenwood, J.A., and Tabor, D. (1958) The friction of hard sliders on lubricated rubber: the importance of deformation losses, *Proc. Phys. Soc.*, **71**, 989. [97]

Greenwood, J.A., and Tripp, J.H. (1967) The elastic contact of rough spheres. *ASME, J. Appl. Mech.*, ser. E, **34**, 1, 153-159. [5]

Greenwood, J.A., and Williamson, J.B.P. (1966) Contact of nominally flat surfaces, *Proc. Roy. Soc., A*, **295**, 1442, 300-319. [5, 15]

Gubenko, V.S., and Mossakovsky, V.I. (1960) , Pressure of an axially symmetric circular die of an elastic half-space, *J. Appl. Math. Mech.*, **24**, 334-340. [216]

Gupta, P.K, and Cook N.M. (1972) Statistical analysis of mechanical interaction of rough surfaces, *ASME, J. Lubr. Tech.*, ser. F, **94**, 1, 14-23. [165]

Hamilton, G.N., and Goodman, L.E. (1966) , The stress field created by a circular sliding contact, *Trans. ASME, J. Appl. Mech.*, ser. E, **33**, 2, 371-376. [6, 174]

Hamrock, B.J. (1994) *Fundamentals of Fluid Film Lubrication*, McGraw-Hill, New York. [152]

Harricks, P.L. (1976) The mechanism of fretting and the influence of temperature, *Industrial Lubrication and Technology*, **Jan/Feb**, 9-17. [197]

Hattori, T., Nakamura, M., Sakata, H., and Watanabe, T. (1988) Fretting fatigue analysis using fracture mechanics, *JSME Int. Journal*, Series 1, **31**, 1, 100-107. [9]

Heathcote, H.L. (1921) The ball bearing: in the making, under test on service, *Proc. Inst. Auto. Engs.*, **15**, 569. [62]

Hertz, H. (1882) Über die Berührung fester elastische Körper, *J. Reine und angewandte Mathematik*, **92**, 156-171. [2]

Hills, D.A., and Nowell, D. (1994) *Mechanics of Fretting Fatigue*, Kluwer Academic Publishers, Dordrecht. [9]

Hirst, W., and Moore, A.J. (1974) Non-Newtonian behavior in elastohydrodynamics lubrication, *Proc. Roy. Soc.*, 33, 101-113. [152]

Hisakado, T. (1969) On the mechanism of contact between solid surfaces, *Bull. JSME*, **12**, 54, 1528-1549. [5, 165]

Hisakado, T. (1970) On the mechanism of contact between solid surfaces: surface roughness effects on dry friction, *Bull. JSME*, **13**, 55, 129-139. [5]

Holm, R. (1946) *Electric Contacts*, Almquist and Wiksells, Stockholm. [197]

Hough, C.L., and Das, B. (1985) Wear characteristics of polycrystalline diamond compact drill bits in small diameter rock drilling. *ASME, J. Energy Resources Technology*, **107**, 4, 534-542. [278]

Hughes, B.D., and White, L.R. (1980) Analytic approximations for the elastic contact of rough spheres, *ASME, J. Appl. Mech.*, **47**, 3, 194-196. [5]

Hunter, S.C. (1960) The Hertz problem for a rigid spherical intenter and a viscoelastic half-space, *J. Mech. and Phys. Solids*, **8**, 3, 219-234. [7]

Hunter, S.C. (1961) The rolling contact of a rigid cylinder with a viscoelastic half-space, *ASME, J. Appl. Mech.*, Ser. E, **28**, 611. [7, 87]

Hutchings, I.M. (1992) *Tribology: Friction and Wear of Engineering Materials*, CRC Press, Boca Raton, Florida. [1]

Il'ushin, A.A., and Pobedrya, B.E. (1970) *Fundamentals of the Mathematical Theory of Thermo-viscoelasticity*, (I.R.) Nauka, Moscow. [7]

Ionov, V.N. and Ogibalov, P.M. (1972) *Strength of Three-Dimensional Structural Elements* (I.R.) Nauka, Moscow. [169]

Ishlinsky, A.Yu. (1940) Theory of rolling friction and close phenomena, in *Friction and Wear in Machines*, (I.R.) Acad. Sci. USSR, Moscow–Leningrad, pp. 255-264. [7]

Ishlinsky, A.Yu. (1956) On partial slip in rolling contact, (I.R.) *Izv. AN SSSR, OTN*, **6**, 3-15. [87]

Janke, E., and Emde, F. (1944) *Tables of Functions with Formulae and Curves*, Dover Publication, New York City. [83, 85]

Johnson, K.L. (1962) Tangential traction and microslip in rolling contact, in *Rolling Contact Phenomena*, Bidwell, Elsevier, London, p. 6. [87]

Johnson, K.L. (1987) *Contact Mechanics*, Cambridge University Press, Cambridge. [4, 7, 76, 87, 128, 143, 169, 185, 226]

Johnson, K.L., Greenwood, J.A., and Higginson, J.G. (1985) The contact of elastic wavy surfaces, *Internat. J. Mech. Sci.*, **27**, 6, 383-396. [15]

Johnson, K.L., and Tevaarwerk, J.L. (1977) Shear behavior of EHD oil films, *Proc. Roy. Soc.*, A352, 215-236. [152]

Kagami, J., Yamada, K., and Hatazawa, T. (1983) Contact between a sphere and rough plates, *Wear*, **87**, 1, 93-105. [5]

Kalandiya, A.I. (1975) *Mathematical Methods of Two-Dimensional Elasticity*, Mir, Moscow. [5, 286, 288, 289, 290]

Kalker, J.J. (1990) *Three-Dimensional Elastic Bodies in Rolling Contact*, Kluwer Academic Publishers, Dordrecht. [4, 87, 277]

Kalker, J.J. (1991) Viscoelastic multilayered cylinders rolling with dry friction, *ASME, J. Appl. Mech.*, **58**, 666-679. [87, 123]

Kantorovich, L.V., and Krylov, V.I. (1952) *Approximate Methods of Higher Analysis*, Gostekhizdat, Moscow. (English translation, Benster, A.D. (Ed.), Interscience, New York, 1958). [53, 233]

Karasik, I.I. (1978) *Running-in Process of Journal Bearings*, (I.R.) Nauka, Moscow. [265]

Keer, L.M., and Goodman, L.E. (1976) Tangential loading of two bodies in contact, *ASME, J. Appl. Mech.*, **42**, 513-514. [6]

Kendall, K., and Tabor, D. (1971) An ultrasonic study of the area of contact between stationary and sliding surfaces, *Proc. Roy. Soc.*, A, **323**, 321-340. [40]

Khrushchov, M.M. (1946) *Investigation of the Running-in of Bearings Alloys and Journals*, (I.R.) Acad. Sci. USSR, Moscow–Leningrad. [167]

Khrushchov, M.M., and Babichev, M.A. (1970) *Abrasive Wear*, (I.R.) Nauka, Moscow. [196]

Kolesnikov, Yu.V., and Morozov, E.V. (1989) *Contact Failure Mechanics*, (I.R.) Nauka, Moscow. [9]

Konvisarov, D.V., and Pokrovskaia, A.A. (1955) Influence of the radii of curvature of cylindrical bodies on their resistance to rolling under different loads, (I.R.) *Proc. Sib. Phys.-Tech. Inst.*, **34**, 62-79. [62]

Korovchinsky, M.V. (1967) Stress distribution in contact of elastic bodies under the normal and tangential load, (I.R.) *Mashinovedenie*, **6**, 85-96. [6]

Korovchinsky, M.V. (1971) Local contact of elastic bodies with wear of their surface, in *Contact Interaction of Solid Bodies and Calculation of Friction Forces and Wear*, (I.R.) Nauka, Moscow, 130-140. [9]

Kostetsky, B.I. (1970) *Friction, Lubrication, and Wear in Machines*, (I.R.) Tekhnika, Kiev. [1]

Kovalenko, E.V. (1982) On the calculation of shaft-sleeve contact wear, *Izv. AN SSSR, Mechanics of Solids*, **6**, 66-72. [277, 280]

Kragelsky, I.V. (1965) *Friction and Wear*, Butterworths, London. [1, 6, 197]

Kragelsky, I.V. Bessonov, L.F., and Shvetsova, E.M. (1953) Contacting of rough surfaces, *Dokl. Akad. Nauk SSSR*, **93**, 1, 43-46. [165]

Kragelsky, I.V., Dobychin, M.N., and Kombalov, V.S. (1982) *Friction and Wear: Calculation Methods*, Pergamon Press, Oxford. [1, 14, 62, 167, 170, 190]

Kragelsky, I.V., and Nepomnyashchy, E.Ph. (1965) Theory of wear of high elastic materials, in *Polymers in Journal Bearings*, (I.R.) Nauka, Moscow, pp. 49-56. [182]

Kravchuk, A.S. (1980) To theory of contact problem with interfacial friction, *J. Appl. Math. Mech.*, **44**, 1, 122-129. [6]

Kravchuk, A.S. (1981) The solution of the 3-D contact problems with interfacial friction, *Soviet Journal of Friction and Wear*, **2**, 4, 589-595. [6]

Kuo, C.H., and Keer, L.M. (1992) Contact stress analysis of a layered transversely isotropic half-space, *ASME Journal of Tribology*, **114**, 2, 253-262. [110]

Kuzmenko, A.G. (1981) Contact problems with account for wear for cylindrical sliding bearings, *Soviet Journal of Friction and Wear*, **2**, 3, 502-512. [277]

Kuznetsov, Ye.A., and Gorokhovsky, G.A. (1978a) , Stress distribution in a polymeric material subjected to the action of a rough–surface indenter, *Wear*, **51**, 299-308. [15, 28, 174]

Kuznetsov, E.A., and Gorokhovsky, G.A. (1978b) , Influence of roughness on stress state of bodies in friction contact, (I.R.) *Prikl. Mekhanika*, **14**, 9, 62-68. [15, 28, 174]

Kuznetsov, E.A., and Gorokhovsky, G.A. (1980) , Friction interaction of rough bodies from the point of view of mechanics of solids, *Soviet Journal of Friction and Wear*, **4**, 638-649. [15]

Lancaster, J.K. (1973) *Dry Rubbing Bearing*, Section A4 in Tribology Handbook, (Ed. Near, M.J.), Butterworths. [196]

Larsen-Basse, J. (1973) Wear of hard metals in rock drilling: a survey of literature, *Powder Metallurgy*, **16**, 31, 1-32. [196]

Lee, E.H., and Radok, J.R.M. (1960) The contact problem for viscoelastic bodies, *Trans. ASME, J. Appl. Mech.*, ser. E, **27**, 3, 438-444. [7]

Lewis, W.D. (1968) Fluorocarbon resin in piston rings – new performance data for reciprocating nonlubrication air compressor, *Lubrication Engineering*, **24**, 112-127. [196]

Ling, F.F. (1973) *Surface Mechanics*, Wiley Interscience Publication, New York–London. [4, 7]

Ling, F.F., and Lai, W.M. (1980) Surface mechanics of layered media, in Cheng, H.S., and Keer, L.M. (Eds.), *Solid Contact and Lubrication*, **39**, Am. Soc. of Mech. Eng., pp. 27-50. [123]

Liu, M., and Farris, T.N. (1994) Effect of three dimensional near surface defects on rolling and sliding contact fatigue, *ASME, J. of Tribology*, **116**, 4, 841-849. [9]

Lur'e, A.I. (1955) *Three-dimensional Problems of the Theory of Elasticity*, (I.R.) Nauka, Moscow. (English translation by Radok, J.R.M., Interscience, 1964). [6]

Majumdar, A., and Bhushan, B. (1990) Role of fractal geometry in characterization and contact mechanics of surfaces, *ASME, J. of Tribology*, **112**, 205-216. [5, 11, 14]

Majumdar, A., and Bhushan, B. (1991) Fractal model of elastic–plastic contact between rough surfaces, *ASME, J. of Tribology*, **113**, 1-11. [5, 14]

Makushkin, A.P. (1990a) Stress–strain state of elastic layer in contact with spherical indenter, Part 1, Contact pressure analysis, *Soviet Journal of Friction and Wear*, **11**, 3, 31-42. [110]

Makushkin, A.P. (1990b) Stress–strain state of elastic layer in contact with spherical indenter, Part 2, Stresses inside polymeric layer, *Soviet Journal of Friction and Wear*, **11**, 4, 20-26. [110]

Marčenko, V.A. (1971) Randwertaufgaben in Gebieten mit feinkörnigem Rand, *Elliptische Differentialgleichungen*, **2**, 177-180. [105]

Marchenko, E.A. (1979) *Mechanism of Surface Fracture of Metals during Friction Process*, (I.R.) Nauka, Moscow. [9]

Marčenko, V.A., and Khruslov, E.Ya. (1974) *Boundary Value Problems in Domains with Fine-Structured Boundaries*, (I.R.) Naukova Dumka, Kiev. [105]

May, W.D., Morris, E.L., and Atack, D. (1959) Rolling friction of a hard cylinder over viscoelastic material, *J. Appl. Phys.*, **30**, 11, 1713-1724. [7]

Menčik, J. (1996) *Mechanics of Components with Treated or Coated Surfaces*, Kluwer Academic Publishers, Dordrecht. [9]

Meng, H.C., and Ludema, K.C. (1995) Wear models and predictive equations: their form and content, *Wear*, **181-183**, 443-457. [193]

Mikhlin, S.G., and Smolitsky, K.L. (1967) *Approximate Methods for Solutions of Differential and Integral Equations*, American Elseiver Pub. Co., New York. [217]

Miller, G.R., Keer, L.M., and Cheng, H.S. (1985) On the mechanics of fatigue crack growth due to contact loading, *Proc. Roy. Soc. London*, **A397**, 197-209. [9]

Mindlin, R.D. (1949) Compliance of elastic bodies in contact, *J. Appl. Mech.*, **16**, 3, 259-268. [6, 6]

Molgaard, J., and Czichos, H. (1977) The application of systems techniques to the study of wear, *ASME, Wear of Materials*, 30-35. [193]

Moor, N.B., Walker, B.H., and Appl, F.C. (1978) A model of performance and life of diamond drill boots, *Trans. ASME, J. Pressure Vessel Tech.*, **100**, 164-171. [196]

Moore, D.F. (1975) *Principles and Applications of Tribology*, Pergamon Press, London. [1, 62, 162]

Morin, A.J. (1853) *Leçons de Mécanique Pratique. Résistance des Matériaux*, L. Hachette, Paris. [7]

Morland, L.W. (1962) A plane problem of rolling contact in linear viscoelasticity theory, *ASME, J. Appl. Mech.*, **29**, ser. E, 2, 345-352. [7, 87]

Morland, L.W. (1967) Exact solution for rolling contact between viscoelastic cylinders, *The Quarterly J. Mech. and Appl. Math.*, **20**, 1, 73-106. [7]

Morland, L.W. (1968) Rolling contact between dissimilar viscoelastic cylinders, *The Quarterly J. Mech. and Appl. Math.*, **25**, 4, 363-376. [7]

Mossakovsky, V.I., Kachalovskaya, N.E., and Golikova, S.S. (1985) *Contact Problems of the Mathematical Theory of Elasticity*, (I.R.) Naukova Dumka, Kiev. [4]

Mossakovsky, V.I., Kachalovskaya, N.E., and Samarsky, V.I. (1986) Contact problem with friction for the axisymmetric punch, in *Mathematical Methods of Mechanics of Solid*, (I.R.) Nauka, Moscow, pp. 96-104. [6]

Mossakovsky, V.I., and Mishchishin, I.I. (1967) Rolling contact of elastic bodies, *J. Appl. Math. Mech.*, **31**, 5, 870-876. [87]

Mossakovsky, V.I., and Petrov, V.V. (1976) Influence of friction on microslip, *Dokl. Akad. Nauk SSSR*, **231**, 3, 603-606. [6]

Mozharovsky, V.V., and Starzhinsky, V.E. (1988) *Applied Mechanics of Composite Layered Bodies*, (I.R.) Nauka i Tekhnika, Minsk. [101]

Muki, R. (1960) Asymmetric problems of the theory of elasticity for a semi-infinite solid and thick plate, in Sneddon, I.N., and Hill, R. (Eds.) *Progress in Solid Mechanics*, North Holland, pp. 401-439. [6]

Muskhelishvili, N.I. (1946) *Singular Integral Equations*, (I.R.) Nauka, Moscow. (English translation by Radok, J.R.M., Noordhoff, 1953). [227]

Muskhelishvili, N.I. (1949) *Some Basic Problems of the Mathematical Theory of Elasticity*, 3rd Edn., (I.R.) Nauka, Moscow. (English translation by Radok, J.R.M., Noordhoff, 1953). [4, 5, 65, 101]

Nayak, P.R. (1971) Random process model of rough surfaces, *ASME J. Lub. Tech. Trans.*, Ser. F, **93**, 398. [13]

Nikishin, V.S., and Shapiro, G.S. (1970) *Spatial Problems of the Elasticity Theory for Multilayered Media*, (I.R.) AN SSSR, Moscow. [8, 110, 114]

Nikishin, V.S., and Shapiro, G.S. (1973) *Elasticity Theory Problems for Multilayered Medium*, (I.R.) Nauka, Moscow. [8, 110]

Peterson, M.B., and Winer, W.O., Eds. (1980) *Wear Control Handbook*, ASME, New York. [1]

Petrishin, V.I., Privarnikov, A.K., Shavalyakov, Yu.A. (1965) On the solution of problems for multilayer foundation, (I.R.) *Izv. AN SSSR, Mechanics of Solids*, **2**, 138-143. [108]

Petrovsky, I.G. (1973) *Ordinary Differential Equations*, Dover Publications, New York. [257]

Pinegin, S.V., and Orlov, A.V. (1961) Resistance to motion for certain types of free rolling, (I.R.) *Izv. AN SSSR, Mekhanika i Mashinostroenie*, **3**, 91-97. [62]

Pisarenko, G.S., and Lebedev, A.A. (1976) , *Deformation and Strength of Materials in the Complex Stress State*, (I.R.) Naukova Dumka, Kiev. [173]

Plemelj, J. (1908) Ein Ergänzungssatz zur Cauchyschen Integraldarstellung analytischer Funktionen, Randwerte betreffend, *Monatsh. Math. Phys.*, **19**, 205-210. [65]

Polzer, G., Ebeling, V., and Firkovsky, A. (1988) External friction of solid bodies, dissipative structures and selforganization, *Soviet Journal of Friction and Wear*, **9**, 2, 12-18. [194]

Popov, G.Ya., and Savchuk, V.V. (1971) Contact problem for a circle contact region taking into account a surface structure of contacting bodies, (I.R.) *Izv. AN SSSR, Mechanics of Solids*, **3**, 80-87. [50]

Popov, G.Ya. (1982) *Concentration of Elastic Stresses Near Punches, Cuts, Thin Inclusions and Stiffening Members*, (I.R.) Nauka, Moscow. [4, 101]

Poritsky, H. (1950) Stresses and deflections of cylindrical bodies in contact with application to contact of gears and of locomotive wheels, *J. Appl. Mech.*, **17**, 2, 191-201. [87]

Pronikov, A.S. (1957) *Wear and Life-time of Machines*, (I.R.) Mashgiz, Moscow. [9]

Pronikov, A.S. (1978) *Engine Reliability*. (I.R.) Mashinostroenie, Moscow. [9]

Rabotnov, Yu.N. (1977) *The Fundamentals of the Hereditary Mechanics of Solids*, (I.R.) Nauka, Moscow. [7]

Rabinowicz, E. (1965) *Friction and Wear of Materials*, John Wiley and Sons, New York. [1, 164, 197]

Rabinowicz, E. (1971) The determination of the compatibility of metals through static friction tests, *ASLE Trans.*, **14**, 198-205. [197]

Regel, R.V., Slutsker, A.I., and Tomashevsky, E.E. (1974) *Kinetics Foundation of Strength of Solids*, (I.R.) Nauka, Moscow. [170, 183]

Reynolds, O. (1875) On rolling friction, *Phil. Trans. Roy. Soc.*, **166**, 155-175. [62, 124]

Rhee, S.K. (1970) Wear equations for polymers sliding against metal surfaces, *Wear*, **16**, 431-445. [196]

Romen, Yu.S. (1969) *Nonlinear Oscillation of Train in an Arbitrary Curved Track*, (I.R.) Transport, Moscow. [311]

Rudzit, I.A. (1975) *Microgeometry and Contact Interaction of Surfaces*, (I.R.) Zinatis, Riga. [5]

Rvachev, V.L., and Protsenko, V.S. (1977) *Contact Problems of the Elasticity Theory for Complicated Regions*, (I.R.) Naukova Dumka, Kiev. [4]

Sadeghi, F., and Dow, T.A. (1987) Thermal effects in rolling/sliding contacts, part 2: analysis of termal effects in fluid film, *Journal of Tribology*, **109**, 512-518. [152]

Sainsot, Ph., Leroy, J.M., and Villechase, B. (1990) Effect of surface coating in a rough normally loaded contact, *Mechanics of Coating (Tribology Series)*, **17**, 151-156. [13, 112]

Sanchez-Palensia, E. (1980) *Non-Homogeneous Media and Vibration Theory*, Springer Verlag, New York. [105]

Sasada, T., and Kando, H. (1973) Formation of wear particle by the mutual transfer and growth process, *Proc. 18-th Jap. Congr. Mater. Reser.*, 32-35. [164]

Sasada, T., and Norose S. (1975) The formation and growth of wear particles through mutual transfer, *Material Transfer. JSLE-ASLE Int. Lubr. Conf.*, 73-80. [164]

Savin, T.N. (1968) *Stress Distribution Near Holes*, (I.R.) Naukova Dumka, Kiev. [101]

Seabra, J., and Berthe, D. (1987) Influence of surface waviness and roughness on the normal pressure distribution in the Hertzian contact, *ASME Journal of Tribology*, **109**, 3, 462-470. [13]

Seal, M. (1957) Friction and wear of diamond, *Conf. on Lubr. and Wear*, **12**, 252-256. [73]

Seifert, W.W., and Westcott, V.C (1972) A method for the study of wear particles in lubricating oil, *Wear*, **21**, 1, 27-42. [164]

Semenyuk, N.F. (1986a) Average values of total and mean curvatures of asperity top, heights of asperities for anisotropic rough surface, *Soviet Journal of Friction and Wear*, **7**, 5, 830-840. [13]

Semenyuk, N.F. (1986b) Probability density of asperity height distribution, asperity top characteristics of anisotropic rough surface, *Soviet Journal of Friction and Wear*, **7**, 6, 1017-1027. [13]

Semenyuk, N.F., and Sirenko, G.A. (1980a) Topography description of anisotropic rough surfaces using the random field model. Part 1, *Soviet Journal of Friction and Wear*, **1**, 3, 465-471. [13]

Semenyuk, N.F., and Sirenko, G.A. (1980b) Topography description of anisotropic rough surfaces using the random field model. Part 2, *Soviet Journal of Friction and Wear*, **1**, 5, 815-823. [13]

Semenyuk, N.F., and Sirenko, G.A. (1980c) Topography description of anisotropic rough surfaces using the random field model. Part 3, *Soviet Journal of Friction and Wear*, **1**, 6, 1010-1020. [13]

Shchedrov, V.S. (1950) Analysis of experimental laws of running-in on the basis of general equations of wear theory, in *Friction and Wear in Machines*, (I.R.) **6**, pp. 3-12. [167]

Sikarskie, D.L., and Altiero, N.J. (1973) The formation of chips in the penetration of elastic-brittle materials (rock), *J. Appl. Mech.*, **40**, Ser. E, 3, 791-798. [188]

Singer, I.L., and Pollock, H.M., Eds. (1992) *Fundamentals of Friction: Macroscopic and Microscopic Processes*, Kluwer Academic Publishers, Dordrecht. [1]

Sneddon, I.N. (1972) *The Use of Integral Transforms*, McGraw–Hill, New York. [108]

Soldatenkov, I.A. (1985) Wear of a thin elastic coating with a varying contact area, *Soviet Journal of Friction and Wear*, **6**, 2, 247-254. [10, 277, 281]

Soldatenkov, I.A. (1987) Wear of coating in elastic junctions with varying contact region, *Soviet Journal of Friction and Wear*, **8**, 2, 206-213. [10, 282]

Soldatenkov, I.A. (1989) Problem of wear of a half-plane by a counteracting disk, *Izv. AN SSSR, Mechanics of Solids*, **24**, 6, 107-110. [225]

Soldatenkov, I.A. (1994) Asymptotic analysis of the elastic problem solution for the strip of the variable width, *Izv. AN SSSR, Mechanics of Solids*, **29**, 6, 57-73. [231]

Specht, W. (1987) New particulars of wear of heavy railway carriage wheels, *Glasers Annalen*, **9**, 271-280. [310]

Spence, D.A. (1973) An eigenvalue problem for elastic contact with finite friction, *Proc. Cambridge Philosophical Society*, **73**, 249-268. [6]

Staierman, I.Ya. (1949) *Contact Problem of the Elasticity Theory*, (I.R.) Gostekhizdat, Moscow. [4, 15, 47, 50, 138, 183]

Sui, P.C., and Sadeghi, F. (1991) Non-Newtonian thermal elastohydrodynamic lubrication, *Journal of Tribology*, **113**, 390-397. [152]

Sviridenok, A.I., and Chijik, S.A. (1992) Contact interaction of smooth surfaces, *Soviet Journal of Friction and Wear*, **13**, 1, 130-137. [12]

Sviridenok, A.I., Chijik, S.A., and Petrokovets, M.I. (1990) *Mechanics of Discrete Friction Contact*, (I.R.) Nauka i Tekhnika, Minsk. [5, 11, 13, 42]

Szolwinski, M.P., and Farris, T.N. (1996) Mechanics of fretting fatigue crack formation, *Wear*, **198**, 93-107. [9]

Tabor, D., and Wynne-Williams, (1961) The effect of orientation on friction of polytetrafluoroethylene, *Wear*, **4**, 5, 391-400. [73]

Tabor, D. (1952) The mechanism of rolling friction, *Phil. Mag.*, Ser. 7, **43**, 345. [62, 97]

Tabor, D. (1955) The mechanism of rolling friction: the elastic range, *Proc. Roy. Soc.*, *A*, **229**, 198. [98]

Teply, M.I. (1983) Estimation of wear in friction pair shaft and bush, *Soviet Journal of Friction and Wear*, **4**, 2, 249-257. [10]

Thomas, T.R., Ed. (1982) *Rough Surfaces*, Longman, London. [5]

Tikhonov, A.N., and Arsenin, V.Ya. (1974) *Method of the Solution of Ill–Posed Problems*, (I.R.) Nauka, Moscow. (English translation by Winston and Sons, Washington, 1977). [275]

Timoshenko, S.P. (1943) *Strength of Materials*, D. Van Nostrand, New York. [206]

Timoshenko, S.P., and Goodier, J.N. (1951) *Theory of Elasticity*, 3rd Edn., McGraw-Hill, New York, London et al. [20]

Ting, T.C.T. (1968) Contact problems in the linear theory of viscoelasticity, *ASME, J. Appl. Mech.*, ser. E, **35**, 248. [7]

Tomlinson, J.A. (1929) A molecular theory of friction, *Phil. Mag.*, Ser. 7, **46**, 905-939. [62]

Tricomi, F. (1957) *Integral Equations*, Interscience, New York–London. [203]

Tsuchiya, K., and Tamai, T. (1970) Fluctuations of contact resistance in sliding contact, *Wear*, **16**, 5, 337-349. [164]

Ufland, Ya.S. (1965) *Survey of Articles on the Applications of Integral Tramsforms in the Theory of Elasticity*, English translation: Sneddon, I.N. (Ed.) Raleigh, North Carolina State University. [108]

Usov, P.P., Drozdov, Yu.N., and Nikolashev, Yu.N. (1979) Theoretical study of shaft-sleeve friction pair stress state with account for wear, (I.R.) *Mashinovedenie*, **2**, 80-87. [277]

Vantorin, V.D. (1962) Motion along a plane with anisotropic friction, in *Friction and Wear in Machines*, (I.R.) **16**. Izdat. Akad. Nauk, Moscow, pp. 81-120. [73]

Verigo, M.F., and Kogan, A.Ya. (1986) *Railway and Train Interaction*, (I.R.) Transport, Moscow. [301]

Vorovich, I.I., Aleksandrov, V.M., and Babeshko, V.A. (1974) *Nonclassical Mixed Problems of the Elasticity Theory*, (I.R.) Nauka, Moscow. [4, 8, 51, 54]

Vorozhtsov, M.I., Gusman, A.M., Draganchuk, O.T., and Reutov, V.A. (1989) Experimental and theoretical investigation of rock fracture by diamond–hard alloy plates, (I.R.) *Trudy VNIIBT*, **67**, 131-146. [319]

Waterhause, R.B. (1981) *Fretting Fatigue*, Applied Science Publishers, London. [9]

Waterhause, R.B. (1992) Fretting fatigue, *Int. Materials Reviews*, **37**, 2, 77-97. [9]

Westergaard, H.M. (1939) Bearing pressure and cracks, *ASME*, Ser. E, J. Appl. Mech., **6**, 1, 49-53. [15]

Westman, R.A. (1965) Axisymmetric mixed boundary-value problems of the elastic half-space, *ASME J. Appl. Mech.*, Ser. E, **32**, 411-417. [6]

Xia-Qiu Wang, (1994) Influence of angle attack and lateral force on rail wear (laboratory study), *Conference on Contact Mechanics and Wear of Rail/Wheel Systems, Preliminary Proceedings*, Vancouver, 404-411. [308]

Zmitrovicz, A. (1990) Liniowe i nieliniove równania konstytutywne tarcia anizotropowego, *Zeszyty Naukowe Akademii Gorniczo-Hutniczei*, Mechanica, **9**, 2, 141-154. [73]

Index

Mechanics

SOLID MECHANICS AND ITS APPLICATIONS

Series Editor: G.M.L. Gladwell

20. W. Schiehlen (ed.): *Advanced Multibody System Dynamics.* Simulation and Software Tools. 1993 ISBN 0-7923-2192-8
21. C.-W. Lee: *Vibration Analysis of Rotors.* 1993 ISBN 0-7923-2300-9
22. D.R. Smith: *An Introduction to Continuum Mechanics.* 1993 ISBN 0-7923-2454-4
23. G.M.L. Gladwell: *Inverse Problems in Scattering.* An Introduction. 1993 ISBN 0-7923-2478-1
24. G. Prathap: *The Finite Element Method in Structural Mechanics.* 1993 ISBN 0-7923-2492-7
25. J. Herskovits (ed.): *Advances in Structural Optimization.* 1995 ISBN 0-7923-2510-9
26. M.A. González-Palacios and J. Angeles: *Cam Synthesis.* 1993 ISBN 0-7923-2536-2
27. W.S. Hall: *The Boundary Element Method.* 1993 ISBN 0-7923-2580-X
28. J. Angeles, G. Hommel and P. Kovács (eds.): *Computational Kinematics.* 1993 ISBN 0-7923-2585-0
29. A. Curnier: *Computational Methods in Solid Mechanics.* 1994 ISBN 0-7923-2761-6
30. D.A. Hills and D. Nowell: *Mechanics of Fretting Fatigue.* 1994 ISBN 0-7923-2866-3
31. B. Tabarrok and F.P.J. Rimrott: *Variational Methods and Complementary Formulations in Dynamics.* 1994 ISBN 0-7923-2923-6
32. E.H. Dowell (ed.), E.F. Crawley, H.C. Curtiss Jr., D.A. Peters, R. H. Scanlan and F. Sisto: *A Modern Course in Aeroelasticity.* Third Revised and Enlarged Edition. 1995 ISBN 0-7923-2788-8; Pb: 0-7923-2789-6
33. A. Preumont: *Random Vibration and Spectral Analysis.* 1994 ISBN 0-7923-3036-6
34. J.N. Reddy (ed.): *Mechanics of Composite Materials.* Selected works of Nicholas J. Pagano. 1994 ISBN 0-7923-3041-2
35. A.P.S. Selvadurai (ed.): *Mechanics of Poroelastic Media.* 1996 ISBN 0-7923-3329-2
36. Z. Mróz, D. Weichert, S. Dorosz (eds.): *Inelastic Behaviour of Structures under Variable Loads.* 1995 ISBN 0-7923-3397-7
37. R. Pyrz (ed.): *IUTAM Symposium on Microstructure-Property Interactions in Composite Materials.* Proceedings of the IUTAM Symposium held in Aalborg, Denmark. 1995 ISBN 0-7923-3427-2
38. M.I. Friswell and J.E. Mottershead: *Finite Element Model Updating in Structural Dynamics.* 1995 ISBN 0-7923-3431-0
39. D.F. Parker and A.H. England (eds.): *IUTAM Symposium on Anisotropy, Inhomogeneity and Nonlinearity in Solid Mechanics.* Proceedings of the IUTAM Symposium held in Nottingham, U.K. 1995 ISBN 0-7923-3594-5
40. J.-P. Merlet and B. Ravani (eds.): *Computational Kinematics '95.* 1995 ISBN 0-7923-3673-9
41. L.P. Lebedev, I.I. Vorovich and G.M.L. Gladwell: *Functional Analysis.* Applications in Mechanics and Inverse Problems. 1996 ISBN 0-7923-3849-9
42. J. Menčik: *Mechanics of Components with Treated or Coated Surfaces.* 1996 ISBN 0-7923-3700-X
43. D. Bestle and W. Schiehlen (eds.): *IUTAM Symposium on Optimization of Mechanical Systems.* Proceedings of the IUTAM Symposium held in Stuttgart, Germany. 1996 ISBN 0-7923-3830-8

Kluwer Academic Publishers – Dordrecht / Boston / London

Mechanics

SOLID MECHANICS AND ITS APPLICATIONS

Series Editor: G.M.L. Gladwell

44. D.A. Hills, P.A. Kelly, D.N. Dai and A.M. Korsunsky: *Solution of Crack Problems*. The Distributed Dislocation Technique. 1996 ISBN 0-7923-3848-0
45. V.A. Squire, R.J. Hosking, A.D. Kerr and P.J. Langhorne: *Moving Loads on Ice Plates*. 1996 ISBN 0-7923-3953-3
46. A. Pineau and A. Zaoui (eds.): *IUTAM Symposium on Micromechanics of Plasticity and Damage of Multiphase Materials*. Proceedings of the IUTAM Symposium held in Sèvres, Paris, France. 1996 ISBN 0-7923-4188-0
47. A. Naess and S. Krenk (eds.): *IUTAM Symposium on Advances in Nonlinear Stochastic Mechanics*. Proceedings of the IUTAM Symposium held in Trondheim, Norway. 1996 ISBN 0-7923-4193-7
48. D. Ieşan and A. Scalia: *Thermoelastic Deformations*. 1996 ISBN 0-7923-4230-5
49. J. R. Willis (ed.): *IUTAM Symposium on Nonlinear Analysis of Fracture*. Proceedings of the IUTAM Symposium held in Cambridge, U.K. 1997 ISBN 0-7923-4378-6
50. A. Preumont: *Vibration Control of Active Structures*. An Introduction. 1997 ISBN 0-7923-4392-1
51. G.P. Cherepanov: *Methods of Fracture Mechanics: Solid Matter Physics*. 1997 ISBN 0-7923-4408-1
52. D.H. van Campen (ed.): *IUTAM Symposium on Interaction between Dynamics and Control in Advanced Mechanical Systems*. Proceedings of the IUTAM Symposium held in Eindhoven, The Netherlands. 1997 ISBN 0-7923-4429-4
53. N.A. Fleck and A.C.F. Cocks (eds.): *IUTAM Symposium on Mechanics of Granular and Porous Materials*. Proceedings of the IUTAM Symposium held in Cambridge, U.K. 1997 ISBN 0-7923-4553-3
54. J. Roorda and N.K. Srivastava (eds.): *Trends in Structural Mechanics*. Theory, Practice, Education. 1997 ISBN 0-7923-4603-3
55. Yu. A. Mitropolskii and N. Van Dao: *Applied Asymptotic Methods in Nonlinear Oscillations*. 1997 ISBN 0-7923-4605-X
56. C. Guedes Soares (ed.): *Probabilistic Methods for Structural Design*. 1997 ISBN 0-7923-4670-X
57. D. François, A. Pineau and A. Zaoui: *Mechanical Behaviour of Materials*. Volume I: Elasticity and Plasticity. 1998 ISBN 0-7923-4894-X
58. D. François, A. Pineau and A. Zaoui: *Mechanical Behaviour of Materials*. Volume II: Viscoplasticity, Damage, Fracture and Contact Mechanics. 1998 ISBN 0-7923-4895-8
59. L. T. Tenek and J. Argyris: *Finite Element Analysis for Composite Structures*. 1998 ISBN 0-7923-4899-0
60. Y.A. Bahei-El-Din and G.J. Dvorak (eds.): *IUTAM Symposium on Transformation Problems in Composite and Active Materials*. Proceedings of the IUTAM Symposium held in Cairo, Egypt. 1998 ISBN 0-7923-5122-3
61. I.G. Goryacheva: *Contact Mechanics in Tribology*. 1998 ISBN 0-7923-5257-2

Kluwer Academic Publishers – Dordrecht / Boston / London

MIX
Papier aus verantwortungsvollen Quellen
Paper from responsible sources
FSC® C105338

If you have any concerns about our products,
you can contact us on
ProductSafety@springernature.com

In case Publisher is established outside the EU,
the EU authorized representative is:
Springer Nature Customer Service Center GmbH
Europaplatz 3, 69115 Heidelberg, Germany

Printed by Libri Plureos GmbH
in Hamburg, Germany